# ORGANIC CHEMISTRY

## *How to Solve It*

The Geometry of Molecules

RUTH A. WALKER

Professor of Chemistry
Herbert H. Lehman College of the
City University of New York

Illustrated by ROGER HAYWARD

FREEMAN, COOPER & COMPANY

1736 Stockton Street,
San Francisco, California 94133

1736 Stockton Street
San Francisco, California 94133



Printed in the United States of America
Library of Congress Catalogue Card Number 70-148030
International Standard Book Number 87735-208-9

# Preface for Teachers

Many students of organic chemistry, those at an advanced level as well as beginners, find it a difficult study. Their difficulties stem not only from the theoretical concepts but also from the mechanics of the discipline, such as the visualization of molecules in three dimensions; the naming and drawing of molecular structures; the methods for handling problems in mechanisms, synthesis, and proof of structure; the organization of facts in the most effective form for their use in problem solving. While there are many good textual presentations of the substance of organic chemistry, there is a scarcity of helpful introductions to the fundamental mechanics of the study and to problem-solving techniques not encountered elsewhere by the student.

ORGANIC CHEMISTRY—HOW TO SOLVE IT aims to fill this gap by providing the student with an organized study pattern, a detailed step-by-step analysis of the thought processes needed to solve each type of problem found in organic chemistry, and an understanding of spatial relationships sufficient to enable him to work effectively with the concepts of molecular structure. This aid to study is organized so that it can be used in conjunction with any standard text.

*The Geometry of Molecules* encompasses subject matter which for the most part would normally fall within the first semester's work; it deals with the concepts of molecular structure and charge distribution that are the foundation of all thought about organic molecules. Part I covers the geometry of atoms in molecules; Part II, the geometry of the electron cloud. Within each Part the presentation starts at an elementary level and builds toward a more sophisticated level of thinking. At any time the student may well be working in both parts at the same level of difficulty.

No effort has been made to include all of the content of a first year course. The expository material presents the essence of concepts rather than their details and emphasizes the application of concepts in problem solving. The objective is to establish a sound foundation on which the student can build and to give him the tools with which to do the building. This objective can best be achieved by teaching him how to study the subject.

Each Part has been subdivided into sections according to topic. Each topic is presented as a unified whole and is followed by its set of problems. Thus, the order in which specific topics are studied can be arranged to allow for differences in texts without producing discontinuity in the problems.

Throughout the book, many examples have been analyzed as they are solved. In each, the main question has been broken down into a series of smaller, readily answerable questions. Through these, the stepwise thought pattern needed to solve the main question has been presented. The thought patterns of a variety of different types of problems have been developed in detail, and drill has been provided starting with a single simple concept and gradually progressing to the use of complex combinations of several concepts. Answers to problems include sufficient detail for the student to follow the thinking that went into getting the answer.

In addition to the set of problems for each section, there is, at the end of each Part, a set of problems designed to test the student's understanding of all the concepts developed within the Part and to provide him with an integrated review of that phase of the subject.

The major emphasis in this book is on the three-dimensionality of molecules and its chemical consequences. The book provides sufficient opportunity for the student to develop for himself the ability to "see" molecules. Sound thought patterns thereby developed in this area of molecular structure will enable him to approach the problems of mechanisms, synthesis, and structure proof with much greater facility. In short an attempt has been made to help the student to help himself over the initial hurdles in his study of organic and to provide him with a sound basis from which to explore this engrossing, significant branch of chemistry.

*Ruth A. Walker*

# Contents

## Part I Geometry of Atoms in Molecules. How to See and Name Molecules

## Tables for Part I



## Part II Geometry of the Electron Cloud. How to Predict and Depict Charge Distribution



## Tables for Part II

# To the Student

You are about to study organic chemistry. At times during the early stages of the study you may feel that the subject is remote or unreal. If so, think.

Ever since you got up this morning you have been intimately involved in organic chemistry. Any move you make, any thought you think, is possible only because a chemical reaction takes place. Everything you touch is either an organic or an inorganic chemical, and there are many more of the former than of the latter.

The clothes you wear are composed of organic molecules. Cotton and wool are natural polymers (a polymer is a molecule in which a particular arrangement of atoms is repeated many times as a unit in a long chain). Dacron and Nylon are man-made polymers. A unique arrangement of the atoms in each produces the particular properties associated with these natural and synthetic textile fibers. Nature controls the reactions that form cotton and wool; man has not yet learned to synthesize those fibers effectively, but he can control the reactions leading to Dacron and Nylon.

The foods you eat are produced in nature as living (organic) substances. Man merely processes them. The structures of the molecules needed for your nourishment are too complex for man to manufacture them at this time. Although extensive research has determined the composition of fats, proteins, and carbohydrates, and has led to an understanding of how to make such complex organic structures in the laboratory, nature can still synthesize foods more easily, efficiently, and economically than man can. The study of nutrition in its broadest sense involves not only a determination of the structure of nutrients and of how they are changed during metabolism, but also a search for new sources of these nutrients.

Consider medicinal products. Many physiologically active substances, such as aspirin, the "sulfa drugs," novocain, and phenobarbital, have been synthesized by chemists and are now manufactured for general use; many others, such as quinine, penicillin, digitalis, and morphine, are isolated from their natural sources, purified, and then made commercially available. If something interferes with the growth of the bacteria, the trees, the flowers, or the other plants that are the sources of these important medicines, the supply is reduced and the organic chemist is challenged.

The chemical reactions that transform the food you eat into flesh, blood, and the energy you need to move, to think, and to live are infinite in number. Only a tiny fraction of them is as yet understood in detail. Those of you who aspire to become doctors, dentists, biologists, and biochemists hope some day to understand and perhaps even to control some of the biochemical reactions of living tissues. Those of you who aspire to become organic chemists hope to be able to assemble atoms into new structures that will help solve some of society's problems with starvation, disease, ecological pollution, decreasing fuel supplies, and increasing needs for energy.

Before you can achieve any of these aims through organic chemistry, you must understand:

- how the structure of a molecule is determined;
- how the atoms are arranged in various molecules, and why they are in these positions;
- how and why molecules react with each other to form new structures; and
- how to control and use these reactions to make new structures with specific properties.

The fundamental theory has to be learned at a level which may seem quite remote from your ultimate objective, but as you gain an understanding of the subject you will be able to relate it to the practical applications which interest you.

* * * * *

ORGANIC CHEMISTRY—HOW TO SOLVE IT was written in answer to the often asked question, How do I study organic chemistry? Our purpose is to show you how to organize your thinking both in learning the theory of organic chemistry and in applying it to the solution of problems. Organic chemistry can be considered as consisting of four major conceptual areas:

- molecular geometry, both atomic and electronic;
- mechanisms of reactions;
- synthesis of compounds; and
- proof of structure.

This book has as its theme the first of these areas. It is a *how to study* book designed to show you:

- how to visualize the structure of a molecule;
- how to figure out the spatial relationships between the atoms in a molecule from the type of bonding;
- how to interpret and draw both two- and three-dimensional structural formulas;
- how to name organic compounds;
- how to figure out the probable distribution of the charge within a molecule;

- how to indicate the charge distribution in both two- and three-dimensional formulas;
- how to determine the relative acidity and basicity of compounds on the basis of charge distribution.

All of the structural concepts that provide the basis for understanding mechanisms, designing syntheses, and elucidating unknown structures are developed here in detail to provide you with a sound foundation for the rest of organic chemistry.

* * * * *

This study guide has been designed for use with any text. It is divided into two parts: Part I deals with the spatial relationships between the atoms in molecules and how these relationships are indicated by the scientific name of a compound. Part II deals with charge distribution in relation to molecular structure and relates this to acidity and basicity.

In order to use this study guide most effectively proceed as follows:

*1*) Read your textbook assignment through fairly quickly. Then turn to the list of section headings at the beginning of this book and locate the sections dealing with the assigned topics. Read the appropriate sections to get a general idea of their content.
*2*) Study the subject in both books simultaneously, referring back and forth between the two until you understand the concepts involved. This book will augment your text by providing some of the practical details (such as how to interpret a formula) that you need to understand the theory.
*3*) As part of this study, work through the details of the illustrative examples; they have been included to show you how the theory may be applied. When you understand the example, close the book and write out the solution in order to make *certain* that you understand how to go about solving that particular type of problem.
*4*) Work out the problems in the section following the discussion. These will assure that you understand the underlying concepts needed to solve the problems in your text. Do these problems in order since they are programmed.
*5*) Work the assigned problems in your text using the thought pattern established for you by this book.

You may find yourself working in both Part I and Part II at the same time. This is as it should be. Both Parts start at the elementary level and proceed to more advanced ones. You will need to understand the elementary level in both Parts before proceeding to the more advanced level in either. You should study the topics in the order in which they appear in your text, and you need not worry about whether you are in Part I or Part II.

As you proceed through this book, you will become aware that certain thought patterns repeat themselves in the solution of problems relating to molecular structure. You will find that these patterns can be applied to additional types of problems not specifically covered in this book but included in your text. Once you have perceived the logic in organic chemistry, you can start to enjoy the intellectual challenge provided by problems in organic chemistry. Whether or not you eventually use organic chemistry in your profession, you will have learned a technique of studying and a pattern of logical thinking that can be applied to other fields. You will also have increased your general knowledge because of the importance of organic chemistry in your daily life.

PART I

# Geometry of Atoms in Molecules. How to See and Name Molecules

A molecule is a space-filling, three-dimensional unit. It is composed of atoms held in position within a framework of electrical charge from the electrons present in each atom. The structure of each different organic substance is unique. The number, the kind, and the arrangement of the atoms determine the chemical and physical properties of a molecule. To fully understand how and why organic reactions occur you must "see" the molecular structures involved.

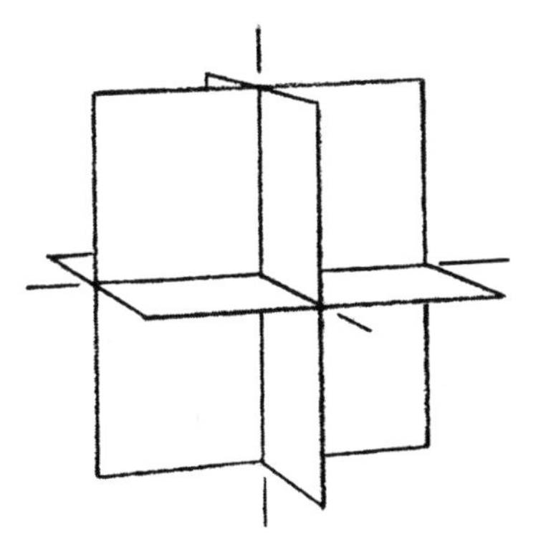

Visualizing the three-dimensionality of an object drawn on a flat surface is difficult for many of us. The process can be made easier by the use of models. Throughout this book, directions are given for constructing models. Take the time to make these models. Each illustrates the specific spatial relationship under discussion at that point in the presentation and will help you to "see" it. After you have made the model, try to draw it; you do not need an artistic drawing, just a series of lines so positioned as to represent for you, on a two-dimensional surface, the three-dimensional relationships of the components of the model.

* * * * *

Each organic substance has its own name. Some of these—such as aspirin, morphine, paraffin, and Nylon—are popular rather than scientific and tell us nothing about the composition of the compound. For that you have to know the chemical formula; it cannot be deduced from popular names. The IUPAC (for "International Union of Pure and Applied Chemistry") system of naming organic compounds goes further; it was developed so that it would be possible to draw the molecular structure of a compound from the IUPAC name even if you had never heard of the compound. Conversely, you can determine the correct IUPAC name for a compound if you know its structure.

The structure of a compound and its IUPAC name are so directly related that learning to name compounds provides a sound introduction to molecular geometry. The study of nomenclature helps you to gain an insight into the many different ways that atoms can be assembled. The presentations of molecular geometry and of nomenclature have been interwoven in this book. First we will study a particular structural relationship between atoms; then we will learn to indicate this structural feature in the names of compounds that contain it.

Only the fundamental rules of nomenclature have been included. Emphasis is on the application of the rules in naming simple compounds. Once you have learned to relate these rules to simple structures, you will be able to name more complicated ones by looking up the specialized IUPAC rules and applying them in the same way. Your procedure is first to learn the broad principles of nomenclature and then to practice them whenever the opportunity arises.

When the textual material you are studying refers to a compound by the name instead of giving the structural formula, take the time to draw a structure for it. This will be time well spent because you will understand what you are reading much better with the formula in front of you. Conversely, consciously identify each structural formula in your text by name. Many students ask questions about "the compound in the upper right-hand corner of the blackboard." Instead, force yourself from the beginning to say the name of the compound. It is only with this kind of practice that you can develop facility in naming compounds and in visualizing structures from names, a facility that will make it easier for you to think more clearly about the rest of organic chemistry.

You will need to learn a new vocabulary with which to describe molecules. Treat this vocabulary as you would that of a new language. When you first learn that *ouvrez la porte* means "open the door," you translate the French words into English ones before the command has meaning for you. However, with repetition you learn to react to the French words themselves without translation. Similarly, translate the new geometric and organic terms into words that are familiar to you until this becomes unnecessary. When you read *enantiomer*, say "exact mirror image"; *n*-pentane becomes "a five-carbon, continuous-chain, saturated hydrocarbon." Soon the words themselves will convey the concept without intermediate translation.

* * * * *

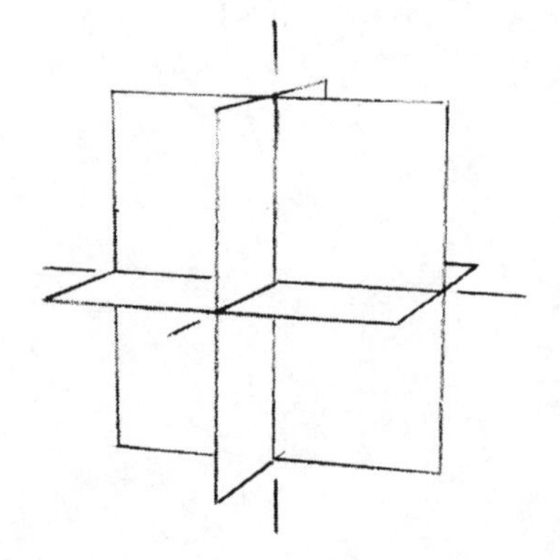

Perhaps the most important study hint for organic chemistry is to use your pencil. Do not just read. Write the name. Draw the formula. Write and draw, write and draw. Practice drawing three-dimensional structures until it becomes easy to visualize them. Then you will understand them and the molecules they represent. Do not just read the illustrative examples. Write down each step as you go through the example. Then close the book and write out the solution again to be sure you understand it. If you list the general directions given at the beginning of each step in the solution to an illustrative example, you will have an outline of the general procedure to be followed in solving all problems of one particular type. The examples have been deliberately set up this way; so take advantage of this to learn how to do problems in organic. As you finish your study of each concept, do the problems relating to it. They provide the drill you need to reinforce the textual material and show how to apply the theory. Do not just read a problem and decide that you know how to do it. Actually write out the answer. Frequently a new perspective takes shape in your mind while you are doing so. This kind of practice also makes it possible for you to write out similar answers more rapidly when you are working under pressure.

Now it is time for you to get the pencils and the models, and find out how well you can learn to "see" and name molecules.

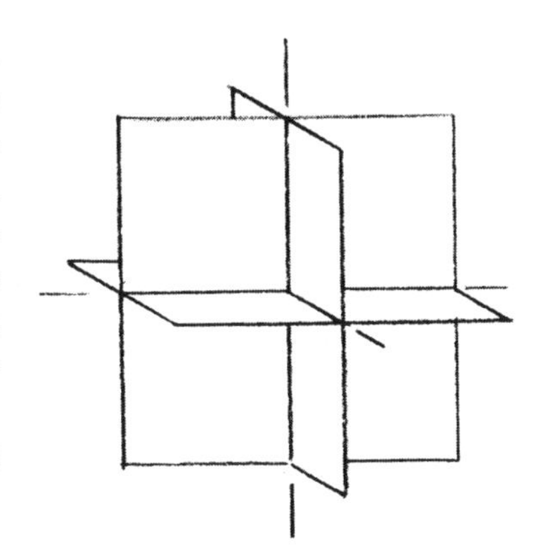

## § 1. The Geometry of Sigma (σ) and Pi (π) Bonding

The neutral atom of an element is associated with a specific number of planetary electrons; the number of planetary electrons varies from element to element but is unique for each element. For instance, hydrogen has one, carbon six, nitrogen seven, and oxygen eight planetary electrons.

Electrons are in constant motion. It is not completely random motion. There is a prescribed volume of space within which there is a ninety percent chance of finding a given electron at any moment in time. This space is referred to as the *orbital* of that electron.

To visualize an orbital, imagine a minute particle moving about within a balloon so rapidly that it effectively occupies the entire balloon. The shape of the balloon then is the shape of the orbital of that electron. If you have trouble imagining how one minute particle could occupy the entire volume of the balloon, consider a bicycle wheel with a single spoke. When the wheel is motionless, you can easily put your hand through it from one side to the other. However, you cannot do so when the wheel is revolved rapidly. The single spoke is then, in effect, occupying the entire space bounded by the rim of the wheel. Somewhat similarly, an electron in motion occupies a three-dimensional space, its orbital.

The size and shape of an orbital, as well as its position relative to the nucleus of an atom, vary according to the energy of the electron. These orbital characteristics can be described mathematically. We do not need to become involved here in these calculations. Our main concern now is the spatial relationships among the atoms as determined by the lengths and directions of the bonds between them. Bond direction depends upon the relative positions of electron orbitals; diagrammatic representations of these relationships will suffice to enable us to understand the role played by electron orbitals in determining molecular geometry.

An atom that is not bonded to any other atom and in which each electron is at its lowest energy level is said to be in the *ground state*. An atom of hydrogen in the ground state has a single electron that moves around within a spherical orbital, with the hydrogen nucleus at the center. An electron in a spherical orbital is designated as an *s* electron.

There can be no more than two electrons in a single orbital. An electron spins on an axis. Electrons that spin in the same direction repel each other; so electrons that share an orbital assume opposite spins. This relationship between the two is described as being *paired*.

*Sigma Bonds*

When two atoms, each with an orbital containing a single electron, come close enough to each other, the orbital from one atom can overlap the orbital of the other, as diagrammed in #1–1 for two hydrogen atoms.

**#1–1**

(Hydrogen nucleus at center of *1s* orbital)

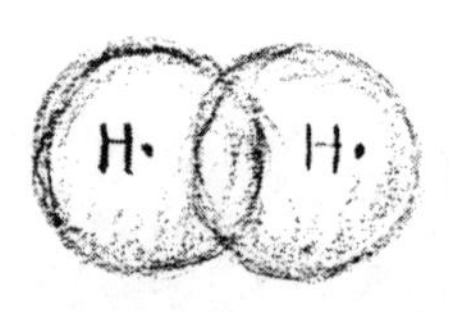

(Overlapping *1s* orbitals)

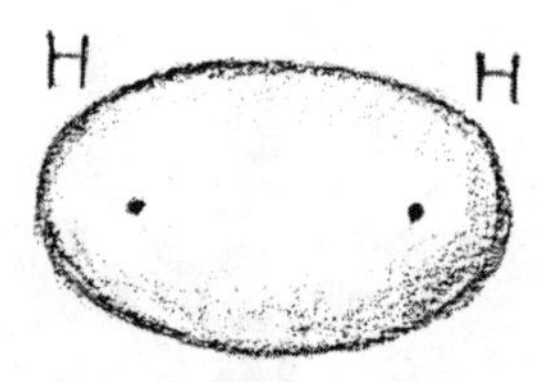

(Molecular orbital from two *1s* electrons)

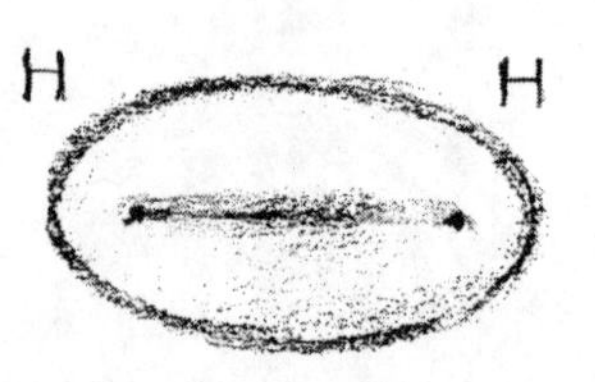

($\sigma$ bond between two hydrogen nuclei)

If you have trouble visualizing how two orbitals can overlap when each is effectively filled by an electron, try the following experiment: Find yourself two pieces of string about two feet long and two keys or similar objects. Tie one to each string. Grasp one string by the end and twirl the key in front of you. The key will describe a circle or an ellipse. Such a circle could be called a flat "orbital" of the key. If you are reasonably dexterous, it is possible to twirl the second key on its string in your other hand so that there is a considerable region of space through which both keys move; the circle described by one key overlaps the circle described by the other (#1–2). Yet, if your timing is right, the two keys do not interfere with each other. An eggbeater's rapidly revolving blades also suggest how small things may "fill" a larger space.

The overlap is analogous to the overlap of two electron orbitals except that, after such overlapping has occurred with electrons, each does not continue to move independently within its original orbital. The electrons interact, their spins become paired, and both electrons occupy a new orbital that they share.

The changes resulting from orbital overlap can be visualized if you picture a hydrogen atom as a balloon with the hydrogen nucleus at its center and the single electron moving about so rapidly that it occupies all of the space within the balloon. Imagine that you could take two such balloons and push them together so that the interface between them disappeared and the two balloons merged into one ellipsoidal unit like the one in #1–1.

As the merger takes place, the two electrons become spin-paired and both of them move about within the entire ellipse. The two hydrogen nuclei are positioned so that there is one near each end of the ellipse. Thus, most of the electrical charge from the two electrons is concentrated between the two nuclei. The region of high electron density between the nuclei tends to hold them in a specific position relative to each other. The two hydrogen atoms are said to be *bonded* together. The region of high electron density is referred to as a *bond*. The distance between the nuclei, which remains constant, is the *bond length*.

The ellipsoidal balloon with the two hydrogen nuclei represents a molecule ($H_2$) of hydrogen, i.e., two hydrogen atoms ($H\cdot$) have come together to form the diatomic hydrogen molecule ($H_2$). The ellipsoidal orbital of the two bonding electrons is a *molecular orbital* whereas the spherical orbital of the electron in a hydrogen atom is an atomic orbital.

Bonding that results from the overlap of two orbitals in the manner described for hydrogen is called *sigma* ($\sigma$) bonding. We refer to the forces holding the atoms together as a bond, and we indicate the bond by a line, H—H, or a pair of dots, H:H, between the atoms. Overlap of orbitals accompanied by the interaction between electrons that leads to bonding can occur only when there is a single electron in each of the two orbitals.

(Space shared in common)

*Geometry of a Plane*

Before considering bonding in the compounds of carbon, we are going to review some of the fundamental geometrical terms and concepts useful in the visualization of molecular structures. The discussion of these geometric concepts includes directions for making models to illustrate the specific spatial relationship being described. To make the models, you will need some roundish objects that can be joined with sticks. The directions are given in terms of a ball and a stick, but apply to whatever objects you wish to use: styrofoam balls and pipe cleaners; small rubber balls and double-ended metal knitting needles; or gumdrops—the roundish objects need not be regular in shape—and toothpicks; or the like.

Only one straight line can be drawn between two points. The length of the line is determined by the distance between the points. Join two balls together with a stick. You now have a single unit that can be moved around in space without producing any change in the distance between the balls. You can roll the model like a cylinder; or turn it end for end; you can hold it in either a vertical or a horizontal position; yet the unit remains unchanged.

Make a second unit using a stick that is half the length of the first one. If all the balls are identical you can differentiate between the two models by indicating that the distance between the two balls in one model is twice that in the other model.

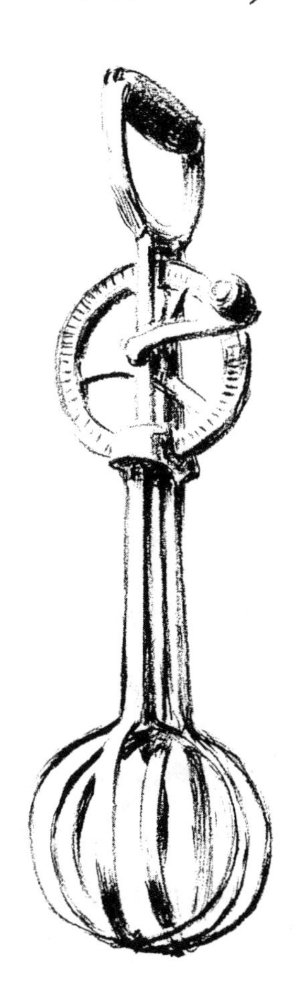

Diatomic molecules such as hydrogen have the geometric properties of two balls on a stick. The balls represent the nuclei of the two bonded atoms and the stick represents the force or the bond that holds them in one position relative to each other. We can visualize the geometry of the molecule from this model even though in reality the nuclei are infinitely small compared to the boundaries of the molecular orbitals and the bond is a broad region of high electron density rather than a narrow line. The model can serve to summarize the overall spatial relationships within the molecule.

The size of an atom is determined by the boundaries of its electron orbitals. The length of a bond between two atoms is determined by the size of the two atoms and the extent to which their respective orbitals overlap. Thus, bond lengths vary according to the kinds of atoms that are bonded together and according to the type of bond (single, double, or triple) that is formed (see Table 1–1). *The shorter the bond, the greater its strength.* You will find tables of bond lengths and bond energies in your textbook.

Other diatomic molecules with the geometric properties of two balls on a stick include:

| $N_2$ | $Br_2$ | HF | NO | CO |
|---|---|---|---|---|
| nitrogen | bromine | hydrogen fluoride | nitrous oxide | carbon monoxide |

* * * * *

The relative positions of three points in space cannot be described by merely indicating the distances between them. The balls A and B are each about one-half inch away from ball D in the three examples in #1–3; yet the relationship between the three balls is different in each example (measured from center to center).

**#1–3**

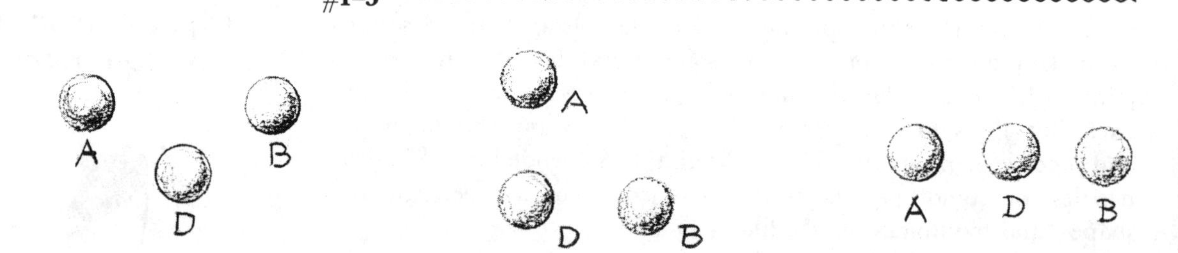


Place three balls of the same diameter on a flat surface. Join the three balls by means of two sticks so that you have a single unit that can be moved about without changing the relationships between the balls. The shape of your model depends upon the positions of the three balls relative to each other. To describe your model so that someone else could reproduce it, you have only to *specify the length* of the two sticks *and the angle* formed where they intersect at the central ball. These two structural features are unique for each different set of three points as indicated in #1–4. There is no relationship between the length of the sticks and the size of the angle they form with each other.

**#1–4**

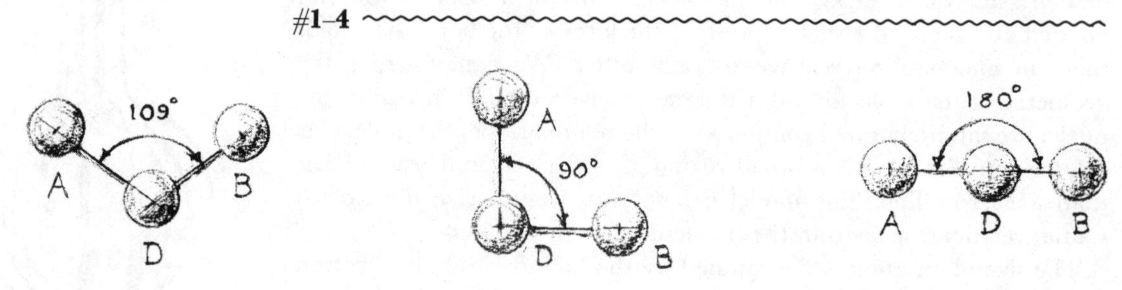

Triatomic molecules have the same geometrical properties as the model you made from three balls and two sticks. The balls represent the tiny nuclei of the three atoms and the sticks represent the bonds holding them in position. The geometry of a triatomic molecule can, therefore, be described by specifying the length of its two bonds and the size of the angle their axes form. Three molecules that you have encountered before in which the angles of bonding* differ are shown in #1–5.

**#1–5**

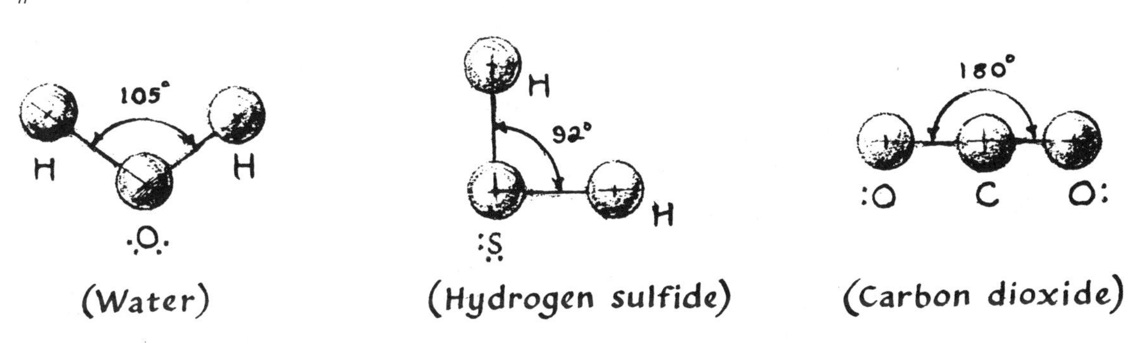


Your three-ball, two-stick model represents a structure that is flat, or planar. That is, the geometrical elements—the points at the centers of the three balls and the two lines connecting them—all lie in one plane. Even though the balls and sticks—or the atoms and orbitals they represent—occupy space and so are three-dimensional, the structure itself is planar if all of its geometrical elements are in one two-dimensional plane, such as the flat page of this book, the surface of a wall, or the top of a table.

You may visualize this relationship by placing your model on a table or other flat surface. The model will touch the table at three points. If all the balls are the same size, then their centers and the axes of the sticks will lie in a plane just above and parallel to the surface of the table. This concept of a plane and of flatness is a useful one in visualizing the molecular geometry of double and triple bonds.

It is always possible to draw a plane through any three given points. It is in fact impossible to position three points so that they do not lie in one plane. Prove this to yourself by trying to arrange three balls so that a single plane cannot pass through the centers of all three. The balls do not need to be of the same size. Additionally there is *only one plane* that can include any particular set of three points. This plane may contain many other points and lines in addition to the three chosen to determine the position of the plane. If all these other geometric elements belong to the same structure, then the structure can be described as "flat."

* The less precise term *bond angle* is frequently used instead of the *angle of bonding*. The first term implies that the bond is bent at an angle, which is not true. The axis of a bond is straight. The angle referred to is that formed at the point where the axes of the two bonds intersect.

The spatial relationships of lines and points within a plane can be designated by the use of only two dimensions. For example, a point P in the plane of this page can be described as being either above or below the horizontal line X–X and as being either to the right or to the left of the vertical line Y–Y, as in #1–6. The straight lines labeled X and Y are called *axes*. They are perpendicular to each other and

**#1–6**

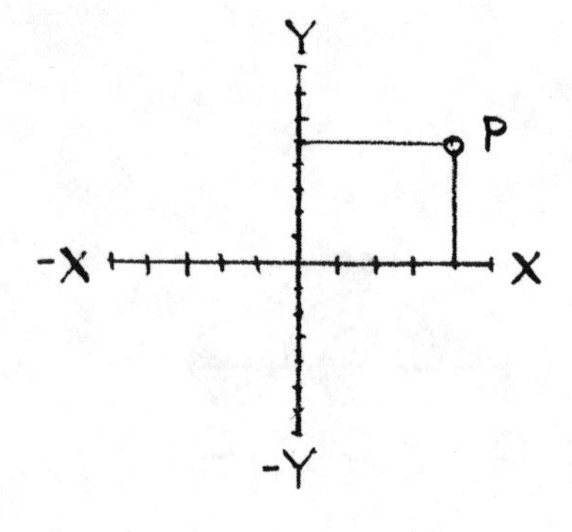


there is only one plane that includes both of them. The distance that P lies to the right of Y can be measured in units of X and is called the X coordinate. The distance that P lies above X can be measured in units of Y and is called the Y coordinate. Thus, the position of P in the plane can be designated by its X and Y coordinates (called Cartesian coordinates).

When we specify the angle at which two lines intersect, as in #1–4, we are indicating their positions relative to each other within a plane. We are doing the same thing when we specify the angles of bonding in a molecule, as in #1–5. All three atoms in a triatomic molecule or ion have to be within a single plane. Thus, the spatial relationships between the atoms in a triatomic molecule involve only two dimensions and we can easily show the relative positions of all three atoms on the flat surface of a page. All three molecules are forms rounded out because of the shape of the electron orbitals; yet the overall shape of the unit may be referred to as flat or planar, just as we described your ball and stick model as flat, because its geometrical units, the points and lines, are all in one plane.

The figures in #1–4 and #1–5 are drawn as though they were positioned in the plane of the page; that is, as though the plane which includes the centers of the three spheres is the plane of the paper. In our work on molecular geometry you will find frequent reference to structures positioned not in the plane of the page but in a horizontal plane perpendicular to the plane of the page.

The position of such a horizontal plane relative to the plane of the page can be visualized if you hold the book as you read this in a vertical position with the cover pressed against the edge of the desk in front of you. The desk top is then a horizontal plane perpendicular to the plane of the page. The two axes of this new plane are designated as X–X and Z–Z, as shown in #1–7*a* and *b*, and the plane is referred to as the XZ plane, to differentiate it from the XY plane (the plane of the page) shown in #1–6. The X–X axis is a horizontal axis common to both planes, while the Z–Z axis is perpendicular to the XY plane and to the X–X and the Y–Y axes, as shown in #1–7*c*, which represents the two planes sharing an X axis. In drawings of this kind the planes are assumed to be translucent and the part of the figure that would normally be hidden from view may be represented by dotted lines.

Place on the desk in front of you the model you made from the three balls and two sticks joined so as to form a 90° angle. Try to diagram what you see as you remain seated at the desk. To indicate that the model is oriented in the XZ plane you have to include some kind of

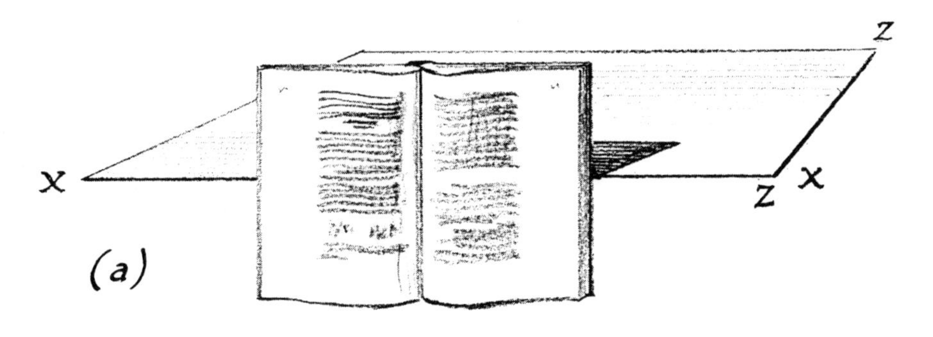
Z
X
Z X
(a)

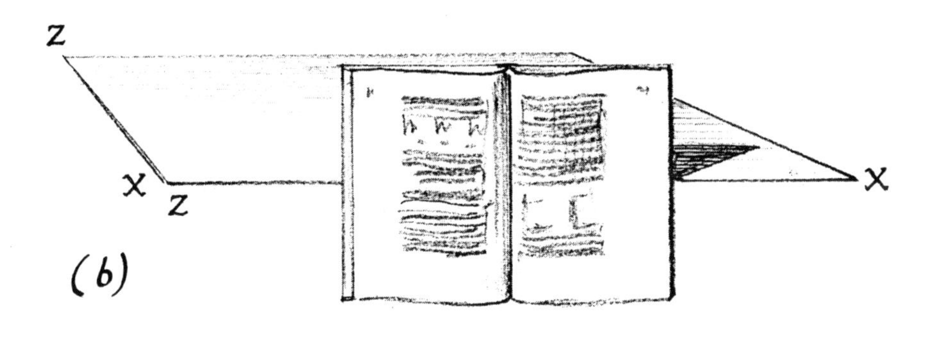
Z
X
Z
X
(b)

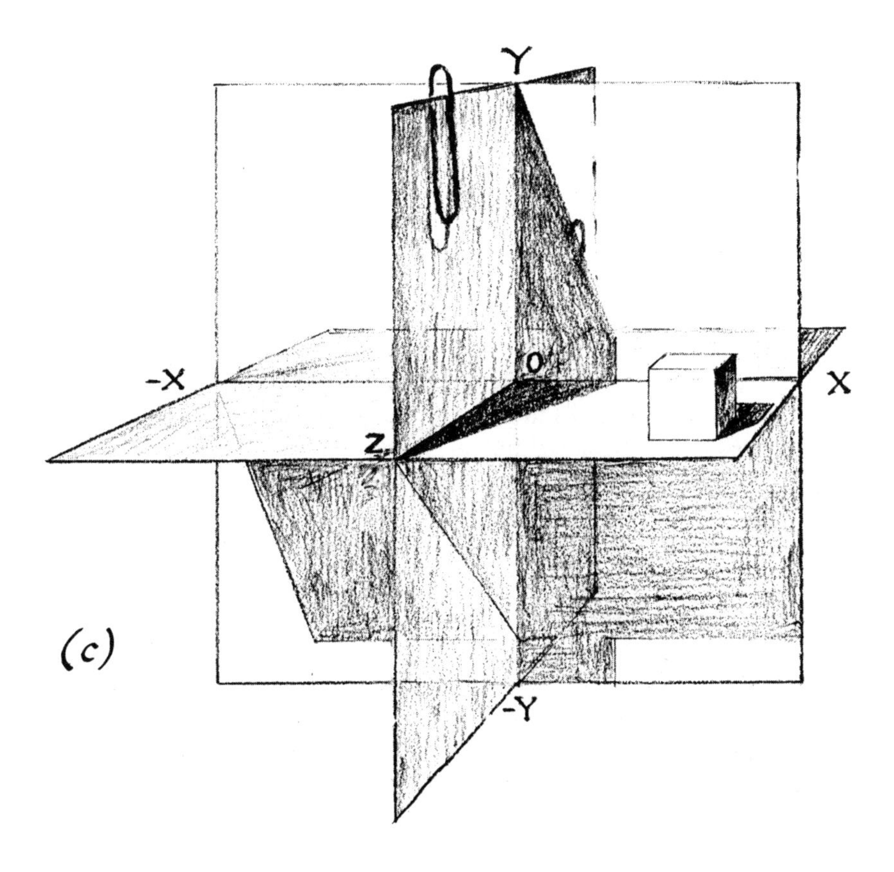
Y
O
-X
X
Z
-Y
(c)

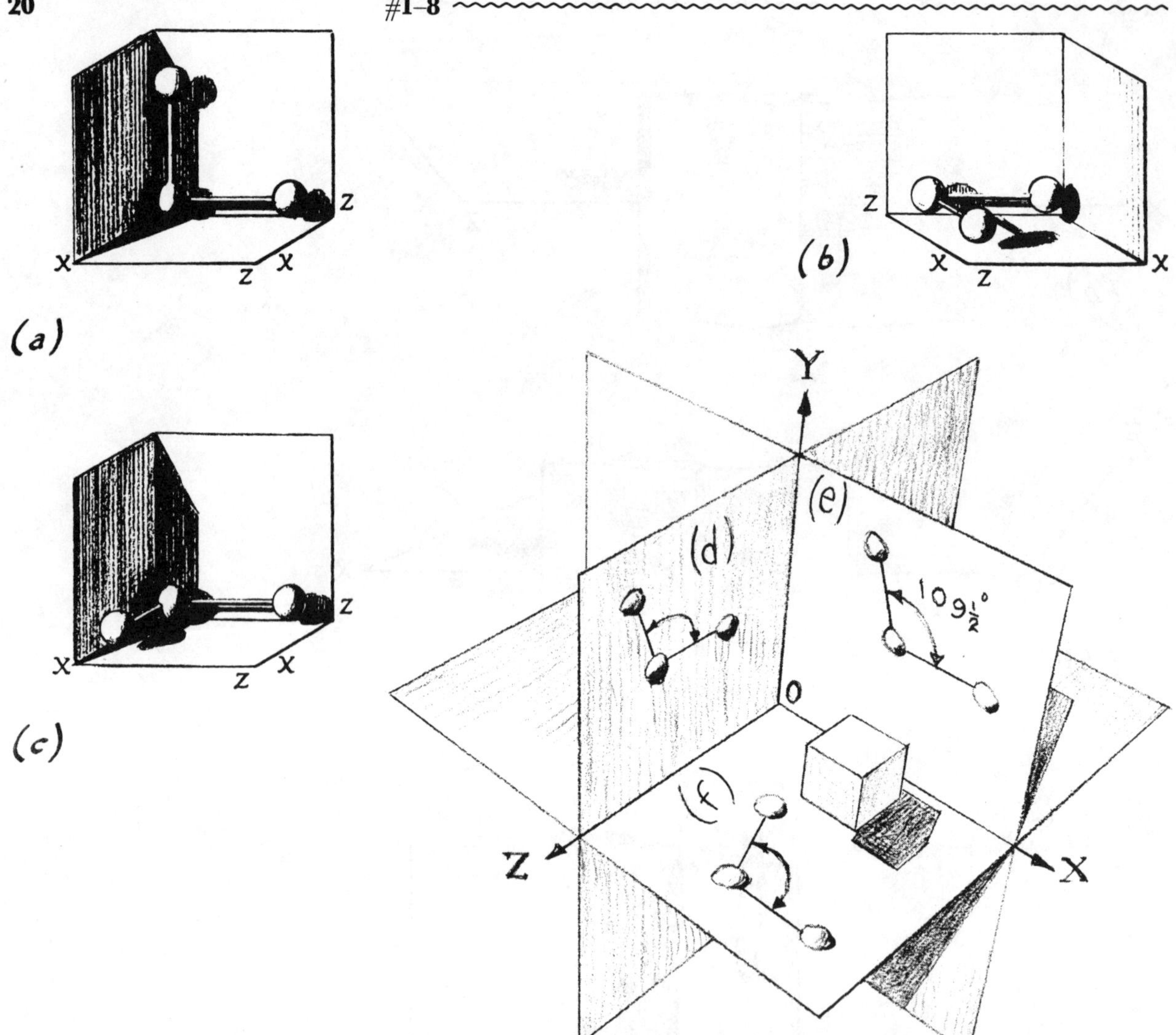


perspective or projection in the drawing. One simple way is to outline the plane XZ, including its axes, as if you were looking at it from either the left or the right side. Projections of the planes are shown in #1–7 from both views, while #1–8*a* and *b* show the model positioned in the XZ plane and #1–8*c* shows the model in the XY plane for comparison.

Although the 90° angles in the projected XZ plane are not actually drawn as 90°, your eye tells you that the axes do form a right angle. Each of the model's "arms" parallels an axis to indicate that the angle of the model is the same as the angle formed by the axes.

If you wish to draw an angle larger than 90°, for example the 109.5° angle common in carbon bonding, draw one arm parallel to one axis of the plane and position the other arm so as to form an angle larger than the projected 90° angle where the axes meet. A 109.5° angle is shown in #1–8 as positioned in plane XZ and in the other planes.

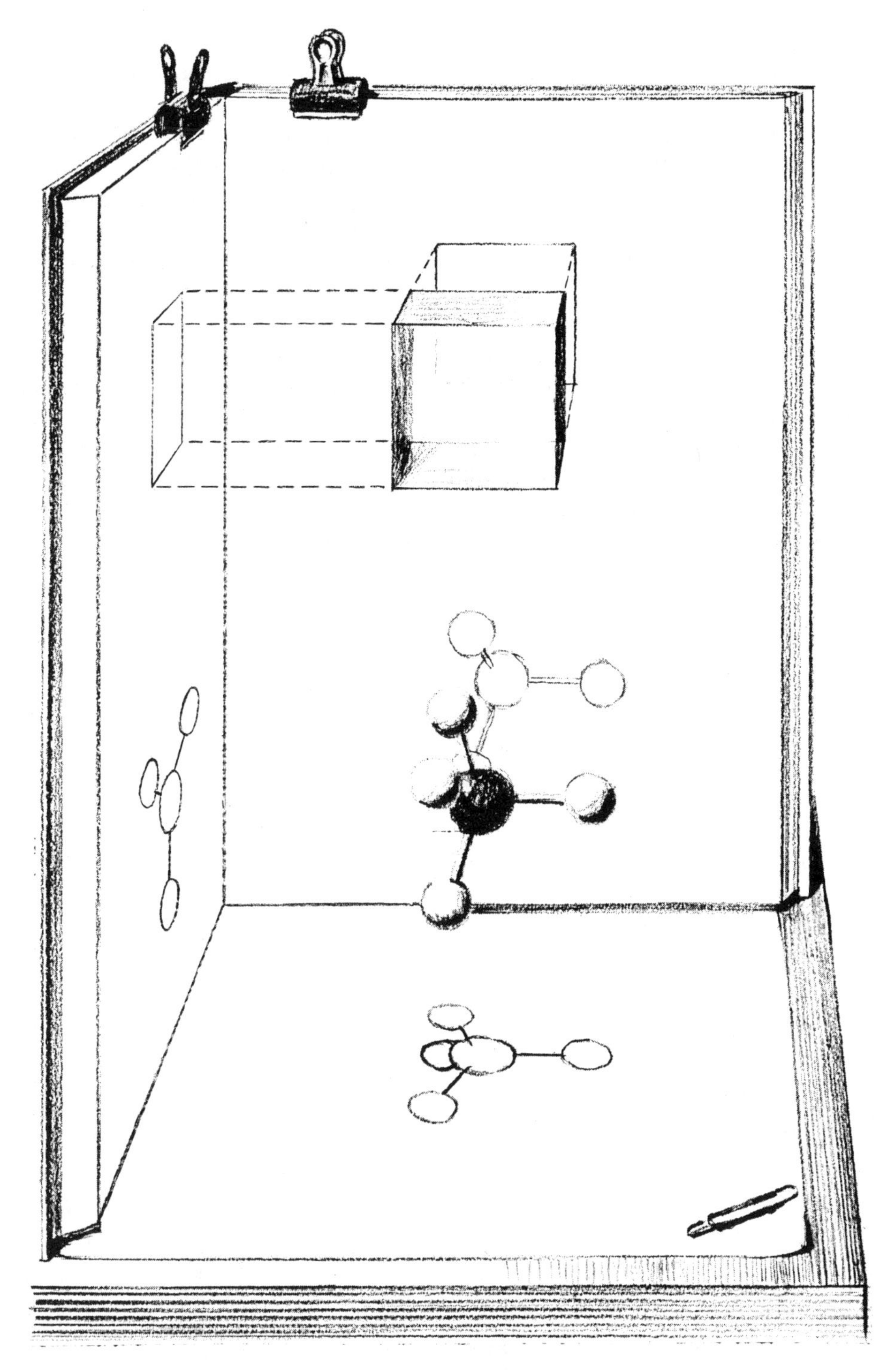

Problems 1 through 6 in §2 provide you with practice in drawing units positioned in both the XY and XZ planes.

*Geometry of Three Dimensions*

Now that you understand the relationship between an XY plane and an XZ plane we can shift our attention from planar structures to those for which we need to consider all three of the dimensions of space. As a guide toward understanding these concepts, convert the two-dimensional model from #1–8 into a three-dimensional structure, as follows.

Place the model on the desk positioned as shown for the XZ plane in either #1–8*a* or *b*. Hold a fourth ball directly over the center one and attach it in this position with a stick. This third stick will form a 90° angle with the top of the desk and with all lines in the plane of the desk. Thus, you have constructed a model in which all three sticks are perpendicular to each other and project in three different directions. All the components of such a structure do not lie in one plane. Therefore, more than one plane or one set of two axes is needed to indicate the spatial relationships between the components of this three-dimensional structure.

The standard set of reference axes used for designating three dimensions consists of three mutually perpendicular axes having the same spatial relationships to each other as the arms of your model except that the axes all extend beyond the central point. Two projections of the reference axes are shown in #1–9. In this book the reference axes, even when only two of them appear in a diagram, are consistently designated as follows:

- X–X axis—the horizontal axis in the plane of the page.
- Y–Y axis—the vertical axis in the plane of the page.
- Z–Z axis—the axis perpendicular to the plane of the page and to both the X–X and Y–Y axes.

Each pair of reference axes determines the position of a reference plane. The three reference planes are designated in this book as follows and are represented in #1–9.

- XY plane—the plane of the page or a plane parallel to the plane of the page.
- XZ plane—a horizontal plane perpendicular to the plane of the page.
- YZ plane—a vertical plane perpendicular to the XY and XZ planes.

(*a*)

#1–9

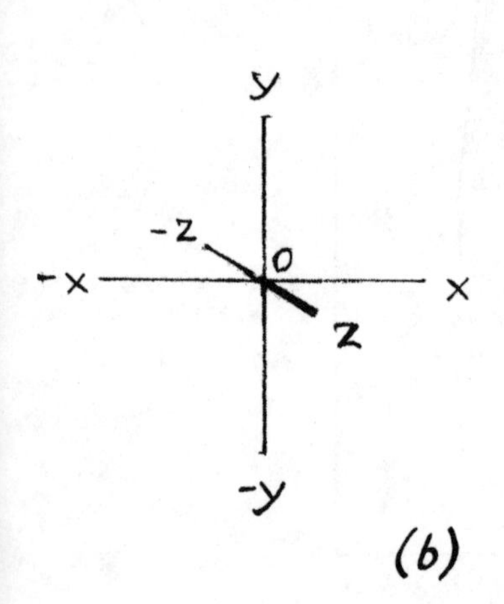


(*b*)

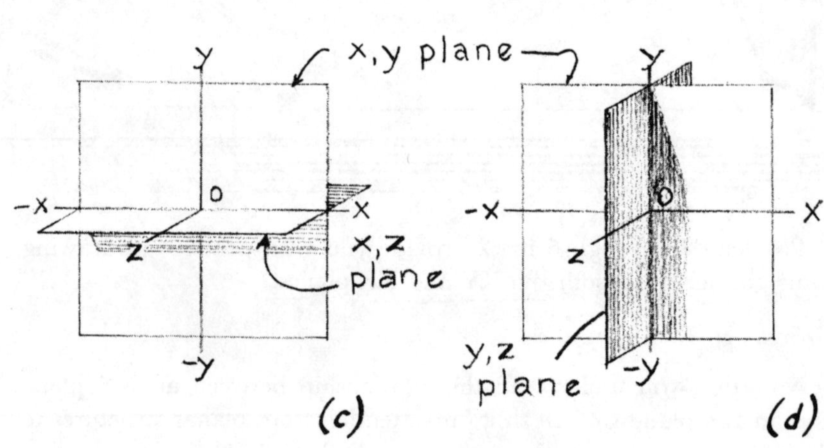


(*c*) (*d*)

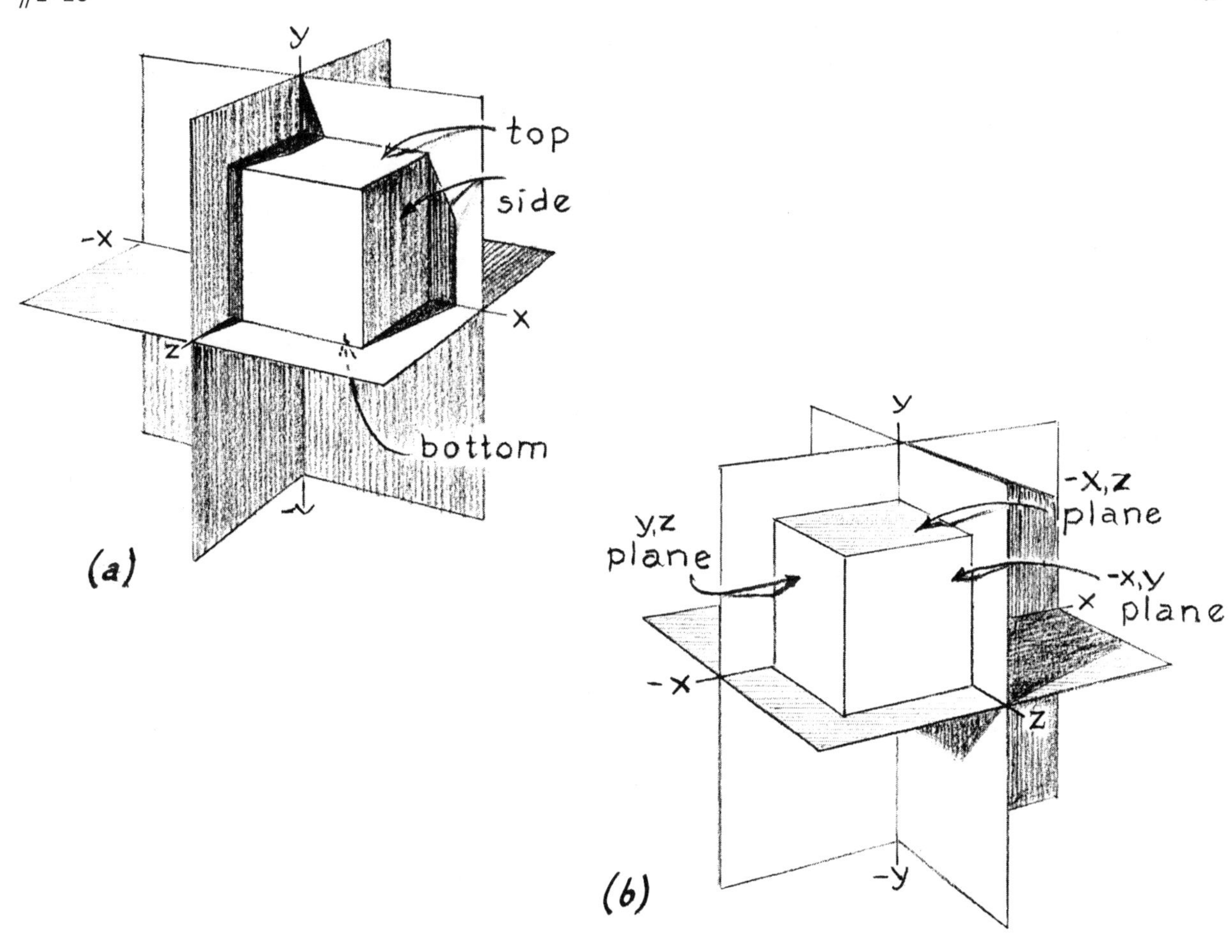
y
top
side
-x
x
z
bottom
y
-x,z
plane
y,z
plane
-x,y
plane
x
-x
z
-y

(a)

(b)

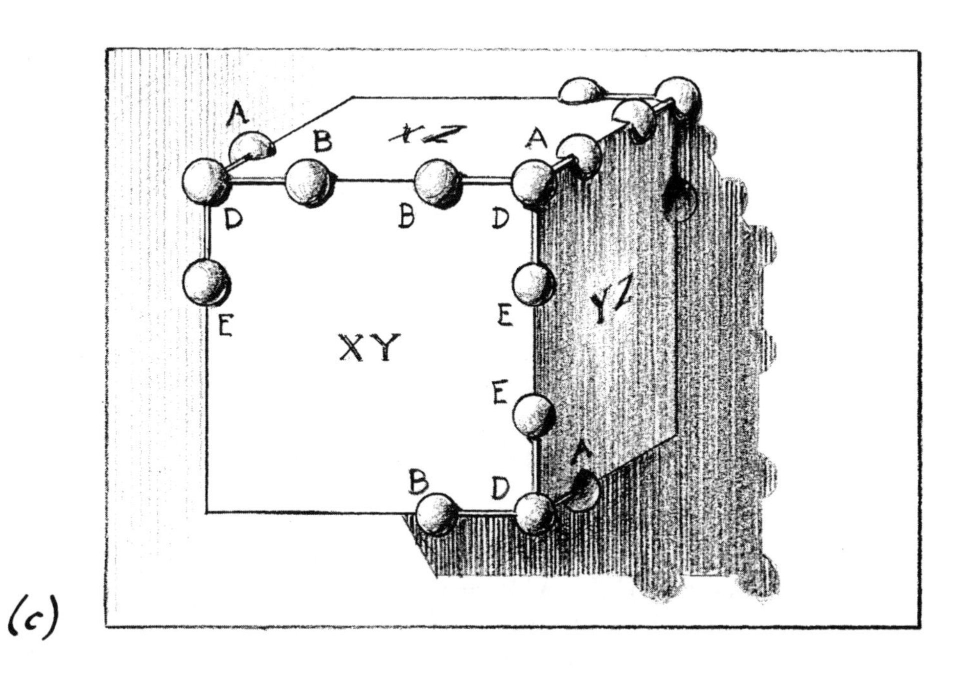
A
B
A
D
B
D
E
E
XY
YZ
E
B
D

(c)

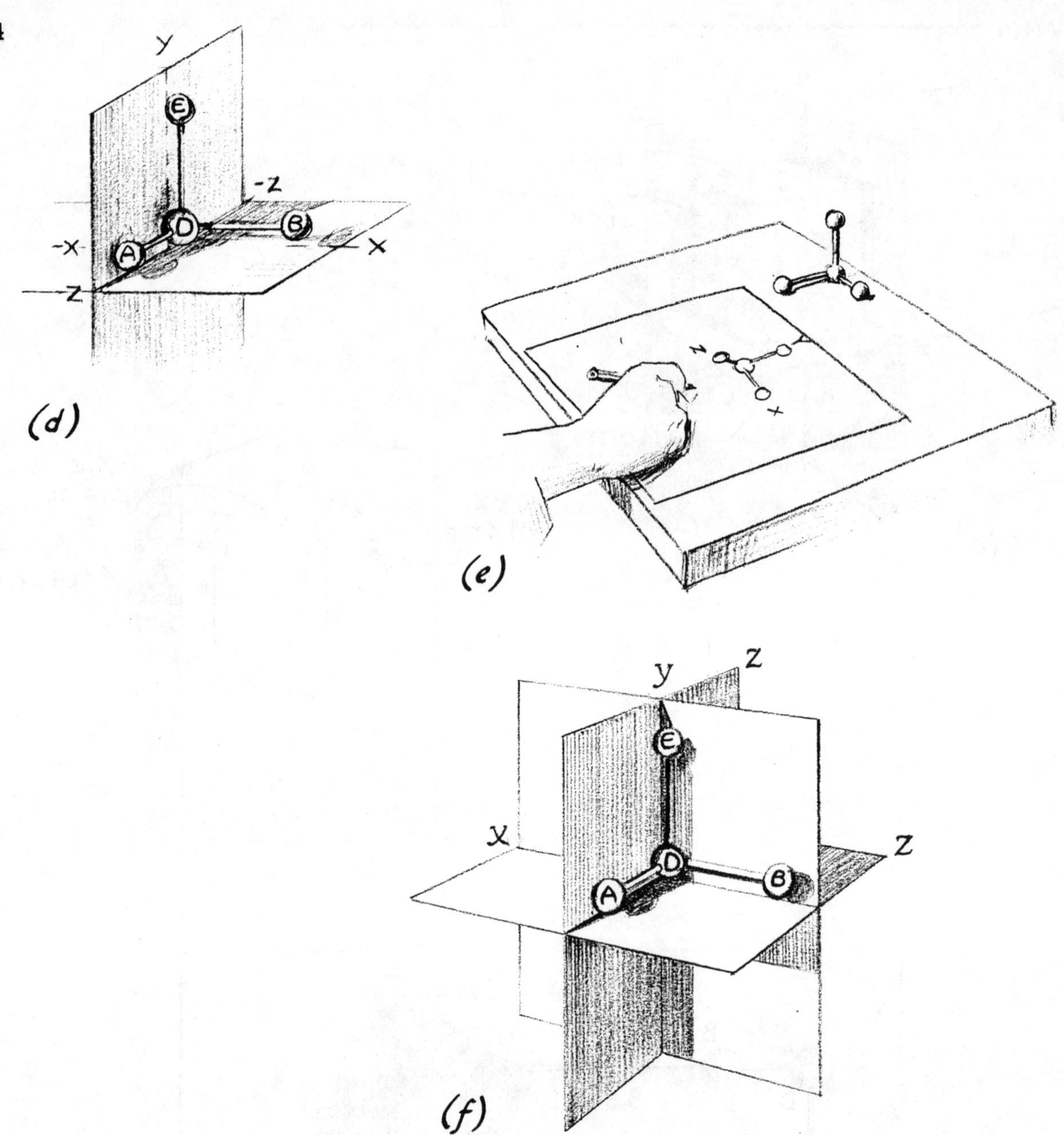


The spatial relationships between these planes can more easily be visualized by considering a cube. The six faces of a cube consist of two XY planes, two XZ planes, and two YZ planes. These are labeled for you in #1–10. The edges of a cube correspond in direction to the reference axes. Your model would fit on the corner of a cube as shown in #1–10.

Reference planes can be utilized to show three-dimensionality in any line drawing. For example, #1–10 depicts your model by indicating that points A, B, and D lie in the XZ plane while points E, D, and A lie in the YZ plane. Alternatively, #1–10 shows A, B, and D in the XZ plane with E, D, and B in the XY plane. This device for showing spatial relationships has been used in #1–11, #1–26, and #9–2. Usually the relationships are clearer if only two of the three reference planes are included along with the reference axes in any one drawing.

It is not always necessary to outline an entire reference plane to indicate position; frequently it is sufficient merely to indicate the reference axis along which a particular part of a structure is aligned or to indicate the angle formed between one part of the structure and the reference axis, as in #1–9.

The reference axes are useful not only for introducing perspective into drawing; they also provide the basis for the verbal description of objects. We refer to the three dimensions they represent when discussing molecular structures, the geometry of bonding, and the spatial relationships between atoms. The terms used for this purpose and their meanings relative to the reference axes are:

- *to the right, to the left:* these terms indicate the position of a point relative to the vertical Y axis and refer to direction parallel to that of the horizontal X axis (#1–11*a*).
- *up, down:* these terms indicate the position of a point relative to the horizontal X axis and refer to direction parallel to that of the vertical Y axis (#1–11*a*).
- *toward you, away from you:* these terms indicate the position of a point relative to the plane of the page (either in front of it or behind it) and refer to direction parallel to that of the Z axis, which is perpendicular to the plane of the page (#1–11*b*).

**#1–11**

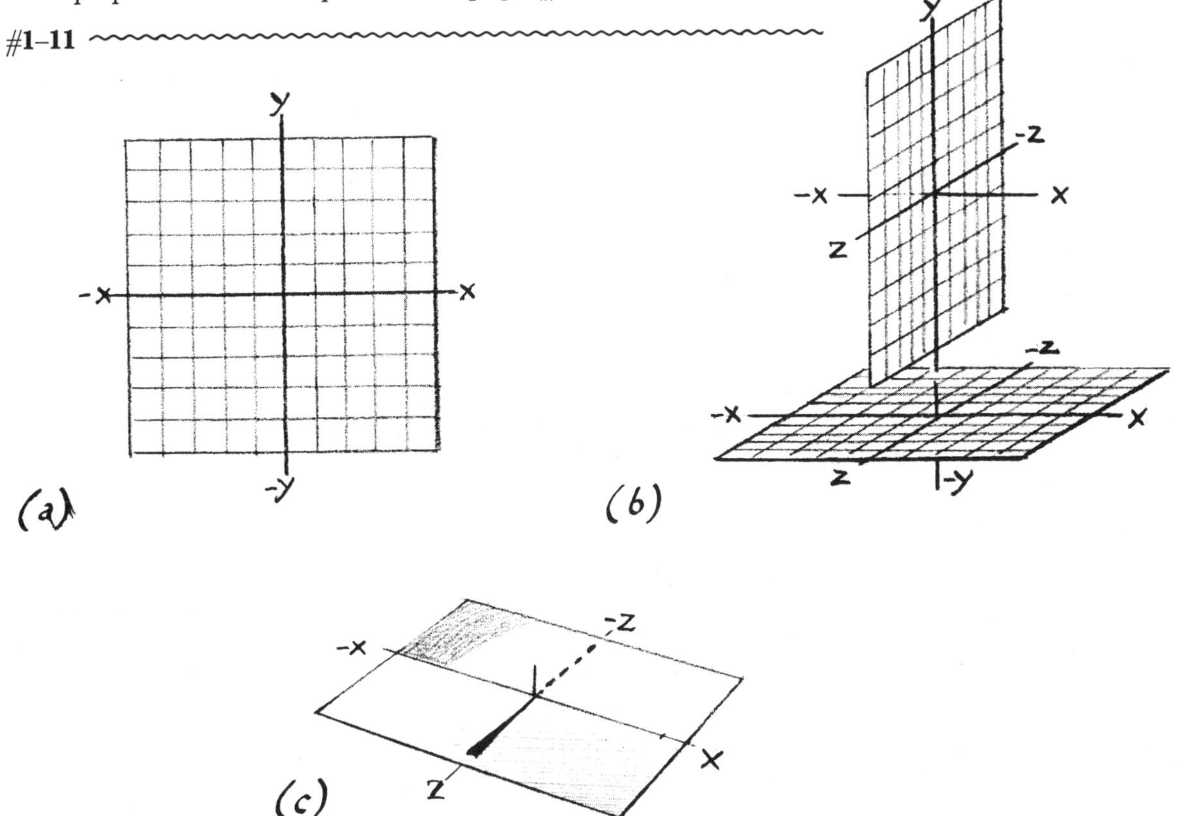

Another way to indicate the relationship between the atoms in a molecule is to position two or more points in the plane of the page (plane XY) and then to indicate the angles at which other elements of the structure project either behind or in front of that plane, or both.

This particular device has been adopted by the organic chemists in the form of projection formulas. A set of conventions has been established for indicating bond directions in projection formulas. Projection formulas for 1,2-dichloropropane are given in #1–12, which

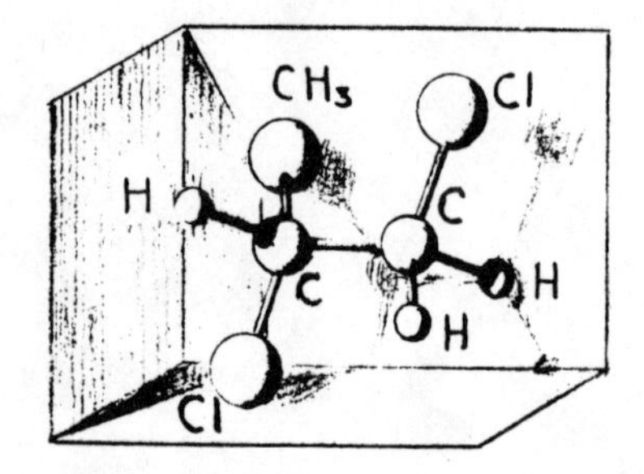


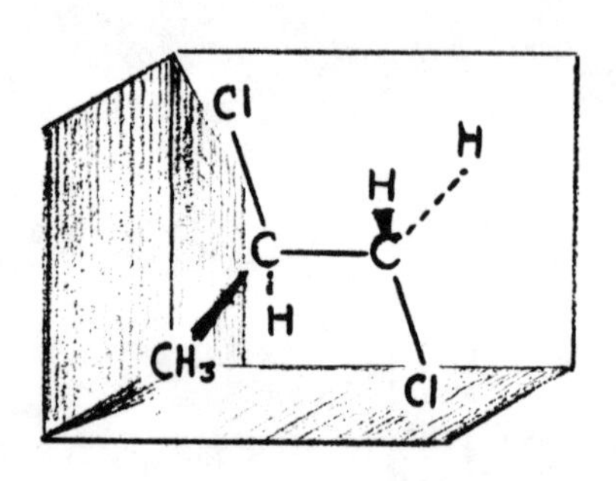


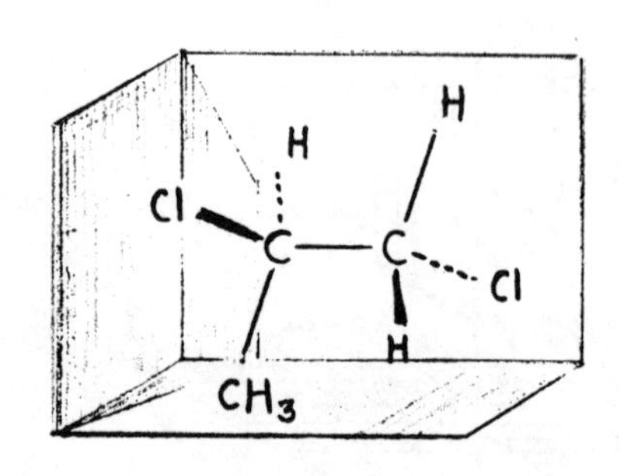


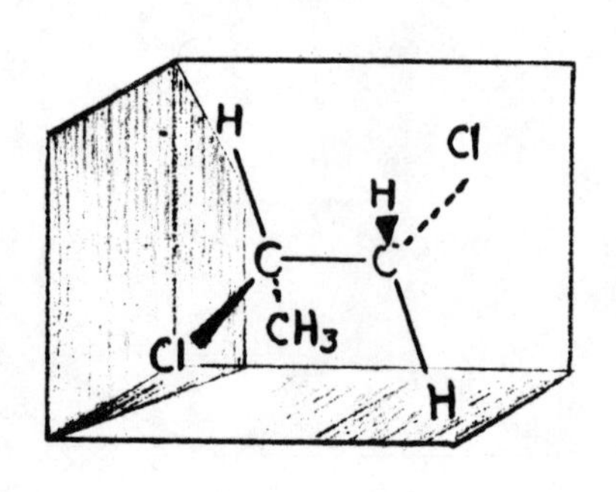


**#1–12**

shows various positions of the atoms, i.e., different conformations of the molecule (see §7). These conventions are:

- A bond in the plane of the page is indicated by a line of even weight or thickness. It may be directed up or down, right or left within this plane.
- A bond that projects out of the plane of the page toward the reader is indicated by a line of increasing weight. The line is wedge-shaped, with the narrow end at the point of origin. The angle that this bond forms with the plane may vary up to and including 90°.
- A bond that projects back of the plane of the page away from the reader is indicated by a dotted line. The angle of projection may have any value up to and including 90°.

The formulas in #1–12 show these conventions as applied to a formula. The application of these conventions to a line drawing of reference axes is illustrated in #1–11*c*.

Before proceeding to study the geometry of bonding at carbon, work problems 7 and 8 in §2.

*Carbon in the Ground State*

To understand the geometry of bonding in the compounds of carbon, examine first the geometry of the electron orbitals in a ground-state carbon atom. There are six electrons around a carbon nucleus:

- Two 1*s* electrons* spin-paired in the spherical orbital closest to the nucleus.
- Two 2*s* electrons spin-paired in a spherical orbital that is larger than the 1*s* orbital.
- Two 2*p* electrons each in a different *p* orbital.

A *p* orbital has two somewhat egg-shaped lobes, both of which are occupied by the single *p* electron. A *p* orbital, unlike a sphere, has a lengthwise direction, which means that, like a stick, it can be oriented with the long axis in any one of the three directions defined by the set of reference axes (#1–13).

* The numbers 1 and 2 designate different energy levels. The two 1*s* electrons are not directly involved in bonding and will be omitted from the rest of the discussion.

#1–13

The two $2p$ orbitals in carbon are perpendicular to each other and are usually represented as $p_y$, drawn vertically, and $p_x$, drawn horizontally. The spatial relationship between the orbitals of the four electrons in the atom are shown in #1–14.

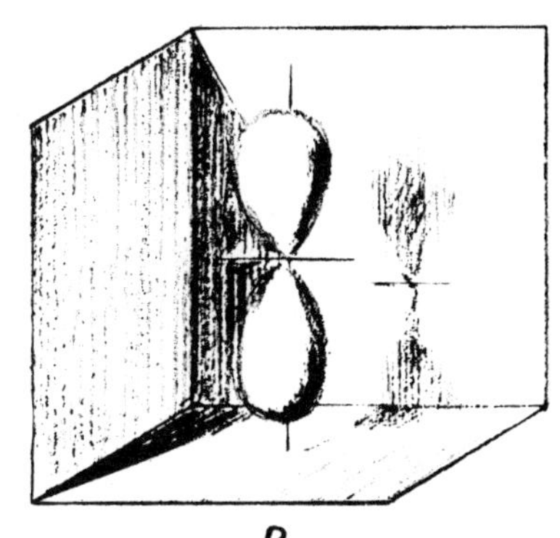

Py

#1–14

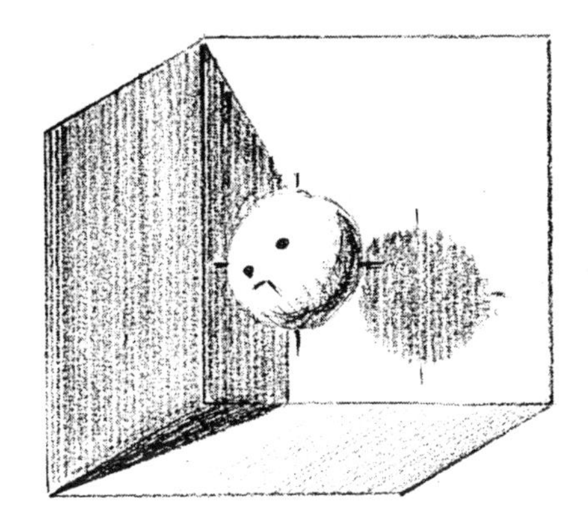

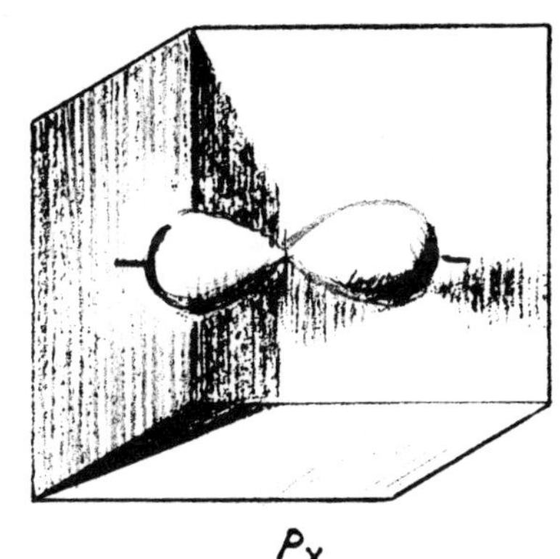

Px

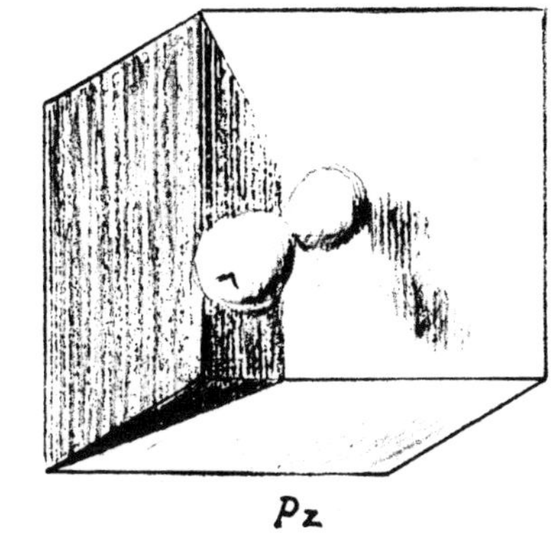

Pz

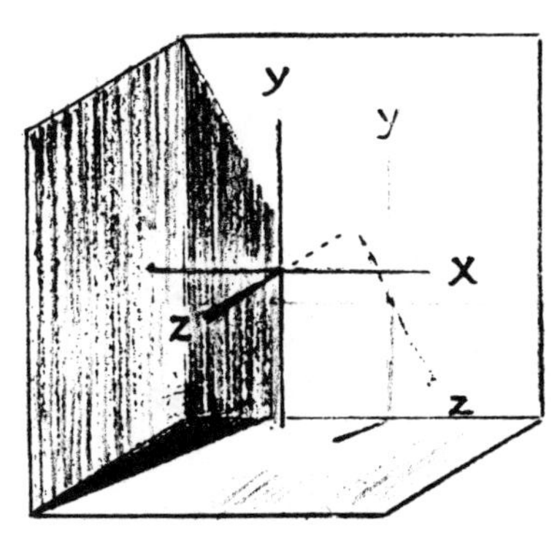


The carbon nucleus is at the center of the sphere. While the $2p$ electrons travel somewhat farther from the nucleus than the $2s$ electrons, the difference has been exaggerated in this diagram to make it easier for you to visualize the relationships. The space occupied by the $p$ orbitals coincides to some extent with the space occupied by the $s$ orbital. To visualize how two orbitals can share the same space, recall that an orbital merely indicates the space within which a particle is moving at a very rapid rate. Remember, also, that we demonstrated (#1–2) how two moving keys could occupy the same space. In this manner two or more electrons can move about within the same space even though their movements describe different patterns.

The overlap of two electron orbitals in the manner described for hydrogen (#1–1) leads to a sigma bond only when each orbital contains a single unpaired electron. Carbon in the ground state has two unpaired electrons, each of which is in a $2p$ orbital. The presence of only two unpaired electrons would seem to indicate that carbon is divalent, i.e., can bond with only two other atoms. If this conclusion were true, the simplest compound between carbon and hydrogen would be $CH_2$.

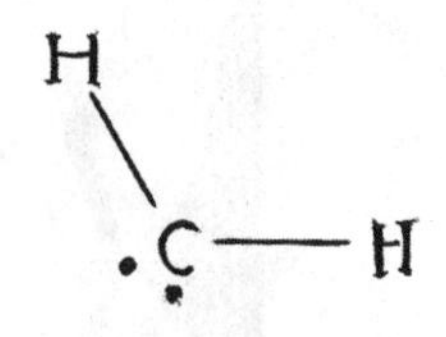


#1–15

However, we find that the simplest molecule formed by carbon and hydrogen is $CH_4$, methane, in which carbon is bonded to *four* other atoms. Methane is frequently represented by the structural formula in

#1–16

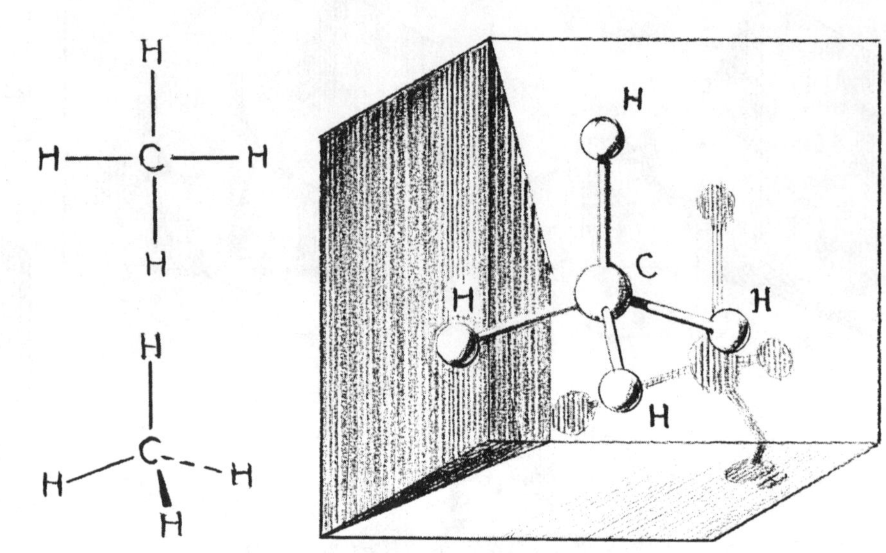


#1–16*a*, but *b* gives a more accurate indication of the spatial relationships between carbon and four hydrogens.

*sp*³ *Hybridization*

We need to explain why, when the ground-state structure of carbon implies divalency, carbon is tetravalent in most of its compounds.

One way for carbon to acquire an electron configuration that includes the four unpaired electrons necessary for the formation of four bonds could be for one of the 2*s* electrons to move into a third 2*p* orbital, giving the configuration 2*s*, $2p_x$, $2p_y$, $2p_z$* as diagrammed in #1–17.

#1–17

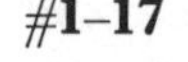

The proposed electron shift would place three single electrons in similar *p* orbitals, while the fourth electron would remain in a spherical orbital. A C—H bond formed by the overlap of an *s* orbital from carbon with an *s* orbital from hydrogen would be shorter than a C—H bond formed by the overlap of a *p* orbital from carbon and an *s* orbital from hydrogen. The circumference of the sphere does not extend as

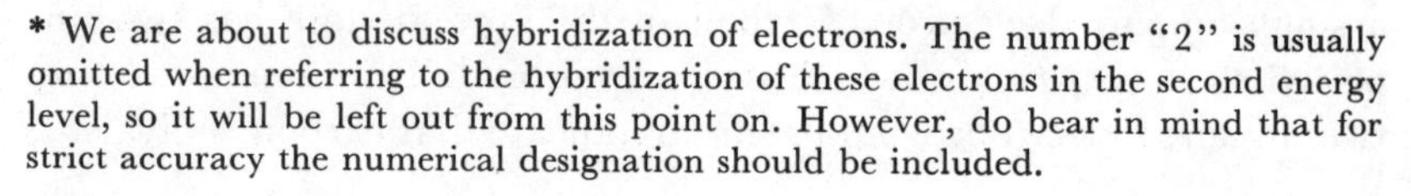

* We are about to discuss hybridization of electrons. The number "2" is usually omitted when referring to the hybridization of these electrons in the second energy level, so it will be left out from this point on. However, do bear in mind that for strict accuracy the numerical designation should be included.

far from the carbon nucleus as do the ends of the lobes of the *p* orbitals. Consequently, overlapping of the two *s* orbitals would place the two nuclei closer together than would overlapping of an *s* and a *p* orbital.

Under these circumstances, three of the bonds in $CH_4$ would be the same length, while the fourth would be shorter. Yet, x-ray determinations show that all four bonds in methane are the same length! Some change other than the mere promotion of an *s* electron to a *p* orbital must have occurred to produce the four equivalent orbitals that make the four equivalent bonds. The additional change is called *electron hybridization* and is understood as follows.

The characteristics of one *s* electron and three *p* electrons blend together to form four equivalent electrons which are called $sp^3$ electrons. The main characteristic of an electron is its energy, which determines the shape of its orbital. Thus, equivalent electrons are at the same energy level and have orbitals that are the same shape and size, but may be oriented differently in space. The shape of an $sp^3$ orbital has been calculated mathematically to be approximately that shown in #1–18.

**#1–18**

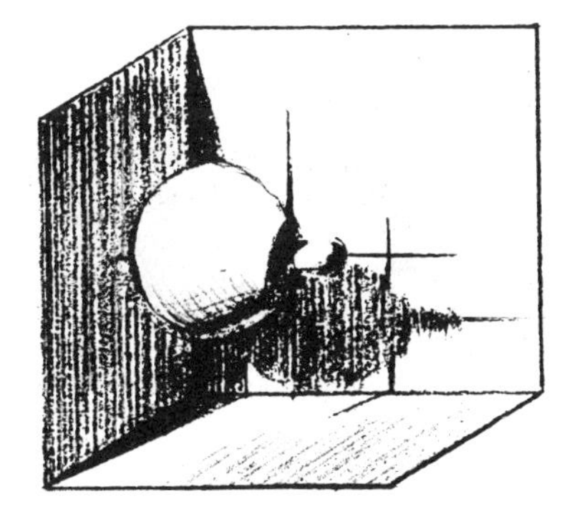

The term $sp^3$ indicates that one *s* and three *p* electrons contributed to the formation of the four identical hybridized electrons. An $sp^3$ orbital is not spherical like an *s* orbital and it does not have two equivalent lobes like a *p* orbital. The new shape combines the properties of both in a ratio of one part *s* to three parts *p*. The carbon to which the four electrons belong is said to be $sp^3$ hybridized and has four $sp^3$ electron orbitals.

The term *hybridization*, used to describe this blending process, comes from biology. A mule is a hybrid of a horse and a donkey; hybrid corn is the product of cross-pollination of two varieties of corn. The following analogy will help you understand hybridization. Suppose that we want to paint four posts and that we have four small jars of paint, one red and three blue. We need all of the paint in the jars to cover all four posts. We want all of the posts to be the same color, so we mix the contents of the four jars together. After they are painted, the posts are neither blue nor red, but a new color with characteristics of both blue and red. We could call the color $rb^3$ but instead we make up a new name for it, purple. The name we use does not really tell us as much about the color as the name $rb^3$.

Each of the four electrons resulting from the blending together of one *s* and three *p* electrons is given a new name, $sp^3$, to distinguish it from the original four electrons. The characteristics of an electron that differentiate it from another electron are the size, shape, and orientation of its orbital, just as color is the distinguishing feature in our analogy.

### *Tetrahedral Bonding*

We now need to determine how the four $sp^3$ orbitals are arranged around the carbon nucleus, because this will determine the positions of the bonds between carbon and other atoms. Our interest need not concentrate on the actual shape of the orbitals, just on their relationship to the carbon nucleus. The small secondary lobe (#1–18) of an $sp^3$ orbital does not play a significant role in bonding so we will leave it out of the discussion. We consider how to place four approximately egg-shaped objects around one point so that their centers are all equidistant from that point.

The spatial or geometrical relationships are essentially the same, whether we represent the orbitals as four spheres or four egg-shaped objects. The concepts may be a bit easier to see with spheres, but both shapes will be used in this discussion.

Place three balls of identical size on a table so that each one touches or is tangent to the other two. Add a fourth ball on top of the original three, forming a pyramidal pile. Now each ball is tangent to the other three and the centers of all four balls are equidistant from one point in the space at the center of the pile.

**#1–19**

Such an arrangement is the *most compact* one for four spheres or sphere-like objects. The "naturalness" of this arrangement even for egg-shaped units can be dramatically illustrated with four egg-shaped balloons.

Prepare two pairs of balloons in which the members of each pair are joined by a rubber band about one inch in length. Place the two pairs on a flat surface in such a position that the rubber bands form a cross (#1–20). You will have to stretch the rubber bands in order to hold all

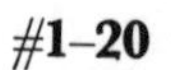

**#1–20**

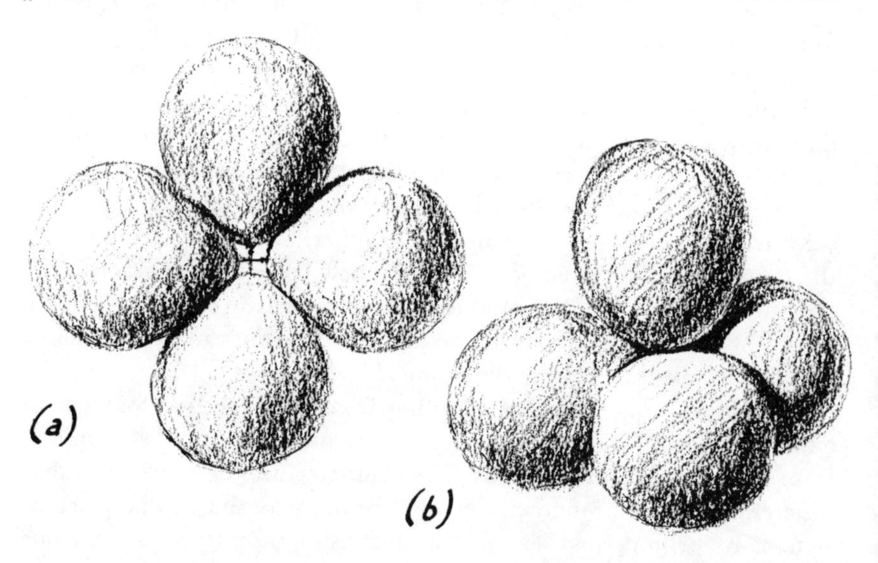


four balloons in one plane with the two pairs at right angles to each other. This is *not a stable arrangement* and when you release the balloons they will regroup in the single, stable, pyramidal unit shown in #1–20*b*. This cluster of balloons provides a good representation of the spatial relationships between the four $sp^3$ orbitals in tetravalent carbon.

The pyramidal arrangement of the four $sp^3$ orbitals is the most important factor determining the angle of bonding in an $sp^3$ hybridized carbon. The point at the center of the pile equidistant from the centers of all four spheres represents the position of the carbon nucleus. In methane each of the four $sp^3$ orbitals overlaps the $s$ orbital from a

**#1–21** 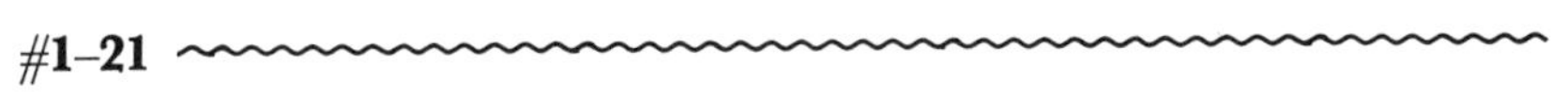

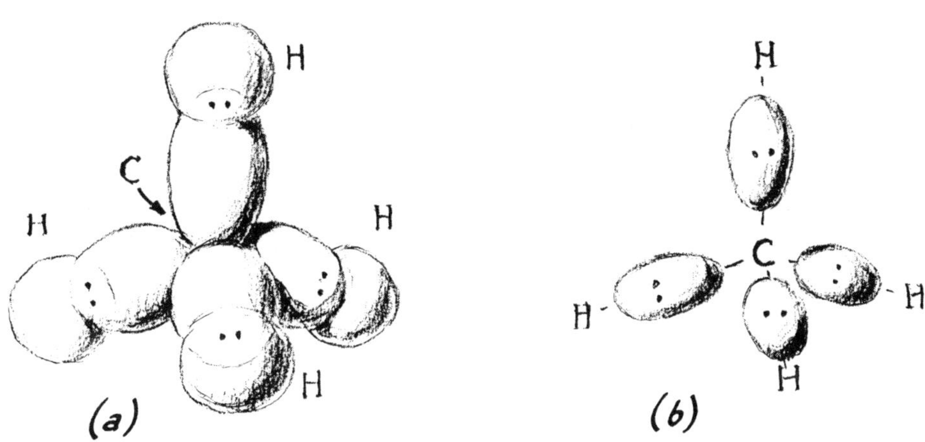


hydrogen atom (#1–21*a*). The unpaired electrons in each set of overlapping orbitals interact. They become spin-paired and form a molecular orbital, thus producing $\sigma$ bonds between carbon and each of the four hydrogens. Each molecular orbital is represented by an ellipsoid in #1–21*b*.

The direction of each $\sigma$ bond is determined by the direction of a line drawn between the nuclei of the two bonded atoms. This line is the main axis of the molecular orbital of the two atoms and coincides with the main axis of the original $sp^3$ orbital. Thus, if we can determine the size of the angles formed between the lines drawn from the center of each sphere to the central point of the pyramid, we will know the angle of bonding in $sp^3$ hybridized carbon. These relationships are diagrammed in #1–22, where the last figure omits all orbitals and

**#1–22**

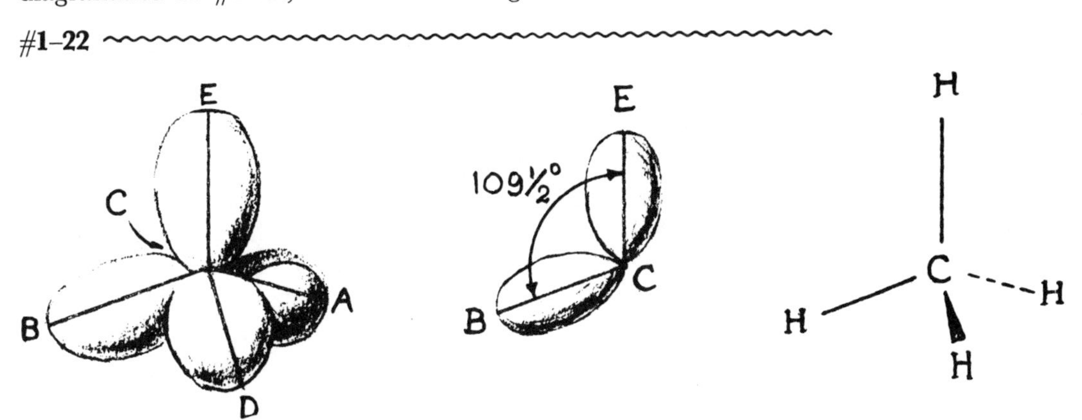


merely indicates the directions in which four $\sigma$ bonds would project from carbon.

The four balls in the pyramid do not lie in one plane. Thus, the lines drawn from the central point of the pyramid to the center of each sphere do not lie in one plane. Therefore, the spatial relationships between these lines (#1–22), which represent the four bonds, have to be described in three-dimensional terms.

There are six angles in this figure, ECA, ECB, ECD, BCA, BCD, and DCA. All the angles are equal, but because of the three-dimensionality of the structure we cannot measure the angles directly from the drawing as we could in #1–5. The value has to be calculated using precepts of solid geometry and trigonometry. Such calculations show that the angle is 109.5°.

While we will not actually do these calculations, leaving you to accept this value without determining it, the following models will help you visualize the spatial relationships that arise because the angle between any two of the four bonds is 109.5°.

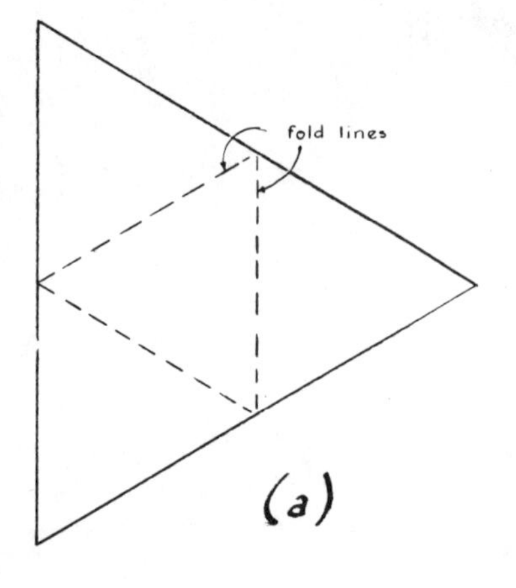


(a)

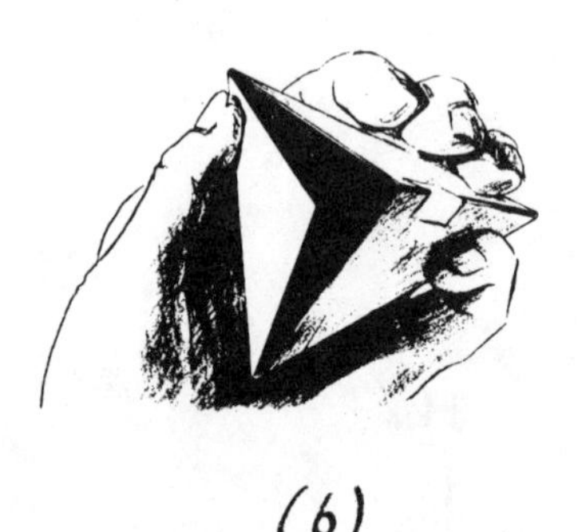

(b)

**#1–23**

On a piece of cardboard draw an equilateral triangle (about two inches on a side). Draw three additional equivalent triangles, attached to the first as shown in #1–23. Cut out this four-triangle figure and bend upward at the lines between the triangles. Where their sides meet, join them with bits of tape. You now have a four-sided figure, a *tetrahedron*, with equivalent faces. Compare the shape of the pyramid of balls (#1–19) with the shape of your tetrahedron. They are similar. If the balls were the correct size you could just fit them inside the tetrahedron, and each ball would touch three sides.

Imagine the carbon nucleus at the center of a tetrahedron. Then each of the four $sp^3$ bonds extends from the carbon nucleus through another nucleus toward a different apex of the tetrahedron. (In geometric terms, the carbon nucleus would be at the point where the four altitudes of the tetrahedron meet.)

To visualize this relationship construct a tetrahedron from six drinking straws (preferably plastic ones) and some string. Make a triangle out of three straws by stringing them like beads, arranging them as in #1–24 and tying the string. Attach a piece of string a little longer than a straw to each apex of the triangle and thread each one through one of the three remaining straws. Raise these three straws and tie them together at the top of the tetrahedron, which has the original triangle as its base. You now have the outline of a tetrahedron.

Tie a piece of small, round elastic between the apexes marked B and E in #1–24. Make this slightly longer than one of the straws so that it is held in place without any tension. Cut a second piece of elastic. Tie one end to apex D. Pass the other end between the first piece of elastic and the straw alongside of it. Tie the other end to apex A. Adjust the elastics so that the point where they cross is in the middle of the tetrahedral figure.

The central point at which the elastics should cross represents the carbon nucleus. The elastics are the four $sp^3$ bonds projecting from the carbon. The angle formed by each pair of bonds has been calculated as 109.5°, frequently called the *tetrahedral angle*. If you take a protractor

#1–24

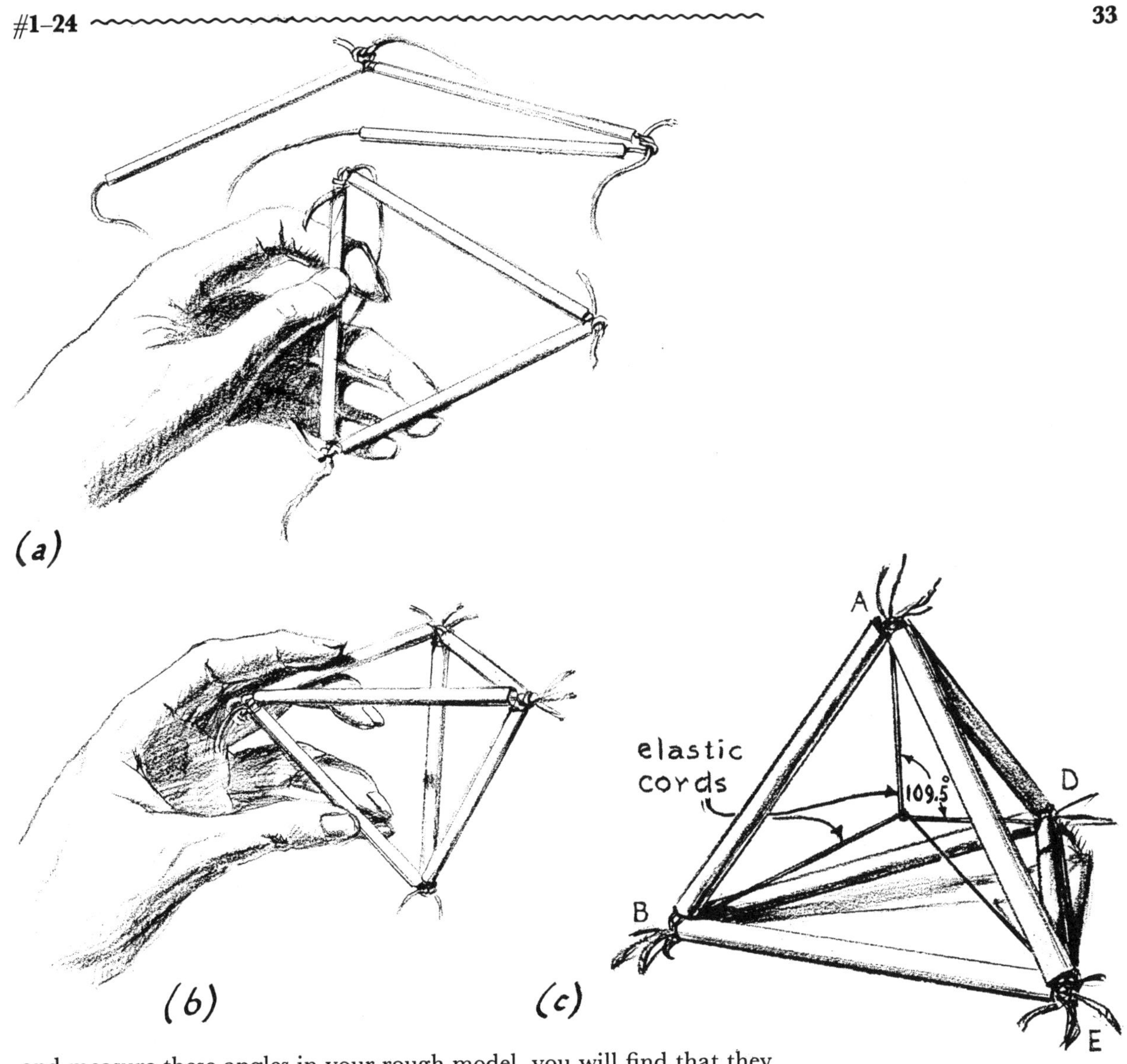


and measure these angles in your rough model, you will find that they are surprisingly close to the correct value.

#1–25

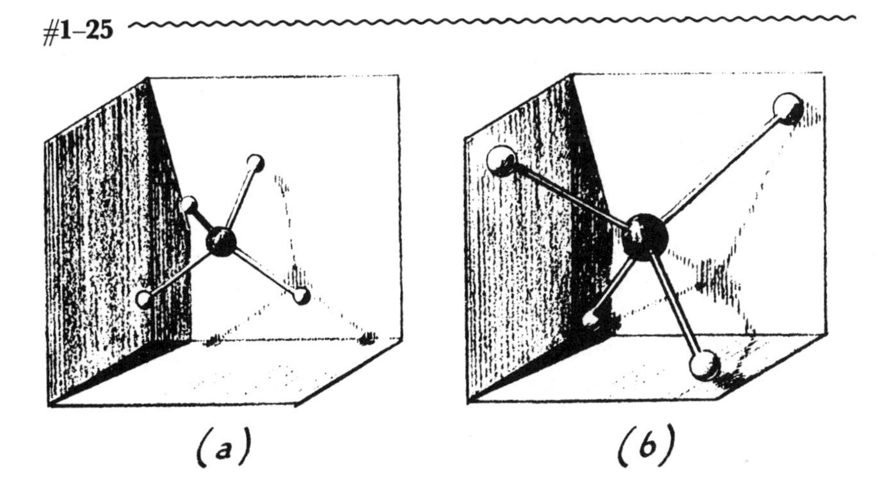

The structure you have made can be analyzed in terms of reference planes. The figure in #1–25 has essentially two components—two 109.5° angles joined together at their apexes so that the plane of one angle is perpendicular to the plane of the other angle. This relationship is shown in #1–26. One of the 109.5° angles has been drawn in the XY plane in such a position that the Y–Y axis bisects the angle. A second 109.5° angle has been drawn in the YZ plane (#1–26*b*). Again the angle has been positioned so that the Y–Y axis bisects it. In *a* the angle is facing down while in *b* it is facing up. The Y–Y reference axis is common to both the XY and the YZ planes. If we position the angles perpendicular to each other in such a way that they have a single Y–Y axis, we obtain the structure shown in #1–26*c*, which shows the four directions of tetrahedral bonding in terms of two reference planes.

**#1–26**

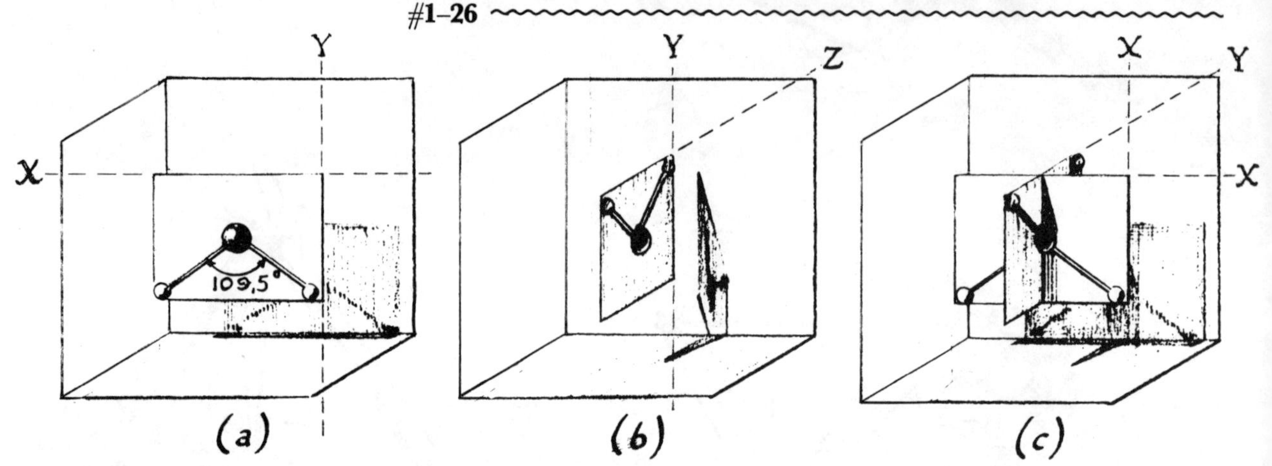


The shape of a molecule of methane, $CH_4$, is such that the unit can be enclosed within a tetrahedron as the pyramid of balls can. The geometry of the tetrahedron determines the angles of bonding for an $sp^3$ hybridized carbon atom. Thus, a carbon so bonded and its angles of bonding are referred to as tetrahedral in arrangement.

A tetrahedron does not necessarily have all four faces (and altitudes) equivalent as your model does. We will consider later a tetrahedral structure in which three faces are equivalent but the fourth has different dimensions. This structure, shown by the bonding of some nitrogen compounds, may be described as pyramidal to indicate that only three sides are equivalent.

The structural similarity between a tetrahedron and a molecule of methane permits the determination of the angles of bonding of $sp^3$ hybridized carbon. The carbon nucleus is at the center of the tetrahedron. That position is determined by the four altitudes of the tetrahedron; an altitude is a line perpendicular to a face and running to the opposite corner or apex of the tetrahedron. In a tetrahedron with equivalent faces, the four altitudes pass through the center of the structure, or a carbon nucleus. From that point toward each of the four corners of the tetrahedron runs the axis of a bond from carbon. The angle between any two such axes in a tetrahedron with equivalent sides is 109.5°.

Thus, when you read and write the chemical formula for methane shown in #1–27 you should keep in mind that it is a short-cut notation that does not show the true geometrical relationship between carbon and the four hydrogens.



**#1–27**

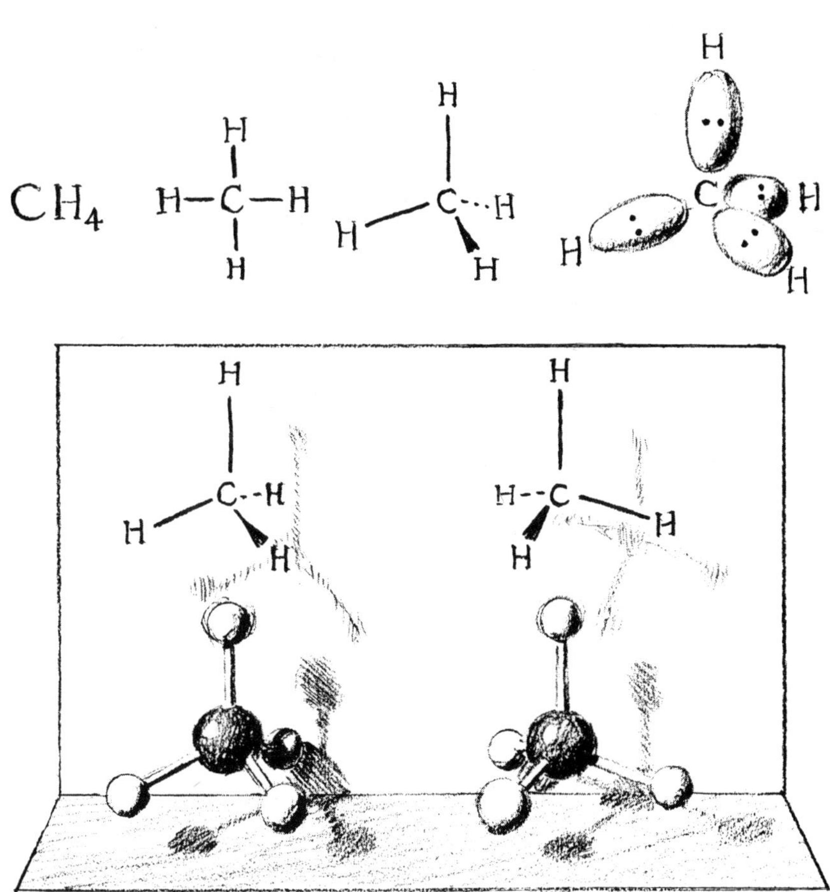


The atoms bonded to the carbon need not be hydrogens. They can be other carbons, or oxygen, nitrogen, chlorine, etc. The bonding will still occur at tetrahedral angles. There may be a slight deviation from 109.5° because of mutual repulsion between atoms and because of crowding. However, the overall form of a bonded $sp^3$ hybridized carbon unit is tetrahedral.

* * * * *

So basic to an understanding of carbon chemistry is visualization of the tetrahedral arrangement that the construction of additional models is advisable. Instructions follow.

The figure #1–28 provides your pattern for a model. You need additionally: some heavy paper or light cardboard—a letter-file folder is of a good weight; some transparent tape; short pieces of wire—pipe cleaners are good; a ruler and some implement such as a penknife to score and cut the cardboard. Proceed as follows:

- Lay out on the cardboard a figure as in #1–28 with four parts. (Each of the four is composed of three triangles, with the measurements given, the triangles so arranged that each part has two points.)
- Rule in the lines—some dotted, some solid, as noted in #1–28.
- With a knife and ruler, score these lines (lightly!), so that the various parts will bend readily.
- Cut or punch out the small round central "spot" in each of the four parts.
- Starting from either end of the four-part piece: (*a*) Bend *up* along the dotted lines of a piece until the two points meet. A bit of tape across the back will hold the points together. (*b*) Bend *back* along the solid line between every two parts (until two such small triangles are back to back).
- Test your folding of the structure by making sure that all four punched-out centers are together. Then use bits of tape to hold the whole together and make the structure a firm one.

2.8284 · 3 · 19½° · 35¼ · 115¾° · 109.5 · 1.06 · a b c d e f g h i j k l

9 units @ 1 each
Cut out on heavy lines.
Fold up on dotted lines.
Fold back on full lines.

A

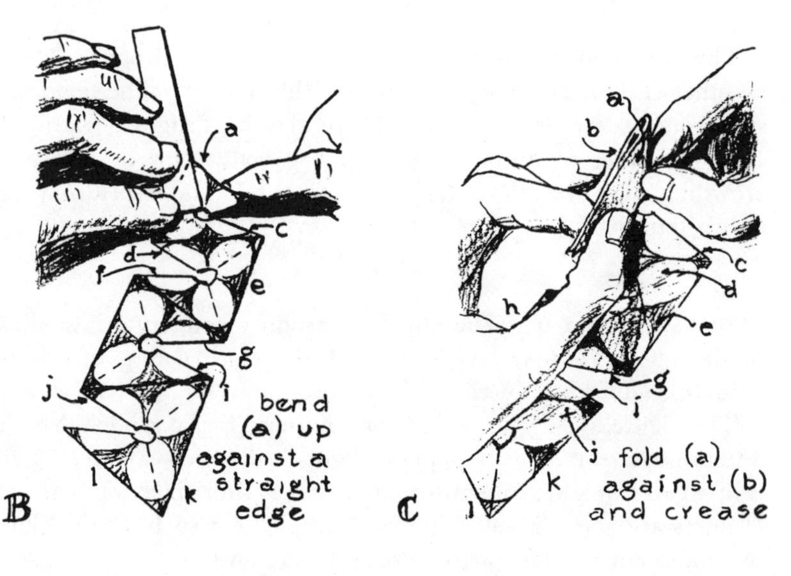


B C

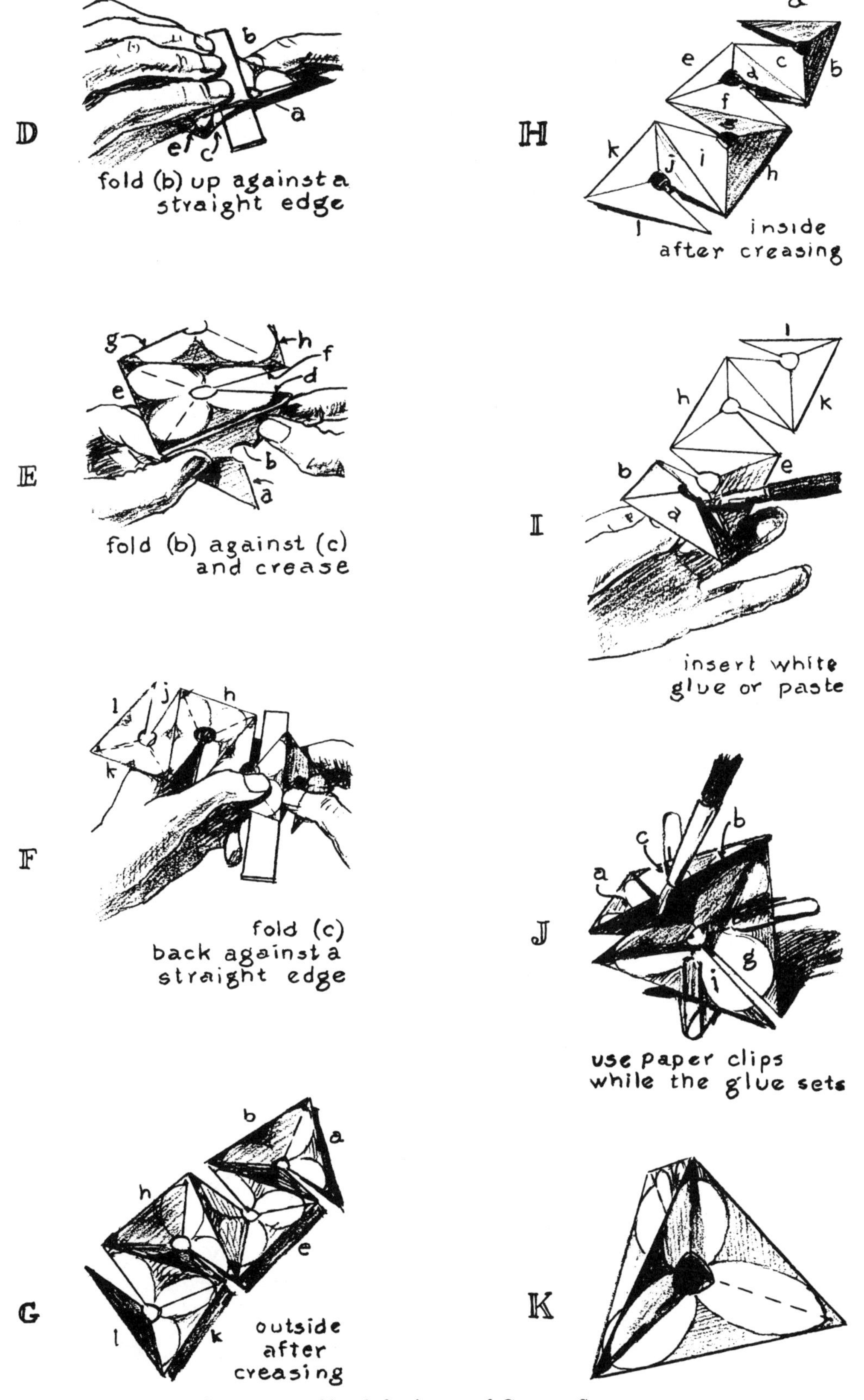


See center of book for insert of *Cooper Structure*.

The form you have constructed (called a *Cooper Structure*) illustrates the geometry of an $sp^3$ hybridized carbon atom. This model with its four recessed sides shows the edges or outline of an equilateral tetrahedron and the axes of four $sp^3$ orbitals projecting from a carbon nucleus at the center towards the four apices of the tetrahedron. Any two of the orbital axes form an angle of 109.5° where they intersect at the nucleus.

Carbon has this electron configuration only when it is bonded to four other atoms as in $CH_4$, $CH_3CH_3$, $CH_3CH_2OH$, $CH_3CH_2CH_2NH_2$, etc. An $sp^3$ orbital overlaps end-to-end with an orbital of another atom to produce a sigma bond. Direct end-to-end overlap causes the axis of the sigma bond to coincide with the axis of the $sp^3$ orbital involved in the bond formation. Therefore, the angle formed by any two of the four sigma bonds is 109.5°.

Two *CooperStructures* can be joined by running a double pipe cleaner from one central hole to the other along the axis of an orbital from each structure (best done if the two orbitals are cut or split at the end). Two or more *Cooper Structures* so joined illustrate the effect of tetrahedral bonding on the molecular geometry of compounds containing more than one carbon. Make a second *CooperStructure* and join it to the first so that you can visualize the arrangement of the atoms in $CH_3CH_3$.

One more model unifies the picture by emphasizing this relationship between the bonds in a tetrahedral carbon. Make a cardboard model of such a carbon structure as follows: Using stiff cardboard, such as you might get from the back of a pad, cut out two pieces with the shape shown in #1–29. Make a cut about an eighth of an inch wide of such length that it extends at least half way across the piece of cardboard at the point indicated in the drawing. The direction of this cut should be such that it would bisect the H–C–H angle at C if it were continued all the way.

**#1–29**

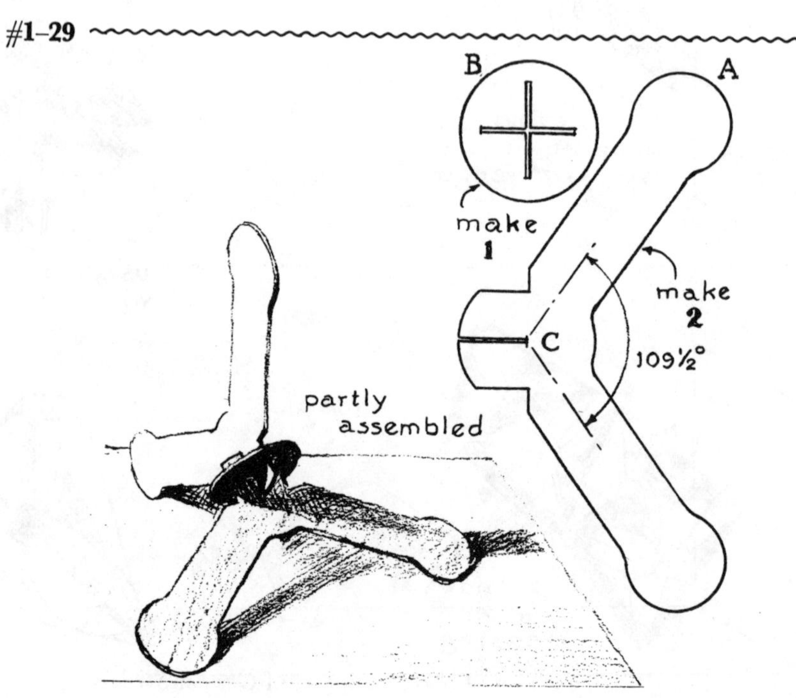

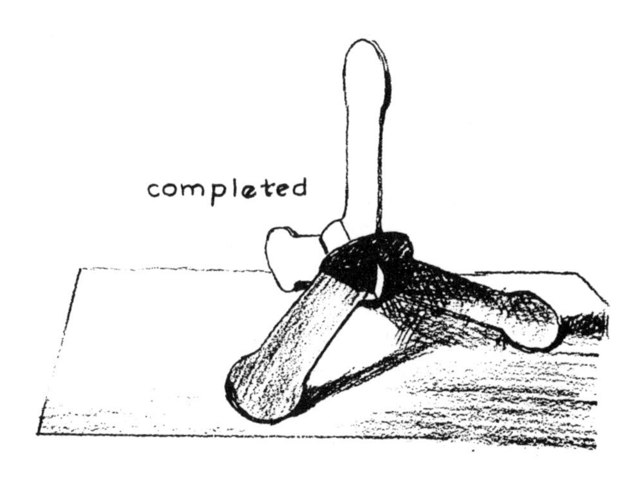


To assemble: Insert the split end of one of the two A pieces into the vertical cut in B. Insert the second A piece into the horizontal cut in B. Push both pieces as far towards each other as possible. Each arm of the structure forms an angle of 109.5° with each other arm. You have a model that shows the orientation of the bonds in an $sp^3$ hybridized carbon. You will frequently have to draw this form, a process simplified if you keep in mind that the model is constructed from two planar pieces at 90° to each other.

The effect of tetrahedral bonding on the structure of molecules with carbon chains is illustrated in #1–30. While the formulas we use for

**#1–30**

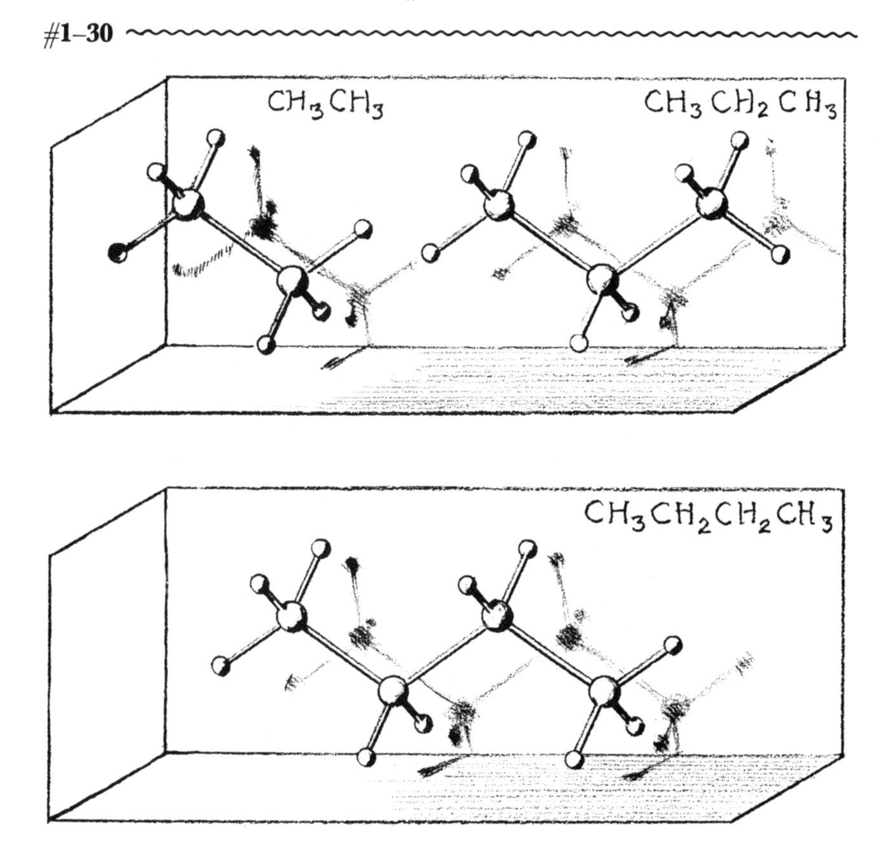

these compounds (shown at the top in #1–30) seem to indicate a straight chain, it is necessary to keep in mind that the actual three-dimensional relationships lead to a zigzag effect.

Do problem 9 in §2, which asks you to draw the model of tetrahedral bonding in sets of reference planes different from those in #1–26. Problem 10 should also be done now.

### $sp^2$ *Hybridization*

Sigma bonding with $sp^3$ hybridized electrons is only one of the three types of bonding that occur between carbon atoms in organic compounds. There is also bonding involving $sp^2$ and $sp$ electrons. The geometry of bonding varies according to the nature of the electrons doing the bonding. We shall first consider the effect of $sp^2$ hybridization.

We learned that hybridization of all four electrons from the excited* configuration $s$, $p_x$, $p_y$, $p_z$ (#1–17) leads to $sp^3$ electrons and tetrahedral bonding. Bonding also occurs during which only three of these four electrons undergo hybridization. The one $s$ electron and two of the $p$ electrons blend their characteristics into three equivalent $sp^2$ electrons, leaving one $p$ electron to continue moving about within its bilobed orbital. It is as if we decided to paint one of the posts in our previous analogy blue and the other three the shade of lavender obtainable from one part of red and two parts of blue paint ($rb^2$).

Calculations reveal that an $sp^2$ orbital is similar in shape to an $sp^3$ orbital but does not extend as far from the carbon nucleus. This difference arises from a ratio of $s$ to $p$ that is 1 to 2 rather than 1 to 3. The arrangement of three $sp^2$ orbitals around a carbon nucleus can be discussed in terms of spheres and then the observed relationships can be applied to the sphere-like, somewhat egg-shaped orbitals.

Place three balls of identical size on a flat surface so that each one is tangent to the other two. These spheres are all in one plane; the centers of all three spheres are equidistant from a point in the middle of the space between the spheres. The distance between this point and the center of each sphere can be represented as a straight line. All the lines are in a single plane so we can measure the angles they form with each other directly on the drawing. The 360° of a full circle has been divided into three equal parts so each angle is 120° as in #1–31.

Similarly, the approximately egg-shaped $sp^2$ orbitals assume a position around the carbon nucleus that leads to angles of 120° between the lines drawn from the carbon nucleus along the main axis of each orbital. These lines represent the direction of the three $\sigma$ bonds formed by an $sp^2$ hybridized carbon.

The fourth electron, which does not become involved in the hybridization, continues to occupy a bilobed $p$ orbital perpendicular to the plane of the three $sp^2$ orbitals and extending both above and below it, as in #1–32.

* The term *excited* indicates that the electron configuration is not that of the ground-state atom.

#1–31

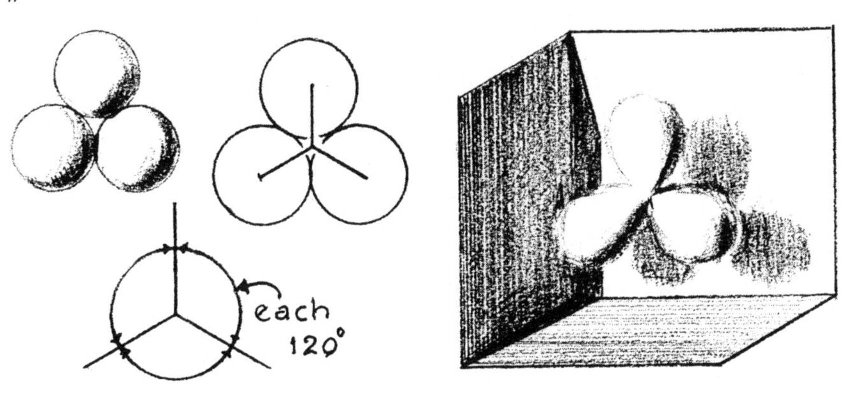


#1–32

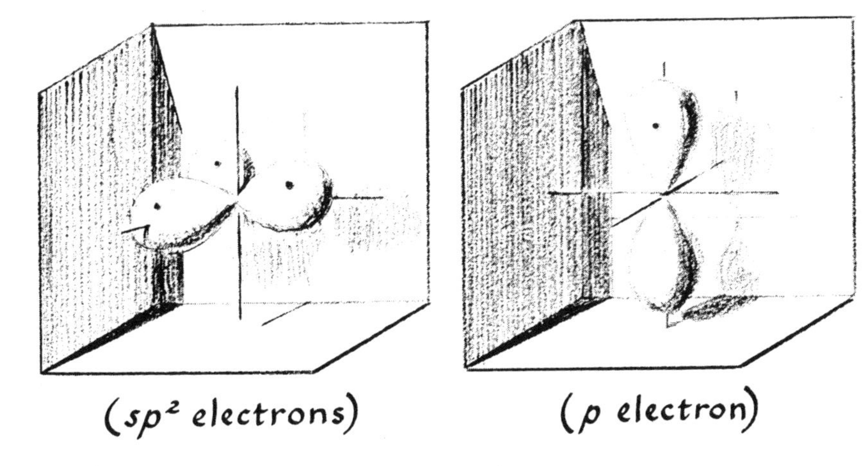


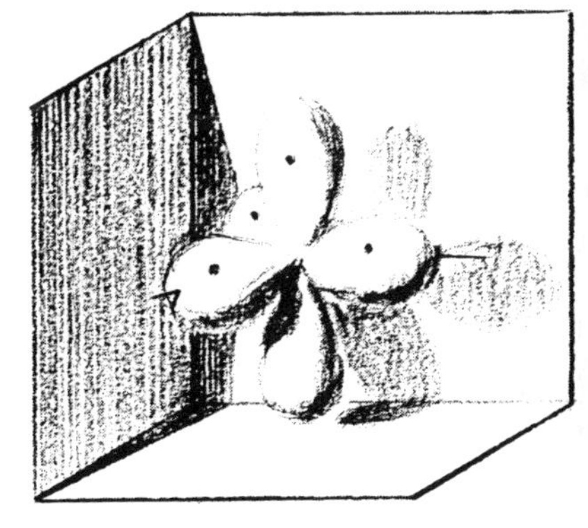

The spatial relationships of the four orbitals can be shown by a plane and four lines as in #1–33. The carbon nucleus is at the point

#1–33

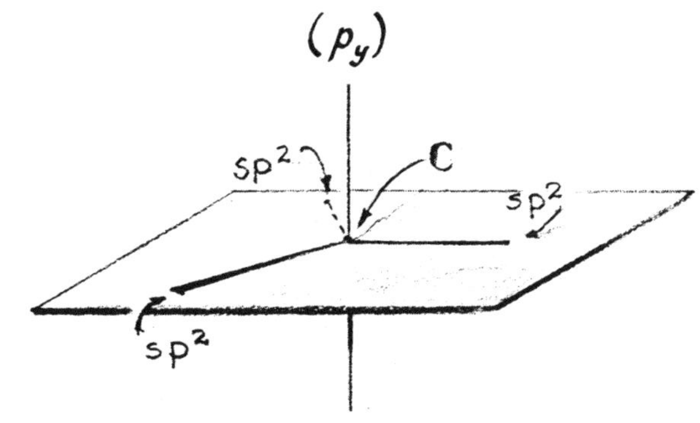


where the lines intersect.

Construct a model of $sp^2$ hybridization so that you can actually *see* the spatial relationships involved. There are several ways in which

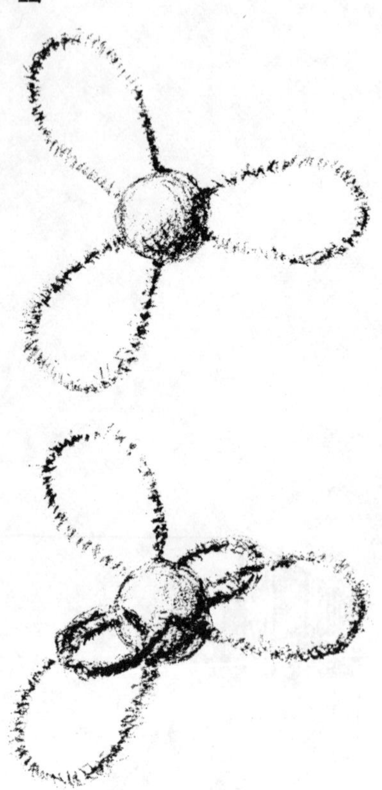

this can be done. If you have a model kit available, use it. Alternatively, styrofoam spheres and pipe cleaners are good materials to use. Let a ball represent the carbon nucleus. Then you can indicate the relative placement of the $sp^2$ orbitals by three horseshoe-shaped pipe cleaners inserted with even spacing around the mid-line of the sphere, as in #1–34.

**#1–34**

The $p$ orbital can be represented by two more pipe cleaners inserted so that they are perpendicular to the plane of the original three.

**#1–35**

You can reproduce #1–33 in a model as follows: Cut the form, as shown in #1–36, out of cardboard. The three arms meet at 120° angles, and represent the three $sp^2$ electron orbitals. Insert another piece vertically (at a 90° angle) through the center. The vertical piece indicates the direction of the $p$ orbital of the unhybridized electron.

**#1–36**

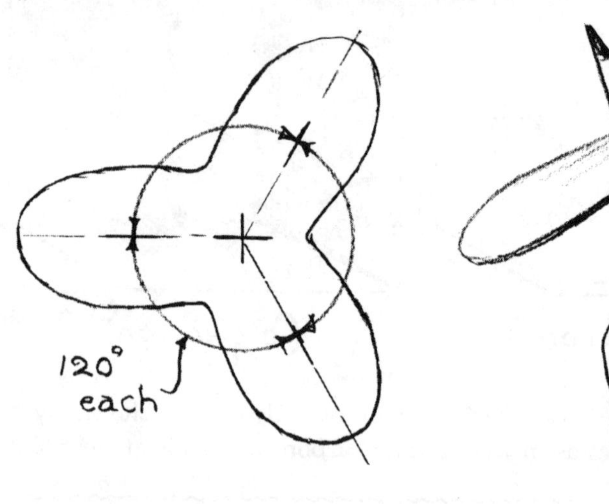


*Pi Bonds*

Hybridization of the $sp^2$ type occurs whenever there are only three atoms bonded to carbon in a neutral molecule. The orbital of each of the three $sp^2$ electrons from one carbon can overlap the orbital of any unpaired electron from another atom. This overlap is represented in #1–37*a* for the two carbon atoms and four hydrogen atoms in ethene ($CH_2{=}CH_2$). The type of electron involved in each overlap is indicated to emphasize for you that $\sigma$ bonds may arise from the overlap of different kinds of electrons. The molecular orbitals formed by the overlap of atomic orbitals in this manner are represented by ellipses in #1–37*b*. Because the relationship between atoms at each carbon is one of three sphere-like objects around a single point (see #1–31), the angles of bonding are 120° and all six atoms lie in a single plane.

The diagram in #1–37 accounts for the orbital overlap of three of the electrons on each carbon but does not reveal the fate of the two

#1–37

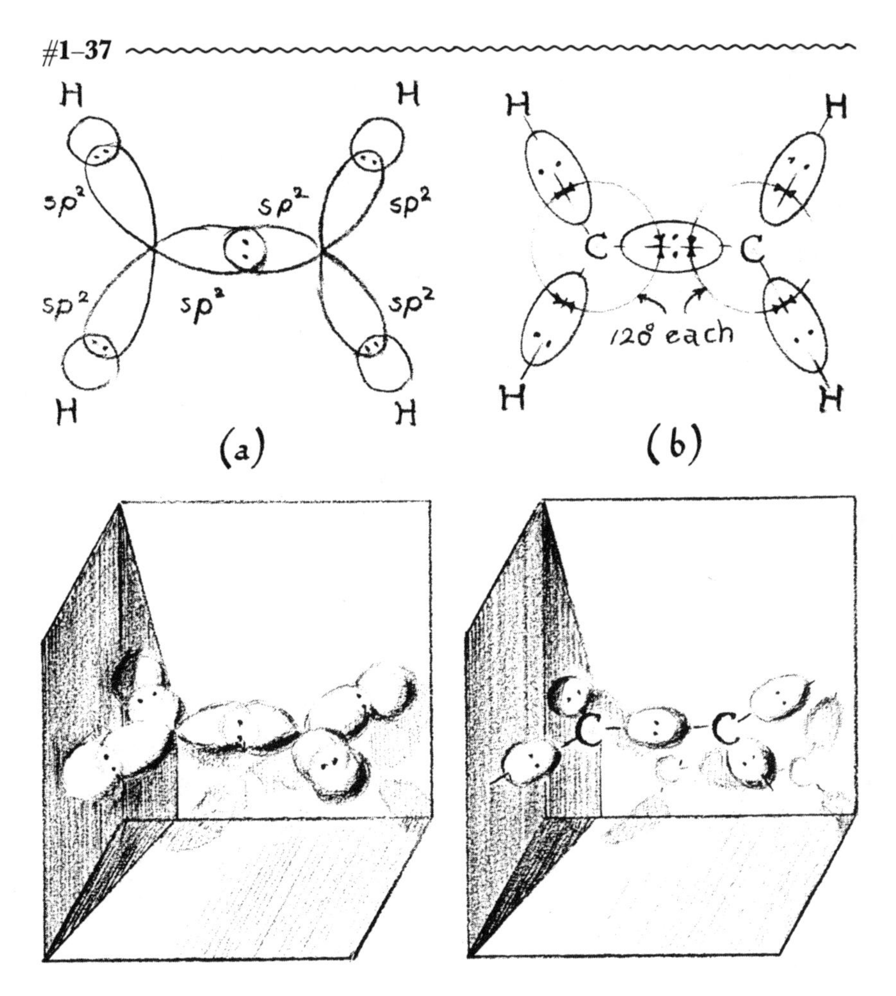


unhybridized $p$ electrons. The axes of the two $p$ orbitals are each perpendicular to the plane through the six atoms and consequently they are parallel to each other as indicated in #1–38.

#1–38

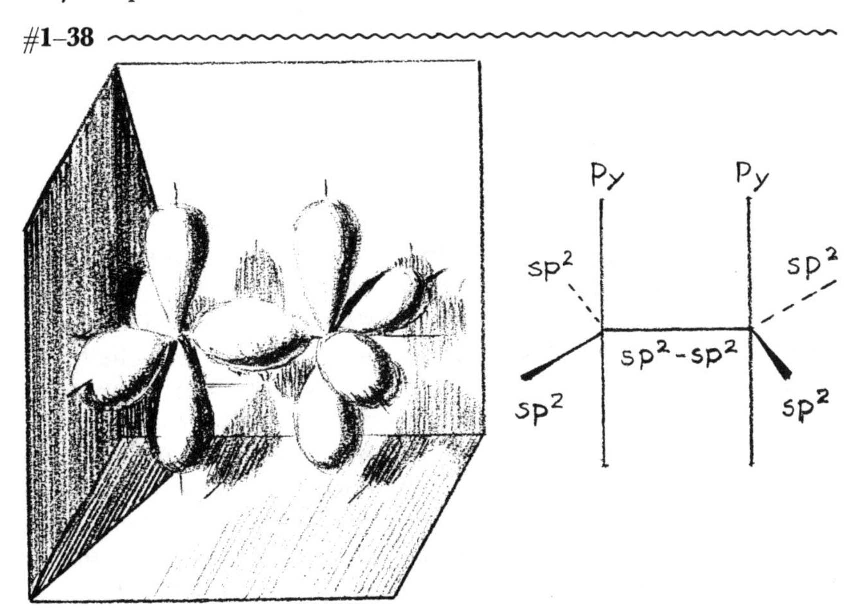

Make a model of the figure in #1–38 so that you can actually *see* the relationship described here. Duplicate whatever model you made of an $sp^2$ hybridized carbon atom and join the two units by overlapping an $sp^2$ orbital from each. This operation places the axes of the two $p$ orbitals parallel to each other and perpendicular to the plane through the six atoms. Although your model effectively shows the alignment of the two $p$ orbitals, it cannot show that the egg-shaped figures are close enough to overlap. This overlap is sideways, as indicated in #1–39*a*, in contrast to the end-to-end overlap of sigma bonding. As

**#1–39**

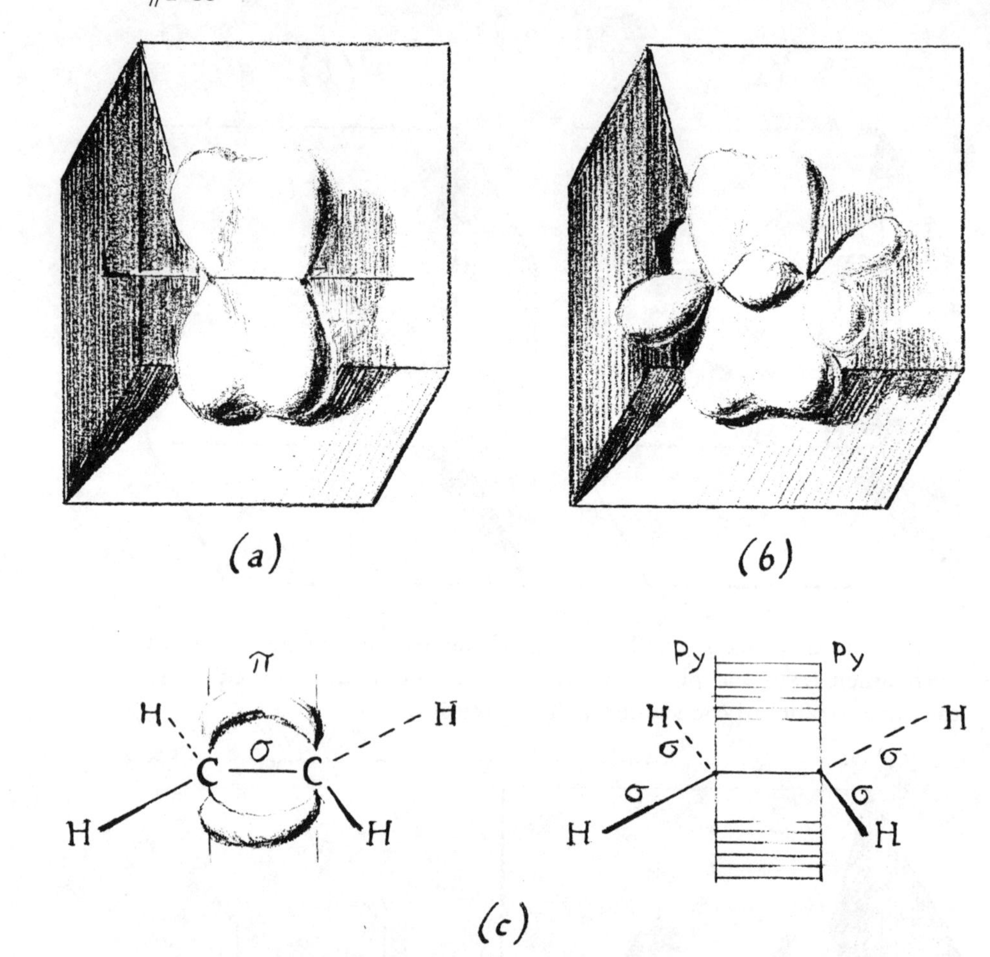


with $\sigma$ bonding, the two unpaired electrons interact when their orbitals overlap. They become spin-paired in a common molecular orbital. Because the atomic orbitals of $p$ electrons are bilobed, the new molecular orbital consists of two parts, one above and one below the axis of the sigma bond between the two carbons as indicated in #1–39*b*.

The increased electron density between the carbon nuclei results in an additional bonding force between the two atoms. This second bond is called a pi ($\pi$) bond. The two-part orbital is a pi ($\pi$) orbital and the electrons that occupy it are pi ($\pi$) electrons. Thus, the two carbons are held together by two bonds: a $\sigma$ bond formed by the end-to-end overlap of the orbitals of two $sp^2$ electrons and a $\pi$ bond formed by the sideways overlap of the orbitals of two *p* electrons. The combined bonding forces are referred to as a *double bond.*

The sideways forces of a $\pi$ bond diagrammed in #1–39*c* tend to pull the two carbons closer together so that the distance between two doubly bonded carbon atoms is less than the distance between two singly bonded carbon atoms, i.e., a double bond is shorter than a single bond (Table 1–1).

A double bond involves four electrons and is represented by two lines or by four dots as in #1–40. A molecule containing a double bond

**#1–40**

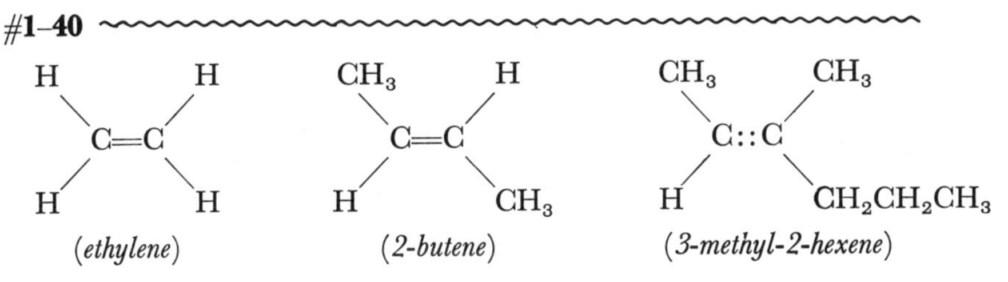


is flat at the site of unsaturation and the angles of bonding are approximately 120° around the double bond. The term "approximately" was included because the angle may vary a few degrees depending upon the relative size and characteristics of the substituents on the carbon.

*sp Hybridization*

The third type of bonding between two carbon atoms produces a triple bond, one $\sigma$ and two $\pi$ bonds. This multiple bonding occurs on carbon when carbon in a neutral molecule is bonded to only two other atoms and hybridization involves only one *s* and one *p* electron from the configuration $s$, $p_x$, $p_y$, $p_z$ (#1–17). The charge from the *s* and a *p* electron is blended to form two equivalent *sp* orbitals, while the remaining *p* electrons continue to occupy bilobed orbitals perpendicular to each other as indicated in #1–41. The two *sp* orbitals bear the same relationship to each other as tangent spheres with the carbon nucleus

**#1–41**

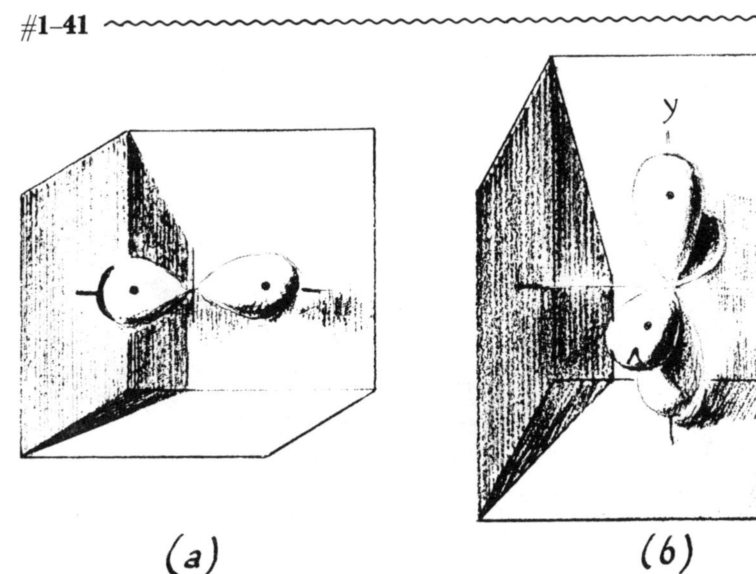

at the point of tangency. The axes of the two *p* orbitals are perpendicular to the line drawn through the centers of the two *sp* orbitals as well as mutually perpendicular. The spatial relationships among the orbitals in only one of the two carbons involved when there is *sp* hybridization are shown in #1–42.

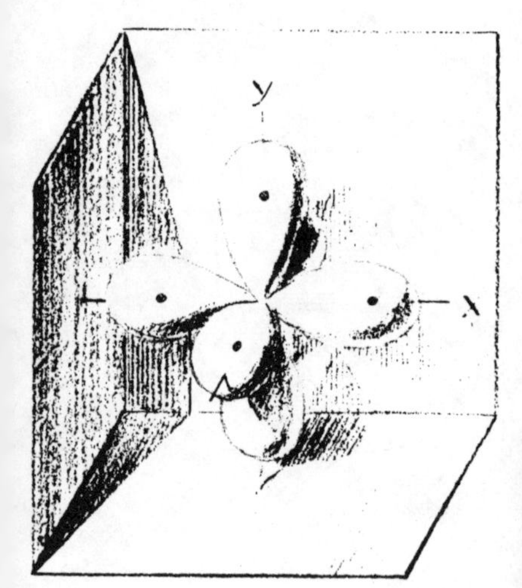


**#1–42**

The end-to-end overlap of the orbital of one *sp* electron from each of the two triply bonded carbon atoms places the carbons so that the axes of the two *p* orbitals of one carbon parallel the axes of the two *p* orbitals on the other carbon, as shown in #1–43. Each set of parallel

**#1–43**

orbitals is close enough for sideways overlap. The electrons from the orbitals that overlap interact and become spin-paired, forming molecular orbitals, one bilobed $\pi$ orbital for each set of overlapping *p* orbitals, as indicated in #1–44. Thus there is a $\pi$ cloud above and

**#1–44**

below the axis of the $\sigma$ bond and also one behind and in front of it. The two molecular orbitals also tend to overlap, with the result that the four pi electrons produce a cylindrical cloud of charge around the $\sigma$ bond, as indicated in #1–44.

The two carbons in acetylene are held together by one $\sigma$ and two $\pi$ bonds, giving a *triple bond.* A triple bond is represented in a formula by three lines or by six dots, as shown in #1–45. A triple bond is shorter than a double bond. The two triply bonded carbons and the atom

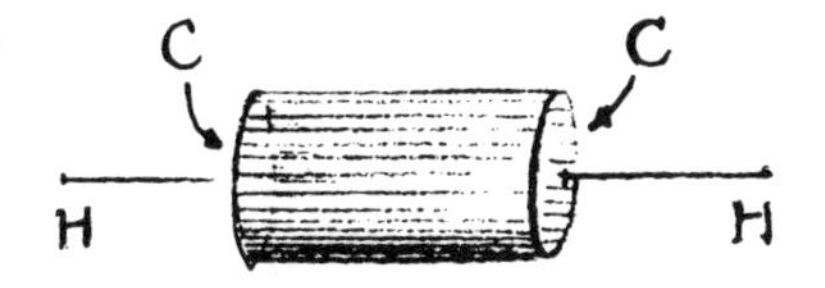


**#1–45**

H—C:::C—H (*acetylene*)

$CH_3C{\equiv}CCH_3$ (*2-butyne*)

$CH_3CH_2C{\equiv}C$—H (*1-butyne*)

substituent on each of them must lie in a straight line. A molecule is flat in the region of a triple bond and the angles of bonding are 180°.

You should be able to look at a formula of an organic compound and recognize immediately the "kinds" of electrons that went into the formation of each bond. The information which provides the basis for your analysis of the nature of the bonding in an organic molecule can be summarized as follows:

- When carbon is bonded to four atoms, four electrons have undergone $sp^3$ hybridization.
- When carbon is bonded to three atoms, three electrons have undergone $sp^2$ hybridization. The fourth electron, a $p$ electron, is involved in a $\pi$ bond.
- When carbon is bonded to two atoms, two electrons have undergone $sp$ hybridization. The other two electrons, both $p$ electrons, are involved in two $\pi$ bonds.
- All the electrons contributed by one carbon atom to the formation of $\sigma$ bonds must be the same. One carbon cannot contribute one $sp^3$ and one $sp^2$ electron to two different $\sigma$ bonds.
- When there is a double bond between two carbons, each carbon contributes one $sp^2$ and one $p$ electron to the double bond.
- When there is a triple bond between two carbons, each carbon contributes one $sp$ and two $p$ electrons to the triple bond.

Example 1 illustrates the application of these concepts to a particular structure.

*Example 1*—Label each bond in the compound in #1–46 as to the type of bond ($\sigma$ or $\pi$) and the kinds of electrons involved. Also indicate the angle at which bonding occurs between each group of three adjacent atoms.

**#1–46**

```
     H
     |
 Cl—C           H
     | \       /
     H   C==C
        /     \
     H          H
  (3)  (2)  (1)
```

*Step 1.* Number each carbon and list the various bonds to be accounted for, including bond type.

- All single bonds are $\sigma$ bonds.
- All double bonds consist of one $\sigma$ bond and one $\pi$ bond.

The bonds in this molecule and their type may be designated as:

**Table 1–1**

**Summary of orbital hybridization in carbon**

| *Electrons involved in hybridization* | *Designation of hybridized electrons* | *Bonds formed by the carbon* | *Angle of bonding* | *Carbon–carbon bond length* |
|---|---|---|---|---|
| $\overbrace{s\,p\,p\,p}$ | $sp^3$ | $4\sigma$ | 109.5° | 1.54 Å |
| $\overbrace{s\,p\,p}\,p$ | $sp^2$ | $3\sigma$<br>$1\pi$ | 120° | 1.35 Å |
| $\overbrace{s\,p}\,p\,p$ | $sp$ | $2\sigma$<br>$2\pi$ | 180° | 1.20 Å |

| | | | | | |
|---|---|---|---|---|---|
| C*1*=C2 | $\sigma + \pi$ | C*2*—H | $\sigma$ | C*3*—H | $\sigma$ |
| C*1*—H | $\sigma$ | C*2*—C*3* | $\sigma$ | C*3*—Cl | $\sigma$ |

*Step 2.* Determine the kinds of electrons contributed by each carbon to the bonding in the molecule.

- Carbon bonded to four different atoms contributes four $sp^3$ electrons, one to each of the four bonds. There is only one tetrahedrally bonded carbon in the molecule. This is C*3*. Each of the bonds to C*3* is formed by at least one $sp^3$ electron as indicated in Table 1–2.
- Whenever carbon forms a double bond, the electron in the $\sigma$ bond portion of the double bond is an $sp^2$ electron; the $\pi$ bond consists of a $p$ electron; and the other two bonds from that carbon each contains at least one $sp^2$ electron, as shown in Table 1–2.

**Table 1–2**

**Bonding in 3-chloropropene** ($CH_2{=}CHCH_2Cl$)

| *Bond* | *Type* | *Electrons* |
|---|---|---|
| C*1*=C*2* | $\sigma$ | $sp^2$—$sp^2$ |
| | $\pi$ | $p$—$p$ |
| C*1*—H | $\sigma$ | $sp^2$—$s$ |
| C*2*—H | $\sigma$ | $sp^2$—$s$ |
| C*2*—C*3* | $\sigma$ | $sp^2$—$sp^3$ |
| C*3*—H | $\sigma$ | $sp^3$—$s$ |
| C*3*—Cl | $\sigma$ | $sp^3$—$p$ |

*Step 3.* Determine the kinds of electrons contributed by the non-carbon atoms.

- A hydrogen contributes an $s$ electron.
- We shall designate the electron from a halogen as a $p$ electron.

These non-carbon electrons are also included in Table 1–2 and we now have designated the kind of each of the electrons involved in the bonding of 3-chloropropene.

*Step 4.* Label the various angles of bonding.

- All the angles at a tetrahedral carbon are 109.5°.
- All the angles at a doubly bonded (trigonal) carbon are 120°.

#1–47
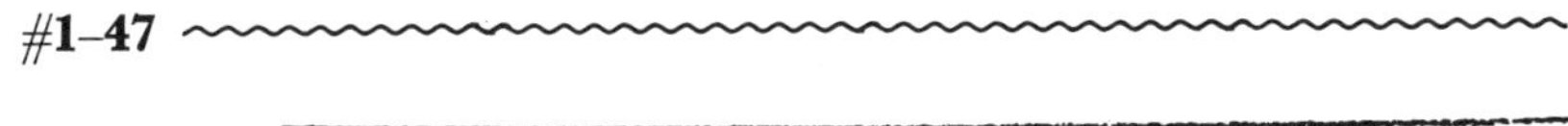

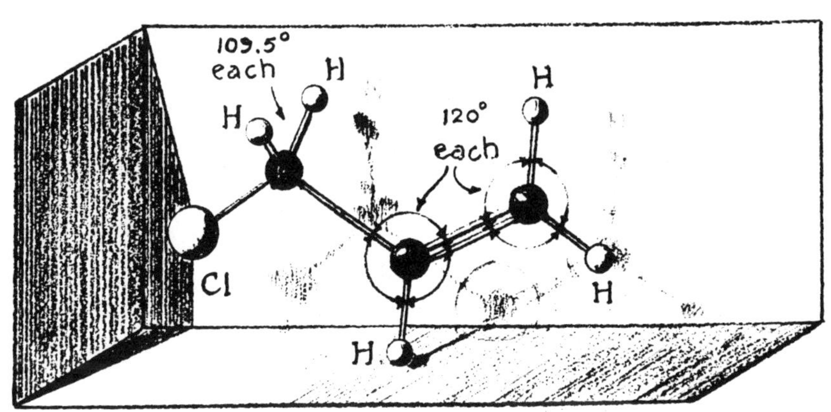


*Bonding of Oxygen and Nitrogen*

Each of the three types of hybridization, $sp^3$, $sp^2$, and $sp$, produces a different molecular geometry for carbon. These three fundamental bond shapes described for carbon also apply to the bonding of other elements.

Of particular interest in organic molecules is the geometry of the bonding of oxygen and nitrogen. When these elements form single, double, or triple bonds, the geometry follows the patterns discussed for carbon if one considers the orbitals of the nonbonding as well as the bonding electrons.

The configuration in the ground state of the second-energy electrons in oxygen is: $2s^2$, $2p_x^2$, $2p_y$, $2p_z$. This notation says that there are two electrons in a spherical orbital, two electrons in a bilobed $p$ orbital, and one electron in each of the other $p$ orbitals (#1–48).

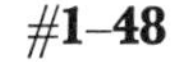
#1–48

The two unpaired $p$ electrons are available for bonding. Oxygen is divalent and forms compounds such as: $H_2\ddot{O}:$, $CH_3\ddot{\underset{..}{O}}H$, and $CH_3\ddot{\underset{..}{O}}CH_3$ (water, methanol, and dimethyl ether). If the electrons maintained their ground-state relationship after bonding, the angle formed by oxygen and the two substituent atoms would be 90°. However, it has been shown to be about 105°, which is closer to the angle in tetrahedral bonding than to a right angle (#1–49).

**#1–49**

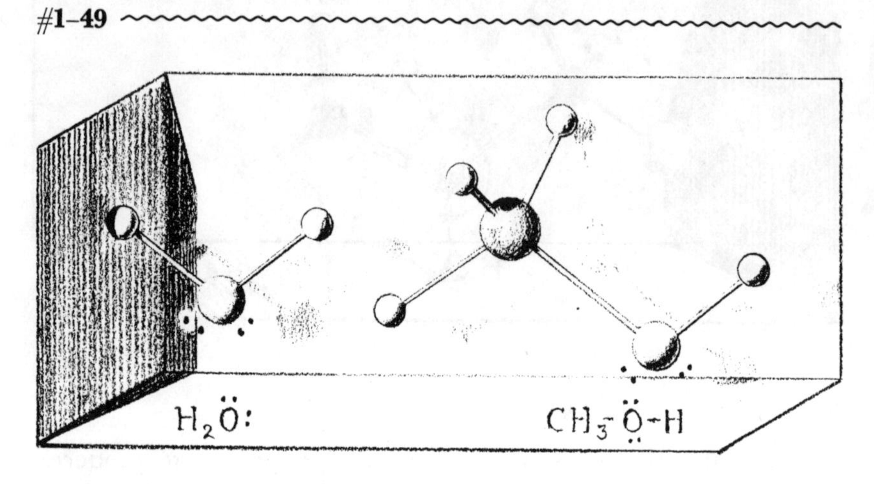


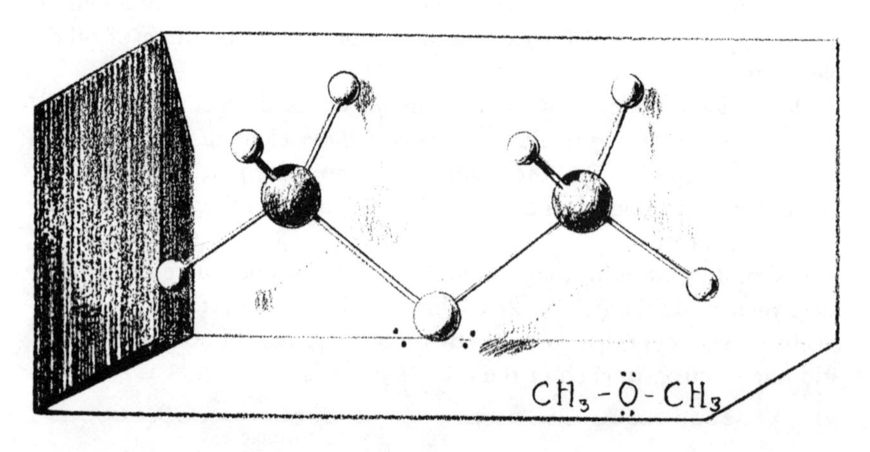


There are four pairs of electrons around the oxygen in each of these molecules, two bonding and two nonbonding. Each pair can be considered as occupying a sphere-like orbital. Thus, the geometry becomes that of four spheres around a point, i.e., the geometry of $sp^3$ hybridization and tetrahedral bonding (#1–50).

There is one major difference between this structure and a tetrahedral carbon. The orbitals here are not equivalent. Nonbonding electrons are less restrained than bonding electrons. The orbitals of the nonbonding electron pairs would tend more toward a spherical shape, resulting in a larger angle between the orbitals of the nonbonding electron pairs than between the more ellipsoidal orbitals of the bonding electrons. In other words, the orbitals of the nonbonding

#1–50

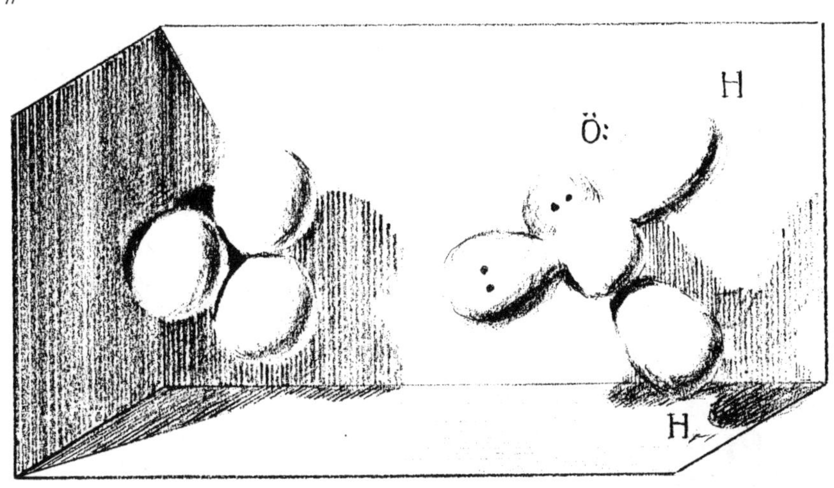


electron pairs tend to force the bonding orbitals a bit closer together so that the angle of bonding is less than 109.5°.

Although the size of the bonding angle is determined by the three-dimensionality of the electron orbitals, the geometry of the molecule at the oxygen atom involves only three atoms. The two-dimensional relationships of the three atoms can be shown in a single plane. You will have drawn many of these molecules from a purely geometric viewpoint in problem 3, §2, without then knowing the reason for the shape.

Nitrogen in the ground state has the configuration $2s^2$, $2p_x$, $2p_y$, $2p_z$. This leaves three unpaired $p$ electrons available for bonding (#1–51).

#1–51

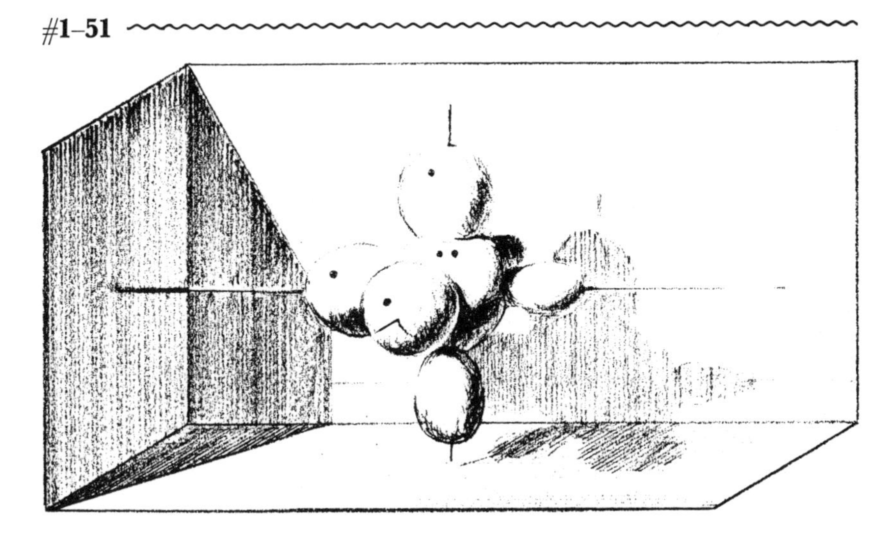

The three hydrogens in ammonia, $NH_3$, do not form angles of 90° with each other and nitrogen. All three angles of bonding in ammonia are 107°, indicating that again we have tetrahedral bonding in which the angle has been decreased somewhat from 109.5° (#1–52).

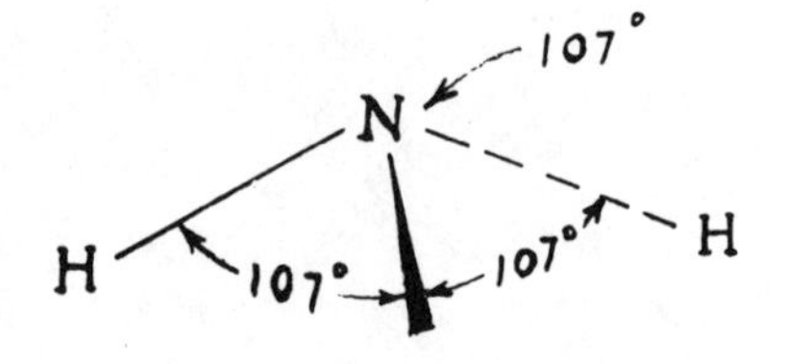


#1–52

In molecules of trivalent nitrogen, there are three pairs of bonding electrons and one pair of nonbonding electrons. This produces the geometry of four spheres around a central point. A single pair of non-bonding electrons does not have as great an effect on the orbitals of the bonding electrons as do two pairs of nonbonding electrons. Therefore, the angle of bonding in the compounds of nitrogen is closer to 109.5° than it is in the compounds of oxygen. Ammonia can be visualized as a pyramid with nitrogen at the apex topped by the orbital of the nonbonding electron pair (#1–53). Organic molecules

#1–53

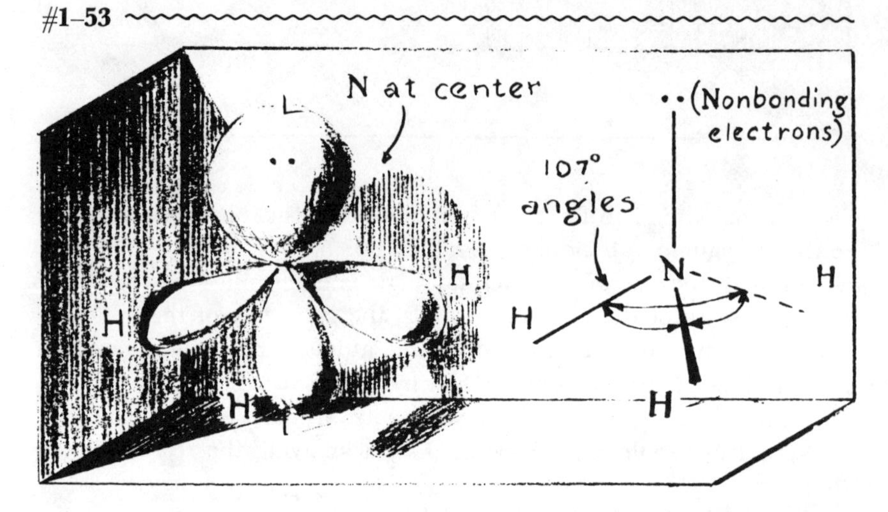


containing nitrogen bonded to three other atoms are pyramidal in shape in the region of the nitrogen atom (#1–54). Do problems 11 and 12 in §2 at this time.

#1–54

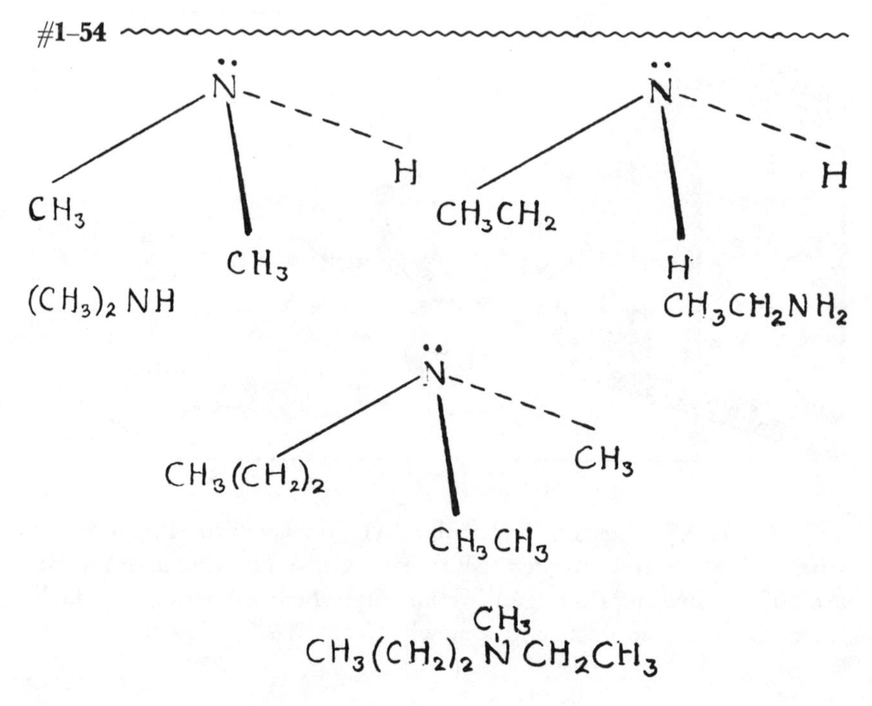

When oxygen forms a double bond with carbon the geometry is that of $sp^2$ hybridization. The orbitals of the two pairs of nonbonding electrons of the oxygen lie in the same plane as the $sp^2$ orbitals of the carbon. The molecule at this point is flat. The angle of bonding at the carbon is 120°. These relationships are depicted for acetaldehyde ($CH_3CHO$), acetone ($CH_3COCH_3$), and acetic acid ($CH_3COOH$) in #1–55.

**#1–55**

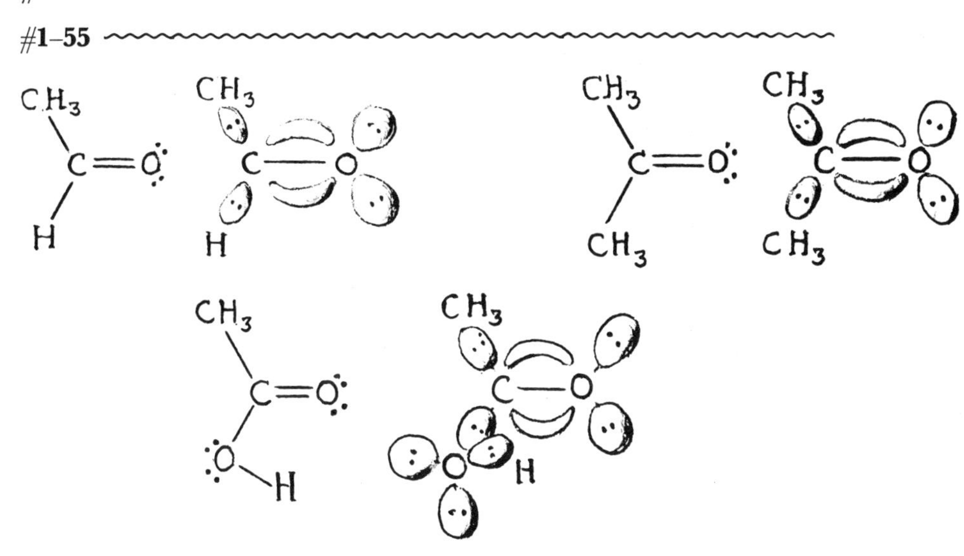


When nitrogen forms a triple bond with carbon, the geometry is that of *sp* hybridization. The orbital of the pair of nonbonding electrons lies in a straight line with the *sp* orbitals of the carbon (#1–56).

**#1–56**

## § 2. Problems. Geometry of Sigma and Pi Bonding

In answering questions 1, 2, 4, 5, and 9, use a dot or a small circle to represent the nucleus of each atom and lines to represent the axes of the bonds. You will find it easier to make the various diagrams if you use a protractor.

**1**) Figure #1–8*c* shows the spatial relationships that exist between three atoms when two atoms are each bonded to a third atom and the angle of bonding is 90°. The unit is shown positioned in plane XY (the plane of the page). Draw similar diagrams for three-atom structures in which the angle of bonding is

*a*) 60° *b*) 105° *c*) 109.5° *d*) 120° *e*) 180°

Include an outline of the plane in your drawing and relate the structure to the X–X and Y–Y axes.

**2**) Figures #1–8*a* and *b* present a three-atom structure with a 90° angle of bonding as though it were positioned in the horizontal reference plane XZ. Draw the structures from problem 1 as though they were positioned in an XZ plane rather than an XY plane. Include a projection of the plane in your drawing with labeled axes.

**3**) An angle of approximately 105° is formed when oxygen is bonded by $\sigma$ bonds to each of two other atoms. Draw a structural formula for each of these compounds which shows the correct angle of bonding at the oxygen. Treat the molecule as a flat unit, positioning oxygen and the two adjacent atoms in the plane of the page. Do not include at this time the geometry of the bonding at these other atoms, e.g., ignore the tetrahedral structure of a methyl group. As examples see #1–5 and #1–49.

*a*) $H_2O$  *b*) HOCl  *c*) $NH_2OH$  *d*) $CH_3OH$  *e*) $CH_3OCH_3$

**4**) Draw diagrams showing the following planar four-atom structures in reference plane XY (see #1–31). Three atoms are each bonded to a central atom in a planar structure wherein:

*a*) the angle formed by the axes of any two adjacent bonds is 120°;

*b*) the angle formed by one pair of adjacent bonds is 100° and the other two bond angles are each 130°;

*c*) the angle formed by each pair of adjacent bonds is different: 100°, 120°, and 140°.

**5**) Draw the structures in problem 4 as though they were positioned in the horizontal reference plane XZ instead of XY (#1–32).

**6**) The following compounds are planar in the region where carbon is bonded to only three other atoms. The angles of bonding at this carbon are approximately 120°. Thus, the spatial relationships of the atoms around this carbon are essentially the same as those you diagrammed in part *a*) of problem 4. Draw each formula in the plane of the page so as to show the relative positions of the atoms bonded to carbon in the "flat" part of the molecule. Do not include the geometry of the tetrahedral carbons.

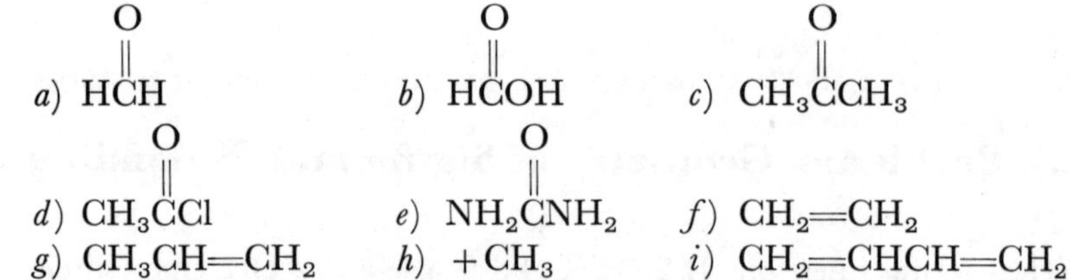


**7**) To strengthen your visualization of the relative positions of the three reference axes and the three reference planes draw a diagram for each of the following situations.

*a*) An XZ plane with a Y axis passing through the midpoint of the plane.

*b*) A YZ plane with an X axis passing through the midpoint of the plane.

*c*) An XY plane with a Z axis passing through the midpoint of the plane.

*d*) Two parallel XZ planes about half an inch apart.

*e*) Two parallel YZ planes about half an inch apart.
*f*) Two parallel XY planes about half an inch apart with the front plane actually in the plane of the page.
*g*) Two parallel XY planes about half an inch apart with the rear plane actually in the plane of the page.
*h*) Two planes sharing an edge in common. The planes are an XY plane and an XZ plane extending behind the plane of the page. The common edge is an X axis.
*i*) Two planes sharing an edge in common. The planes are an XY plane and an XZ plane extending in front of the plane of the page. The common edge is an X axis.
*j*) Two planes sharing an edge in common. The planes are an XY plane and a YZ plane extending behind the plane of the page. The common edge is a Y axis.
*k*) Two planes sharing an edge in common. The planes are an XY plane and a YZ plane extending in front of the plane of the page. The common edge is a Y axis.

**8**) Draw a cube so as to indicate that its face closest to you is in the plane of the page (the X Y plane), and all other faces project behind the X Y plane. Then draw a cube with its back face in the X Y plane and the others projecting in front of the page.

**9**) Draw a diagram to show the relationship between the components in a five-atom structure when each of four atoms is bonded to a central atom and all the angles of bonding are 109.5°. Hold a model of the tetrahedral carbon in front of you as you make this drawing. Use two different devices for showing the spatial relationships:

*a*) Use two perpendicular reference planes, drawing one component angle on each plane as in #1–28. Try this with the following sets of planes: XY and YZ; XY and XZ; YZ and XZ.
*b*) Draw the correct angle in the XY plane and indicate the projection of the other two bonds in front of and behind the plane of the page, as in #1–16*b*. Place this figure in various positions.

**10**) The tetrahedral framework you drew for 9*b* provides the basis for projection formulas of molecules with tetrahedral bonding. Some structures with tetrahedral bonding are listed here. Draw projection formulas showing the spatial relationships among the atoms in these substances.

*a*) $CH_4$ *b*) $CCl_4$ *c*) $CHCl_3$
*d*) $CH_2Cl_2$ *e*) $SO_4^=$ *f*) $ClO_4^=$
*g*) $NH_4^+$ *h*) $CH_3CH_3$ *i*) $CH_3CH_2CH_3$
*j*) $CH_3CH_2CH_2CH_3$

**11**) Problems 4, 5, and 6 dealt with planar four-atom structures. A four-atom unit, with three atoms each bonded to a central atom, can also exist in the form of a pyramid. There are three bonding angles in this structure which may or may not be equivalent. We shall concentrate on structures where they are essentially equivalent. Some pyramidal four-atom molecules and ions with their angles of bonding are the following:

*a*) $NH_3$ (107°) *b*) $PH_3$ (94°) *c*) $P_4$ (60°)

*d*) $NF_3$ (102°) *e*) $PBr_3$ (100°) *f*) $ClO_3^-$ (106°)
*g*) $BrO_3^-$ (111°) *h*) $NH_2CH_3$ (107°) *i*) $NH(CH_3)_2$ (108°)
*j*) $N(CH_3)_3$ (108°)

Draw projection formulas for each compound, ignoring the tetrahedral structure at carbon and concentrating on the pyramidal structure at the hetero atom. As an example see #1–52, which places N and one H in the plane of the page and shows the projection of the other two N—H bonds relative to the plane of the page.

**12**) There is a pair of nonbonding electrons at the apex of each pyramidal structure in problem 11. Add the orbital of these nonbonding pairs to the structures you drew in answer to problem 11 (see #1–53). The presence of this orbital means that the geometry of these four-atom units is essentially that of four spheres around a central point. The nonbonding pair of electrons has the effect of reducing the bonding angle below 109.5°.

By contrast, the planar four-atom units of problem 11*b* do not have nonbonding electrons. Except for $+CH_3$, each of the carbons at the point of planarity has three $\sigma$ bonds and one $\pi$ bond. The geometry is essentially that of three spheres around a central point (#1–31). Indicate in outline form as in #1–38 the overlapping atomic orbitals that lead to the double bond in each compound in problem 6. Then diagram as in #1–39*b* the molecular orbitals, both $\sigma$ and $\pi$, for each.

**13**) The angle of bonding formed by two triply bonded atoms and a substituent on one of these atoms is 180°. Draw a structural formula to show this relationship in

*a*) acetylene *b*) hydrogen cyanide *c*) 2-butyne
*d*) 3-hexyne *e*) 4-chloro-2-pentyne

**14**) Draw projection formulas to show the geometric relationships between the atoms in each of the following compounds. Be sure that the tetrahedral, the pyramidal, and the planar relationships are all clearly indicated.

*a*) HI *b*) $H_2\ddot{O}:$ *c*) $BeCl_2$ (linear)
*d*) $BF_3$ (flat) *e*) $(CH_3)_3\ddot{N}$ *f*) $CH_3Br$
*g*) $PF_3$ (about 100°) *h*) $CH_2Br_2$ *i*) $^-:CH_3$
*j*) $CH_3\underset{\cdot\cdot}{\ddot{O}}CH_3$ *k*) $CH_3CH_2\underset{\cdot\cdot}{\ddot{O}}H$ *l*) $CH_3CH_2\ddot{N}H_2$
*m*) $(CH_3)_2CHCH_3$ *n*) $CH_3\overset{H}{\overset{|}{C}}{=}\overset{H}{\overset{|}{C}}CH_3$ *o*) $CH_3C{\equiv}CH$
*p*) $CH_3\overset{:\ddot{O}}{\overset{\|}{C}}H$ *q*) $CH_3\overset{:\ddot{O}}{\overset{\|}{C}}CH_3$ *r*) $CH_3\overset{:\ddot{O}}{\overset{\|}{C}}\underset{\cdot\cdot}{\ddot{O}}H$
*s*) $CH_3\overset{:\ddot{O}}{\overset{\|}{C}}\underset{\cdot\cdot}{\ddot{O}}CH_3$ *t*) $CH_3\overset{:\ddot{O}}{\overset{\|}{C}}CH_2\underset{\cdot\cdot}{\ddot{O}}CH_3$

**15**) Indicate the type ($\sigma$ or $\pi$) of each bond in the following compounds and state the kinds of electrons ($sp^3$, $sp^2$, etc.) that went into the formation of each bond. Give a value for each angle formed by three adjacent atoms. (The angles shown in the formulas are not the correct ones.)

*a*) $CH_3CH(Br)CH_2\ddot{\underset{..}{O}}H$ *b*) $CH_3C(H){=}C(H)Br$ *c*) $CH_3CH_2C{\equiv}CH$

*d*) $CH_2{=}CHC{\equiv}CCH_3$ *e*) $CH_3\ddot{\underset{..}{O}}CH_2C{\equiv}N$ *f*) $CH_2{=}CHCH({=}\ddot{O}:)$

## § 3. Two-Dimensional Structural Formulas. Position Isomerism

Carbon can bond to many elements, such as hydrogen, nitrogen, phosphorus, oxygen, sulfur, the halogens, sodium, copper, etc. It can also bond to other carbons, forming long continuous chains, branched chains, or cyclic structures, as illustrated in #3–1.

**#3–1**

$CH_3CH_2CH_2CH_2CH_2CH_2CH_3$
(*heptane*)

$CH_3CH(CH_3)CH_2C(CH_3)_2CH_2CH_3$
(3,3,5-*trimethylhexane*)

$CH_2{-}CH_2$, $CH_2$ $CH_2$, $CH_2$ (ring)
(*cyclopentane*)

Silicon can bond with other atoms of silicon—or *catenate*, as it is called—but not to the degree that carbon can. Carbon is unique in this respect. If it did not catenate in this manner, organic chemistry would not be so important to us.

* * * * *

There are several different kinds of formulas that are used to designate the nature of a molecule. The choice of the kind of formula depends entirely upon how much information is needed about a particular compound. The three main types of formula are:

- A *molecular formula* such as $C_3H_8O$ which gives only the number and kind of atoms present. A molecular formula reveals nothing about the way in which the atoms are arranged.
- A *two-dimensional structural formula* indicating the sequence in which the atoms occur in the molecule. Two-dimensional structural formulas for all molecules with the molecular formula $C_3H_8O$ are illustrated by #3–2.

**#3–2**

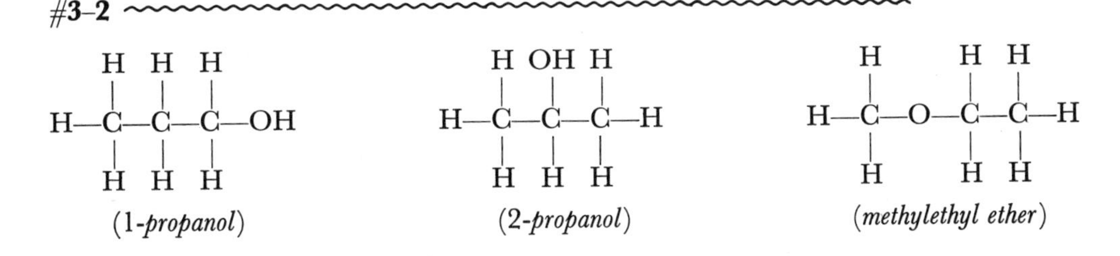
(1-*propanol*) (2-*propanol*) (*methylethyl ether*)

- A *three-dimensional structural formula* such as we used in the previous sections, which is actually a picture showing as accurately as possible on a flat page the spatial relationships between atoms.

Different molecular structures formed from the same numbers and kinds of atoms are called *isomers*. Put the other way, isomers are compounds with the same molecular formula but different structural formulas. Compounds such as those in #3–2, which differ only because of the sequence in which atoms appear in the molecular structure, are called *position isomers*.

Frequently, to save space and the time involved in writing structural formulas like those in #3–2, a condensed form is used. The condensed formulas in #3–3 indicate the bonding between atoms, but not as

**#3–3**

$CH_3CH_2CH_2OH$ $\qquad$ $CH_3CHOHCH_3$ $\qquad$ $CH_3OCH_2CH_3$

explicitly as do the structures in #3–2. In these formulas (#3–3) the substituents on each carbon atom are written after the symbol for the carbon atom. A $—CH_3$ means that this carbon is bonded to three hydrogens. The unit —CHOH in $CH_3CHOHCH_3$ means that the central carbon is bonded to an H and to an O which in turn is bonded to an H. The central carbon is also bonded to two other carbons. A group, such as a $—CH_3$, substituent on a carbon in a chain is placed in parentheses in this system: $CH_3CH_2CH(CH_3)CH_2CH_3$. The central carbon in this molecule has a $—CH_3$ group and a hydrogen atom as substituents in addition to being bonded to the two carbons forming the chain. These relationships are determined by the geometry of the $sp^3$, $sp^2$, and $sp$ electron orbitals, arising from hybridization. Three-dimensional structural formulas for the isomers of $C_3H_8O$ are shown in #3–4. Take note of the effect of tetrahedral bonding.

(1-propanol)

**#3–4**

(Methylethyl ether)

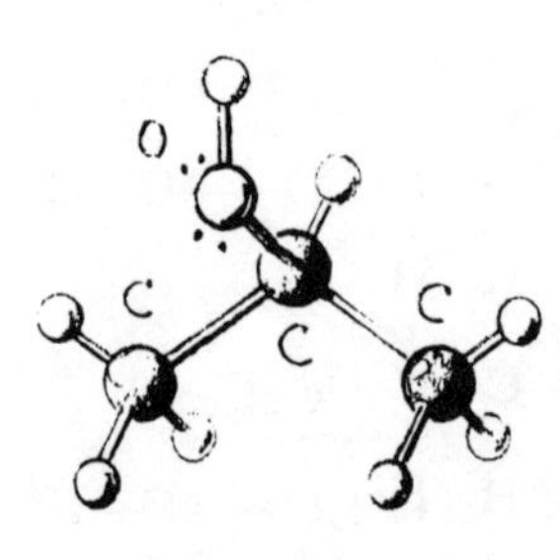

(2-propanol)

Two-dimensional formulas are the ones you will encounter most frequently. They give the most information in the least space. Various conventions have been developed whereby the two-dimensional formula can convey a great deal of information about three-dimensional relationships. However, you must always keep in mind that the formula represents a three-dimensional object. For example, a quick glance at the formulas in #3–5 might lead you to the conclusion that they represent different compounds. However, if one makes a three-

**#3–5**

```
                         CH3                 CH3     CH3
                         |                   |       |
CH3CH2CH2CH2CH3          CH2CH2CH2           CH2CH2CH2
                               |
                               CH3
```

dimensional drawing of each formula, one structure emerges, that shown in #3–6. The implied difference in terminal bonding has no

**#3–6**

```
H   H   H   H   H   H
 \ /     \ /     \ /
  C       C       C
H/ \     / \     / \H
    C         C
   / \       / \
  H   H     H   H
```

reality in space because of tetrahedral bonding and because of free rotation at all single bonds. There is no significance to the terminal ups and downs in the formulas of #3–5. All three of them represent only one molecular structure.

The point for you to consider when looking at the formulas in #3–5 is this: There are five carbon atoms joined in a continuous chain. The ups and downs shown in the formula do not represent branching of the chain. You can take a pencil, trace a line from C*1* to C*5*, and the line will pass through all five carbons, which means that the chain is continuous.

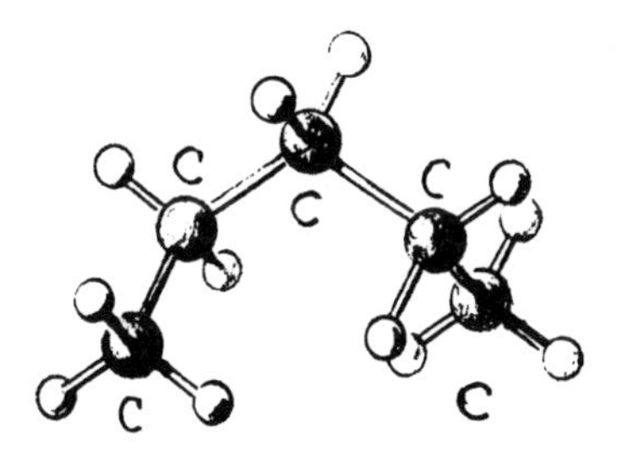


By contrast, consider the formula in #3–7. There are five carbons

**#3–7**

```
    CH3
    |
CH3CHCH2CH3
```

here also. However, you cannot draw a line through all five carbons without taking your pencil off the paper or backtracking. This is a *branched* five-carbon *chain*. The compound in #3–5 is *n*-pentane or normal pentane, $CH_3CH_2CH_2CH_2CH_3$. The compound in #3–7 is a position isomer of *n*-pentane and is called isopentane, $CH_3CH(CH_3)CH_2CH_3$. Compare its three-dimensional formula with that of *n*-pentane (#3–8).

**#3–8**

```
H   H   H   H
 \ /     \ /      H
  C       C      /
 / \     / \    /
H   C   H   C
   / |      / \
  H  |     H   H
     C
    / \
   H   H
```

As you work, remember that the molecules are not glued to the page. They can be picked up and turned around mentally. No matter how the formulas are written on the page, if the same atoms are bonded together in the same order, the formulas represent the same compound. For example, all the formulas in #3–9 represent isobutyl chloride, $(CH_3)_2CHCH_2Cl$.

**#3–9**

```
                      Cl H                                 CH3  H
                      |  |                                 |    |
CH2ClCH(CH3)2       H—C—C—CH3        (CH3)2CHCH2Cl    CH3C————C—Cl
                      |  |                                 |    |
                      H  CH3                               H    H

                      H  CH3                               CH3  Cl
                      |  |                                 |    |
ClCH2CH(CH3)2      Cl—C—C—CH3        (CH3)2CHCH2Cl    CH3C————C—H
                      |  |                                 |    |
                      H  H                                 H    H
```

To make it easier to discuss variations in the bonding of carbon atoms in a molecule, a system has been set up which classifies a carbon atom according to the number of other carbon atoms to which it is bonded. A carbon atom bonded to one other carbon is called a *1° carbon*. The hydrogens on a *1°* carbon are referred to as *1° hydrogens*. A carbon bonded to two other carbons is a *2° carbon* and its hydrogens are *2° hydrogens*. A carbon bonded to three other carbons is a *3° carbon* and its hydrogens are *3° hydrogens* (#3–10). Do problems 3 and 4, §4, and you will understand this classification.

**#3–10**

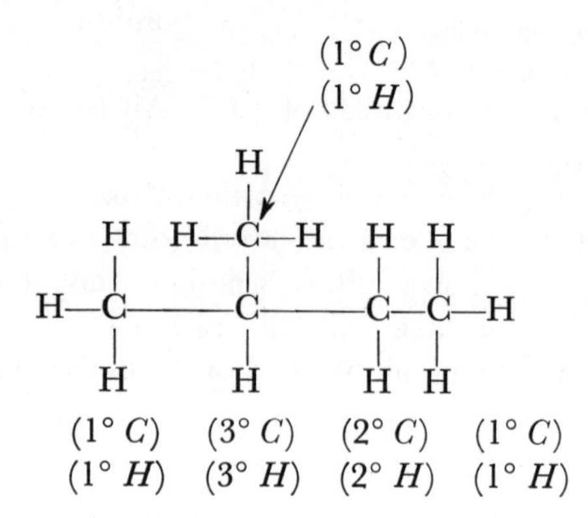


Problem 5 in §4 provides you with practice in differentiating between formulas that represent the same compound and those that represent different compounds. In order to do this differentiation, you have to examine each formula and determine whether or not all the atoms are bonded together in the same sequence. For example, consider the series in #3–11:

**#3–11**

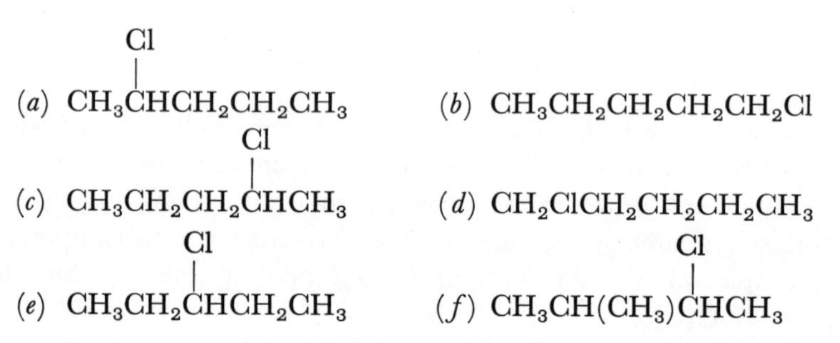


- Structures *a* and *c* are the same. Each is a continuous chain of five carbons with a chlorine on carbon two.
- Structures *b* and *d* are the same. Each is a continuous chain of five carbons with a chlorine on the 1° carbon at the end.
- In structure *e* the chlorine is on a 2° carbon, as it is in *a* and *c*. However, this is a different 2° carbon. It is in the middle of the chain rather than next to the end of the chain.
- The parentheses around the $—CH_3$ in structure *f* means that a methyl group is a substituent on the carbon immediately before the $(—CH_3)$ in the written sequence. Thus, there is branching at carbon two and this structure is unlike any of the others.

Undertaking to draw all the different structures that can be represented by a given molecular formula is another exercise that helps create a feeling for and an understanding of how atoms may be joined to form molecules. Example 1 gives a systematic approach to the problem of determining isomeric structures.

*Example 1*—Draw two-dimensional structural formulas for all the position isomers of $C_6H_{13}Cl$.

*Step 1.* Draw the longest continuous carbon skeleton and determine how many "different kinds" of carbons are present.

**#3–12**

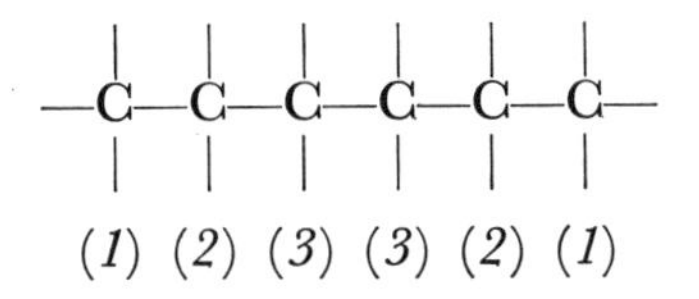


The six-carbon chain has three "different" carbons:

- The carbons numbered *1* are alike. They are 1° carbons and each is a terminal carbon.
- The carbons numbered *2* are alike. They are 2° carbons and each is bonded to a single carbon and to a chain of four carbons.
- The carbons numbered *3* are alike. They are 2° carbons and each is bonded to a chain of two carbons and a chain of three carbons.

*Step 2.* Draw structures with the halogen on each of the "different" carbons and fill in the rest of the bonds with hydrogens.

**#3–13**

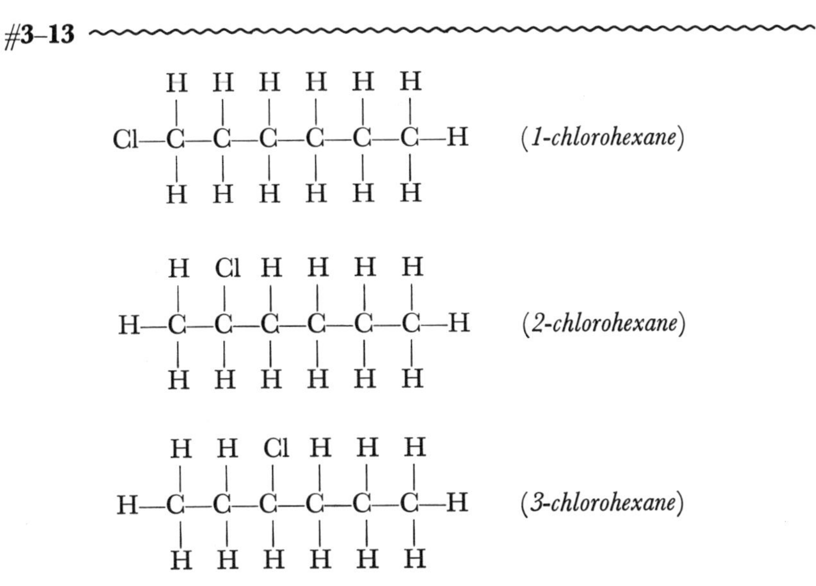


*Step 3.* Draw the next-to-longest carbon chain and locate the "different" carbons in it.

**#3–14**

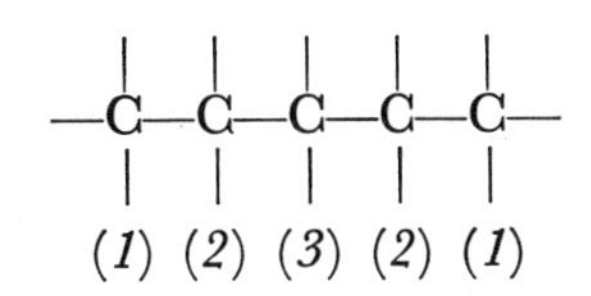


The five-carbon chain also has three "different" carbons.

*Step 4.* Position the extra carbon on each of the "different" carbons, excluding C*1*.

**#3–15**

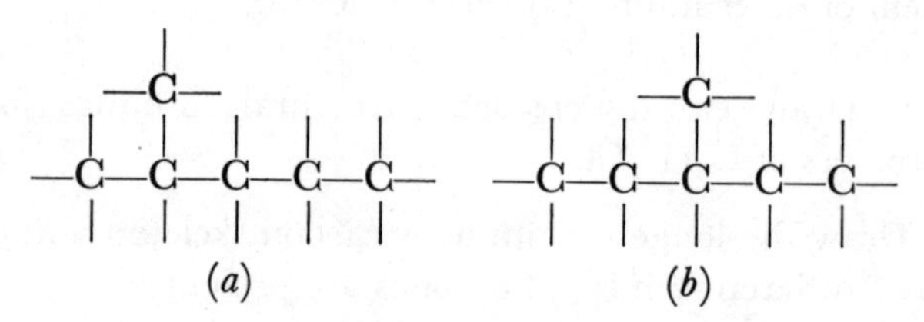

*Step 5.* Locate the "different" carbons in these new carbon skeletons.

**#3–16**

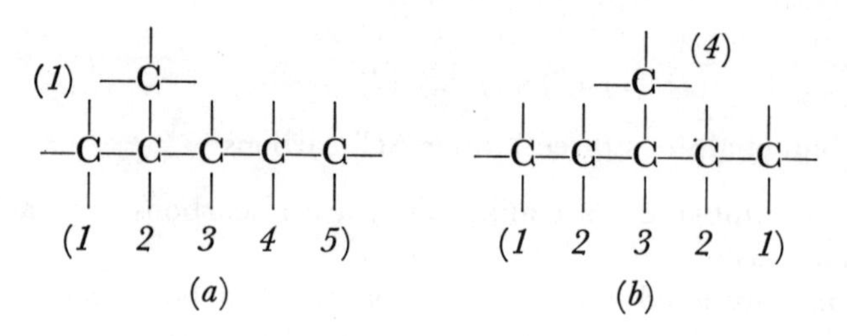

There are five "different" carbons in structure *a*. Those marked *1* are 1° carbons, each of which is bonded to the same 3° carbon. C*5* is also a 1° carbon, but it is bonded to a 2° carbon and so is "different" from the carbons numbered *1*.

Structure *b* has four "different" carbons, two kinds of 1° carbon and two kinds of 2° carbon, as indicated by the numbers.

*Step 6.* Position a Cl on each of the "different" carbons and fill in the rest of the bonds with hydrogens. Remember the carbon is tetravalent.

**#3–17**

(*a*) $CH_2ClCH(CH_3)CH_2CH_2CH_3$ (*1-chloro-2-methylpentane*)

$CH_3CCl(CH_3)CH_2CH_2CH_3$ (*2-chloro-2-methylpentane*)

$CH_3CH(CH_3)CHClCH_2CH_3$ (*3-chloro-2-methylpentane*)

$CH_3CH(CH_3)CH_2CHClCH_3$ (*4-chloro-2-methylpentane*)

$CH_3CH(CH_3)CH_2CH_2CH_2Cl$ (*5-chloro-2-methylpentane*)

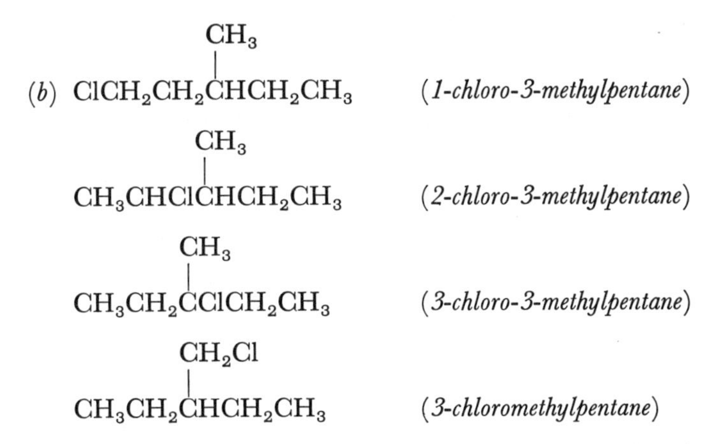


*Step 7.* Draw a four-carbon skeleton and determine how many positions are possible for the two extra carbons.

**#3–18**

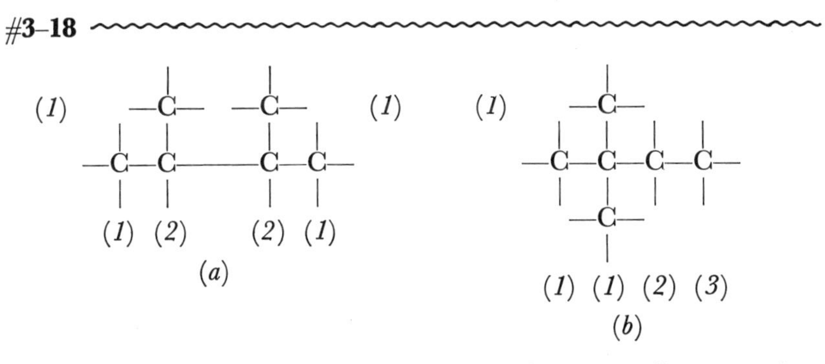


There is one structure with the extra carbons on adjacent carbons of the main chain, *a*, and one structure where they are on opposite sides of the same carbon, *b*. If the two extra carbons are joined and bonded to a nonterminal carbon in the chain, the structure in #3–16*b* results.

*Step 8.* Position a halogen on each of the different carbons in the two structures shown in #3–19. If we were to reduce the main chain to three carbons and to try to position the three remaining carbons, we would start repeating structures. This means that we have completed the problem and have drawn all seventeen position isomers of $C_6H_{13}Cl$.

**#3–19**

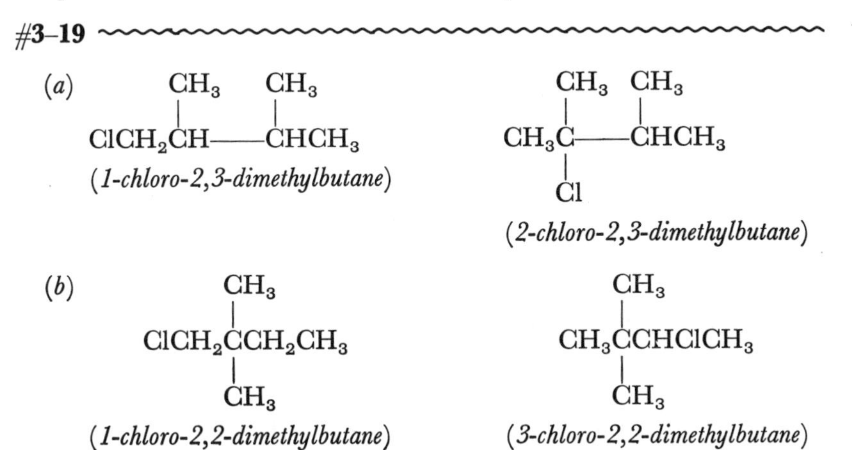


$$CH_3\overset{CH_3}{\underset{CH_3}{C}}CH_2CH_2Cl$$

(*4-chloro-2,2-dimethylbutane*)

## § 4. Problems. Two-Dimensional Structural Formulas. Position Isomerism

**1**) Write "condensed" formulas for:

```
      H  Cl  H
      |  |   |
a) H—C—C—C—OH
      |  |   |
      H  H   H
```

```
         H                   H
         |                   |
      H—C—H    H    H—C—H    H
         |     |         |       |
b) H—————C—————C—————————C———————C—H
         |     |         |       |
         H     CN        H       H
```

```
      Cl    OH   CH3   H    H
      |     |    |     |    |
c) Cl—C—————C————C—————C————C—H
      |     |    |     |    |
      H     H    H     H    H
```

```
              H                          H
              |                          |
      H   H—C—H    H  H           H—C—H    H
      |       |     |  |                |      |
d) H—C———————C—————C—C———O————————C——————C—H
      |       |     |  |                |      |
      H       OH    H  H           H—C—H    H
                                        |
                                        H
```

```
      H  Br     H     O
      |  |      |     ‖
e) H—C—C—————C—————C—H
      |  |      |
      H  H   H—C—H
                |
                H
```

**2**) Write two-dimensional structural formulas for:

*a*) $(CH_3)_2CHCHClCH_2CH_2OH$
*b*) $(CH_3)_3CCH(CH_3)CHO$
*c*) $CH_3CHOHCHBrC(CH_3)_2CH_2OCH_2Cl$
*d*) $CH_2ClCH_2C(CH_3)_2CH_2OH$
*e*) $CH_2OHCH_2CCl_2CH(CH_3)CH_2CHOHCH_3$

**3**) Label each carbon and each hydrogen as to whether it is 1°, 2°, or 3°.

*a)* $CH_3\overset{\overset{CH_3}{|}}{C}HCH_2CH_3$  *b)* $CH_3\underset{\underset{CH_3}{|}}{\overset{\overset{CH_3}{|}}{C}}CH_2CH_2CH_3$

*c)* $CH_3CH_2\overset{\overset{CH_3}{|}}{C}HCH_2CH_3$

**4**) A hydrogen atom is referred to as "different" from another hydrogen atom when the relationship of the first hydrogen to all the other atoms in the molecule differs from that of the second hydrogen atom. There may be different 1° hydrogens in a given molecule and different 2° hydrogens, etc. How many different kinds of hydrogens are there in each of the three structures in question 3?

How many different kinds of hydrogens are there in each of the structures given in question 5?

**5**) Indicate which of the structural formulas in each series represent the same compound.

*a)* *1)* $CH_3CH_2CH_2Cl$  *2)* $CH_3CHClCH_3$  *3)* $H{-}\underset{\underset{H}{|}}{\overset{\overset{Cl}{|}}{C}}CH_2CH_3$

*4)* $CH_3\underset{\underset{H}{|}}{\overset{\overset{Cl}{|}}{C}}CH_3$  *5)* $CH_2ClCH_2CH_3$  *6)* $\underset{\underset{CH_2CH_3}{|}}{CH_2Cl}$

*b)* *1)* $CH_3CH_2\overset{\overset{CN}{|}}{C}HCH_2CH_2CH_3$  *2)* $\overset{\overset{CH_3}{|}}{C}H_2CH\underset{\underset{CN}{|}}{\overset{\overset{CH_3}{|}}{C}}HCH_3$

*3)* $CH_3CH_2CH_2\overset{\overset{CN}{|}}{C}HCH_2CH_3$  *4)* $\underset{\underset{\underset{\underset{CH_3}{|}}{CH_2}}{|}}{CH_2}{-}\underset{\underset{CN}{|}}{\overset{\overset{H}{|}}{C}}{-}\overset{\overset{CH_3}{|}}{C}H_2$

*5)* $\overset{\overset{CH_3}{|}}{C}H_2\underset{\underset{CN}{|}}{C}HCH_2\overset{\overset{CH_3}{|}}{C}H_2$  *6)* $\overset{\overset{CH_3}{|}}{C}H_2CH_2\underset{\underset{CN}{|}}{\overset{\overset{CH_3}{|}}{C}}CH_3$

*c)* *1)* $CH_3\overset{\overset{CH_3}{|}}{C}HCH_2\overset{\overset{CH_3}{|}}{C}HCH_2CH_3$  *2)* $CH_3CH_2\overset{\overset{CH_3}{|}}{C}H{-\!-}\overset{\overset{CH_3}{|}}{C}H\underset{\underset{CH_3}{|}}{C}H_2$

```
      CH3  CH3                              CH3     CH3
      |    |                                |       |
3) H—CCH2CHCH2                 4) CH3CH2CHCH2CHCH3
      |         |
      CH3       CH3

      CH3  CH3                              CH3
      |    |                                |
5) CH3CCH2C—H                  6) CH3CH2CCH2CH2
      |    |                                |   |
      CH3  CH3                              CH3 CH3
```

```
d) 1) CH2————CH2               2) CH2————CHCH3
      |       |                      |       |
      CH2     CHCH2Br                CH2     CH2
        \    /                         \    /
         CH2                             CH
                                         |
                                         Br

               Br
               |
   3) CH2————CCH3              4) CH2————CH2
      |       |                      |       |
      CH2     CH2                    CH2     CHCH2Br
        \    /                         \    /
         CH2                             CH2

   5) CH2————CH2               6) CH2—CH2
      |       |                      |      \
      CH2     CH2                    |       >CH2
        \    /                       |      /
         CH                          CH2—CHCH2Br
         |
         CH2Br
```

*e*) *1*) $CH_3CHOHCH_2CH_3$ *2*) $CH_3CH_2OCH_2CH_3$
*3*) $CH_3CH_2CHOHCH_3$ *4*) $CH_3OCH_2CH_2CH_3$
*5*) $CH_2OHCH_2CH_2CH_3$ *6*) $CH_3CH_2CH_2CH_2OH$

**6**) Draw the structure of all the position isomers that can be represented by each molecular formula. Keep your answers and after you have studied nomenclature write IUPAC names for each structure.

*a*) $C_4H_9Br$ *b*) $C_6H_{14}$ *c*) $C_5H_{10}Cl_2$ *d*) $C_7H_{16}O$ *e*) $C_8H_{18}$

## § 5. The Alkanes. Nomenclature

The *precise* chemical name of a substance provides sufficient information so that any chemist can draw an exact structural formula for the compound without having to look it up in a reference book. This is not true for most of the commercial names, such as

Dacron $(\sim\!\overset{\overset{\displaystyle O}{\|}}{C}C_6H_4\overset{\overset{\displaystyle O}{\|}}{C}OCH_2CH_2O\!\sim)_n$,

aspirin ($o$-$CH_3COOC_6H_4COO^-Na^+$), or

Novocain ($p$-$H_2NC_6H_4COOCH_2CH_2N(C_2H_5)_2$).

Structures cannot be deduced from the commercial names.

Rules governing the systematic naming of organic compounds have been established by the International Union of Pure and Applied Chemistry. These rules are internationally accepted and used, so that all chemists everywhere refer to a particular structure by the same name. Only the more fundamental of the IUPAC naming rules are presented in this book. Our objective here is to introduce you to the underlying "thought pattern" used to translate a name into a structural formula and vice versa. Once you understand the basic approach, you can extend it logically to cover complex situations.

In the IUPAC system each name gives enough information so that you can draw an accurate structural formula of the compound. The IUPAC name tells us precisely:

- the kinds of atom that are present,
- the number of each kind, and
- how they are arranged relative to each other.

Organic compounds composed of only carbon and hydrogen are called *hydrocarbons*. A hydrocarbon without any double or triple bonds is a *saturated hydrocarbon*. The generic name for a saturated hydrocarbon is *alkane*.

Starting with the alkanes we will gradually develop for you the set of rules that makes it possible for a name to convey the information enumerated above. The formula and the IUPAC name for each of the first fifteen continuous chain alkanes are listed in Table 5–1.

Analysis of these names shows that they are each composed of two parts:

- The first part of the name varies according to the number of carbon atoms present: *prop* (three) ane, *hex* (six) ane, *dec* (ten) ane, *dodec* (twelve) ane. These word stems will be found in many different combinations, but they always carry the same numerical significance. Learn to recognize the first ten without having to look them up. Starting with five, the word roots come from Greek. The syllable *alk* is used to denote all the possible numbers of carbon atoms. Thus, each compound in Table 5–1 is an alkane.
- The ending of each name in Table 5–1 is the same, *ane*. The syllable *ane* is used to indicate that the compound is a saturated hydrocarbon.

**Table 5–1**

**Alkanes**

| *Structural formula* | *Molecular formula* | *IUPAC name* |
|---|---|---|
| $CH_4$ | $CH_4$ | methane |
| $CH_3CH_3$ | $C_2H_6$ | ethane |
| $CH_3CH_2CH_3$ | $C_3H_8$ | propane |
| $CH_3(CH_2)_2CH_3$ | $C_4H_{10}$ | butane |
| $CH_3(CH_2)_3CH_3$ | $C_5H_{12}$ | pentane |
| $CH_3(CH_2)_4CH_3$ | $C_6H_{14}$ | hexane |
| $CH_3(CH_2)_5CH_3$ | $C_7H_{16}$ | heptane |
| $CH_3(CH_2)_6CH_3$ | $C_8H_{18}$ | octane |
| $CH_3(CH_2)_7CH_3$ | $C_9H_{20}$ | nonane |
| $CH_3(CH_2)_8CH_3$ | $C_{10}H_{22}$ | decane |
| $CH_3(CH_2)_9CH_3$ | $C_{11}H_{24}$ | undecane |
| $CH_3(CH_2)_{10}CH_3$ | $C_{12}H_{26}$ | dodecane |
| $CH_3(CH_2)_{11}CH_3$ | $C_{13}H_{28}$ | tridecane |
| $CH_3(CH_2)_{12}CH_3$ | $C_{14}H_{30}$ | tetradecane |
| $CH_3(CH_2)_{13}CH_3$ | $C_{15}H_{32}$ | pentadecane |

Each compound in Table 5–1 differs from the one above it by one carbon and two hydrogens. A group of compounds with such a structural relationship is called an *homologous series*. In an homologous series, the numerical relationship between the atoms in each member of the series can be expressed by a single *generic formula*, in which "n" is the number of carbon atoms. This formula for the continuous chain alkanes in Table 5–1 is $C_nH_{2n+2}$. For example: in hexane, n is 6, so the number of hydrogen atoms is $(2 \times 6 + 2)$ 14; in decane, n is 10, so the number of hydrogen atoms is 22.

The generic formula is the same whether the carbon chain is continuous, as for the compounds in Table 5–1, or branched, as in the compound in #5–6.

The carbon skeleton of a compound may also be in the form of a ring. The term *cyclo* is used to indicate the presence of a ring. Thus, a cyclic saturated hydrocarbon is a *cycloalkane*. The syllable *alk* is replaced by the one for the correct number of carbons to indicate a specific compound as in #5–1. The generic formula for the cyclic alkanes is $C_nH_{2n}$. There are two less hydrogens because of ring formation.

**#5–1**

```
CH2—CH2                         CH2—CH2
 |    |     ≡  □               /       \
CH2—CH2                      CH2         CH2  ≡  ⬠
                                \       /
                                  CH2
```

(*cyclobutane*, $C_4H_8$) (*cyclopentane*, $C_5H_{10}$)

Note the two ways of drawing cyclic structures. The one on the right in each set in #5–1 is used most frequently. Be sure to keep in mind, when you are counting hydrogens in a formula, that these line drawings imply two hydrogens on each carbon.

Certain combinations of atoms recur frequently and are named as a unit. For example: $—CH_3$ = methyl, $—CH_2CH_3$ = ethyl, $—CH_2CH_2CH_3$ = propyl. These units are referred to as groups or as *substituents* when they are bonded to a carbon chain or to some atom other than carbon, as shown in #5–2. They are called *radicals* when discussed as entities not bonded to another structure.

**#5–2**

$-CH_2CH_2CH_3$ (on cyclopentane ring)

(*propylcyclopentane*)

$-CH_2CH_3$ (on benzene ring)

(*ethyl benzene*)

$$\begin{array}{c} \quad\;\; CH_3 \\ \quad\;\; | \\ CH_3CH_2CHCH_2CH_3 \end{array}$$

(*3-methylpentane*)

$(CH_3CH_2)_2NH$

(*diethyl amine*)

Each of these groups can be considered as derived from a hydrocarbon in which one hydrogen has been removed. The name of the group is obtained by replacing the *ane* ending of the parent hydrocarbon with *yl*. The generic term is *alkyl group*.

An example of a compound with alkyl groups substituent on a hydrocarbon chain is given in #5–3.

**#5–3**

$$\begin{array}{l} \quad\;\; CH_3 \qquad\quad CH_3 \\ \quad\;\;\; | \qquad\qquad\quad | \\ CH_3CCH_2CHCHCH_2CH_3 \\ \quad\;\;\; | \qquad\;\; | \\ \quad\;\; CH_3 \;\; CH_2 \\ \qquad\qquad\;\; | \\ \qquad\qquad CH_3 \end{array}$$

Such compounds are named by indicating through numbers the position of each substituent on the longest continuous carbon chain. The two main points initially to be determined when naming a compound are:

- The number of carbons in the longest continuous chain, and
- the order in which this carbon chain should be numbered.

These two operations are accomplished as follows:

*1*) A carbon chain is continuous if you can draw a line through each of its carbons without raising your pencil from the paper and without backtracking over any part of the line already drawn. In #5–3 the seven-carbon chain drawn horizontally across the page is the longest one.

*2*) The carbon atoms in the chain are numbered in sequence starting from each end of the chain, as shown in #5–4. The order of numbering

**#5–4**

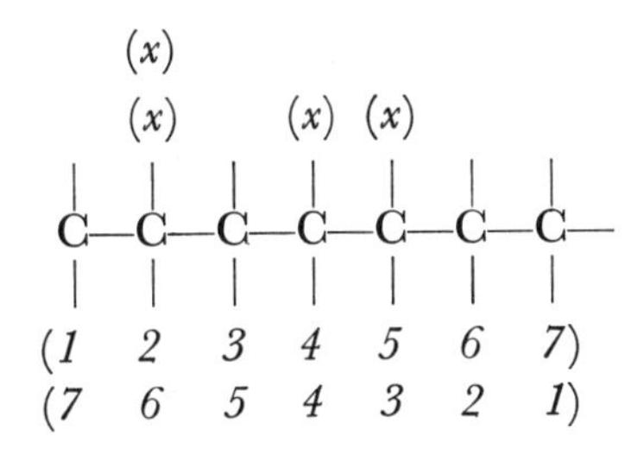


has to be such that the sum of the numbers assigned to substituents has the lowest possible value.

To determine the correct order of numbering, add the numbers that would be assigned to substituents by each order. In #5–4 an *x* indicates the position of the substituent alkyl groups. There are two

$x$'s on one carbon because there are two substituents on that carbon, and that number will appear in the name twice.

The sum of the numbers marked by $x$ is $(6 + 6 + 4 + 3)$ 19 when the chain is numbered from right to left. The sum is $(2 + 2 + 4 + 5)$ 13 when the chain is numbered from left to right. Therefore, the correct order of numbering is from left to right.

We are now ready to name the compound in #5–3 according to IUPAC rules:

- The longest chain has seven (*hept*) carbons and is saturated (*ane*). The compound is named as a derivative of heptane.
- When a group appears more than once, a prefix is used to indicate how many times it occurs: twice is represented by *di*, three times by *tri*, four times by *tetra*, five times by *penta*. In this molecule there are three methyl groups, giving *trimethyl*.
- A number placed before the name of a substituent locates its position on the chain. There is an ethyl group on the fourth carbon, indicated as 4-ethyl. When one group occurs several times, all numbers are listed in ascending order. In this molecule there are two methyl groups on carbon two and one methyl group on carbon five, giving a total of three (*tri*) methyl groups. This information is shown as: 2,2,5-trimethyl. *Note*: There are commas between numbers, and a hyphen between a number and a group name.
- For reference purposes each substituent is listed alphabetically according to the first letter of its name. For the purpose of alphabetizing, the prefix is ignored. Thus, *e*thyl precedes both di*m*ethyl and tri*m*ethyl in any name.

Putting all this information together we get the IUPAC name for the compound in #5–3 as: 4-ethyl-2,2,5-trimethylheptane.

The purpose of all these rules, particularly the one on numbering, is to make certain that every chemist, whether he is in India, Peru, Sweden, or North America, gives the same name to a particular structure no matter from which side he happens to draw it. The compound in #5–3 is drawn in four additional ways in #5–5. As you glance at these formulas, it is not immediately evident that they are

**#5–5**

```
     CH3
     |
     CH2       CH3                       CH3    CH3
     |         |                         |      |
CH3CHCHCH2CCH3                      CH3CCH3   CH2
         |         |                         |      |
         CH2     CH3                       CH2CHCHCH3
         |                                   |
         CH3                                 CH2
                                             |
                                             CH3
```

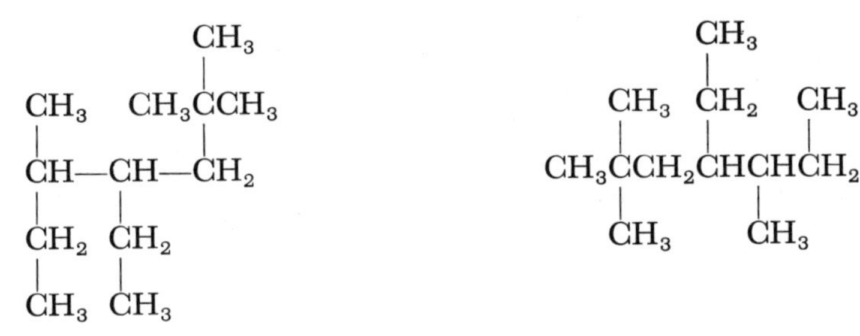

all the same compound as #5–3. Yet if you name each one according to IUPAC rules you will get the same name for each one. Try it.

The alkyl substituent is not always the simple continuous chain used in the above example. The substituent may be branched as in the compound #5–6.

**#5–6**

```
                    CH3
                    |
                 H—C—CH3
                    |
                 H—C—CH3
                    |
CH3CH2CH2CH2CHCH2CH2CH2CH3
```

The IUPAC rules for naming such a substituent are:

- The longest carbon chain in the substituent group is identified and the substituent is named as a derivative of an alkyl group with this number of carbons. For example, in #5–6 there are three carbons in the longest chain of the group, so the substituent is named as a derivative of a propyl group.
- The longest chain in the group is numbered by assigning the number 1 to the carbon bonded to the main chain, as indicated in #5–7.

**#5–7**

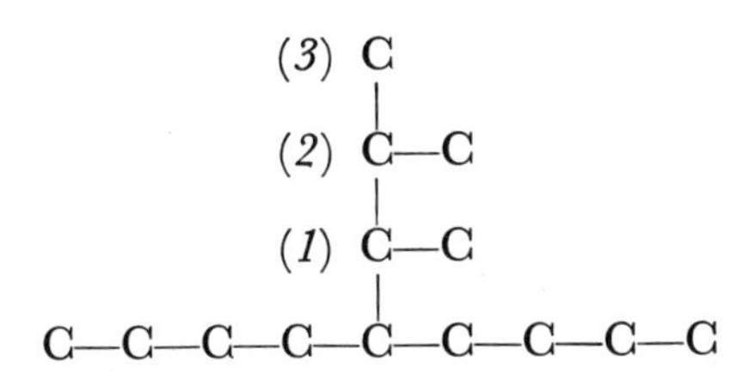


- Each group on the substituent chain is designated by name and number. In this instance there is a methyl group on carbon one and another on carbon two, giving 1,2-dimethyl.
  Thus the entire substituent is a 1,2-dimethylpropyl group.
- The name of the substituent is enclosed in parentheses when it is a complex name such as this one. The position of the substituent group on the main carbon chain is indicated by a number in front of the parenthesis.

Thus, the compound in #5–6 is 5(1,2-dimethylpropyl)nonane.

* * * * *

Before the IUPAC naming system was developed, many names had been coined for compounds and were in common use. Many of these have remained in the vernacular and are referred to as *common names*. These common names are used frequently for structures with four carbons or less and for continuous-chain hydrocarbons with a methyl group on the next-to-last carbon of the chain.

You need to be acquainted with the common names for the radicals arising from alkanes with up to four carbons. The one- and two-carbon radicals, methyl and ethyl, are the same in both naming systems. The difference comes in the three- and four-carbon radicals. Table 5–2 lists these radicals, giving both the IUPAC name obtained as discussed above and the common name, which we will now discuss.

Consider the compound propane, $CH_3CH_2CH_3$. A propyl group is one in which a hydrogen has been removed from the parent hydrocarbon, $CH_3CH_2CH_2$—. As discussed in §3, hydrogens are classified as 1°, 2°, and 3° depending upon whether they are bonded to 1°, 2°, or 3° carbons. A 1° hydrogen behaves differently chemically from the way a 2° hydrogen behaves; so we refer to 1° and 2° hydrogens as different kinds of hydrogens. Propane contains both 1° and 2° hydrogens; so two different propyl groups can be formed, as in #5–8. These are differentiated in the common naming system by the prefix *n* (normal) for the continuous chain group and *iso* (isomer) for the branched chain. The IUPAC designation for these two groups is *propyl* and *1-methylethyl*, respectively.

**#5–8**

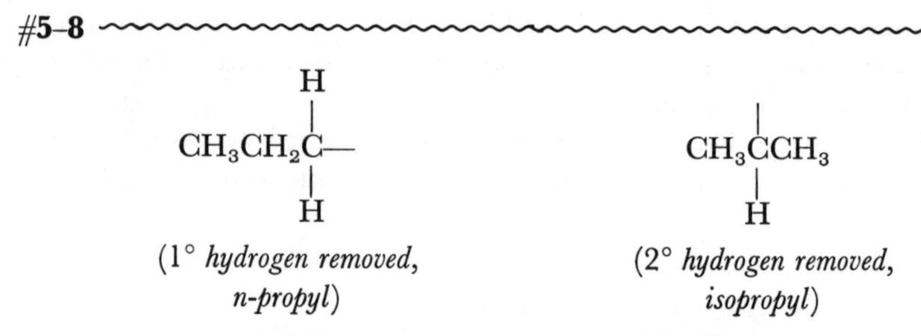


There are two arrangements for four carbons in an alkane. The chain may be continuous or it may be branched, as in #5–9. Note that

**#5–9**

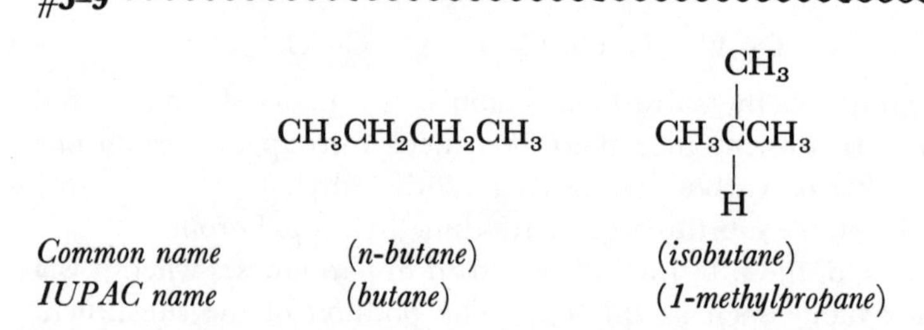


the common name is based on the total number of carbon atoms in the molecule, while the IUPAC name is based directly on only the number of carbons in the longest chain.

There are both 1° and 2° hydrogens in *n*-butane, which leads to two butyl radicals as shown in #5–10. The *n*-butyl radical has a

**Table 5–2**

**Common Names of Simple Alkyl Groups**

| *Alkane* | *Alkyl group* | *Common name* | *IUPAC name* |
| --- | --- | --- | --- |
| $CH_4$ (*methane*) | $—CH_3$ | methyl | methyl |
| $CH_3CH_3$ (*ethane*) | $—CH_2CH_3$ | ethyl | ethyl |
| $CH_3CH_2CH_3$ (*propane*) | $—CH_2CH_2CH_3$ | *n*-propyl | propyl |
| | $CH_3\overset{|}{\underset{H}{C}}CH_3$ | isopropyl | 1-methylethyl |
| $CH_3CH_2CH_2CH_3$ (*butane*) | $—CH_2CH_2CH_2CH_3$ | *n*-butyl | butyl |
| | $CH_3\overset{|}{\underset{H}{C}}CH_2CH_3$ | *sec*.-butyl | 1-methylpropyl |
| $CH_3\overset{CH_3}{C}HCH_3$ (*isobutane*) | $—CH_2\overset{CH_3}{C}HCH_3$ | isobutyl | 2-methylpropyl |
| | $CH_3\overset{|}{\underset{CH_3}{C}}CH_3$ | *tert*-butyl | 1,1-dimethylethyl |
| $CH_3CH_2\overset{CH_3}{C}HCH_3$ (*isopentane*) | $—CH_2CH_2\overset{CH_3}{C}HCH_3$ | isopentyl | 3-methylbutyl |
| $CH_3—\overset{CH_3}{\underset{CH_3}{C}}—CH_3$ (*neopentane*) | $—CH_2\overset{CH_3}{\underset{CH_3}{C}}—CH_3$ | neopentyl | 2,2-dimethylpropyl |

**#5–10**

$$CH_3CH_2CH_2\overset{H}{\underset{H}{C}}— \qquad CH_3CH_2\overset{H}{\underset{|}{C}}CH_3$$

| | | |
| --- | --- | --- |
| *Common name* | (*n-butyl*) | (*sec-butyl*) |
| *IUPAC name* | (*butyl*) | (*1-methylpropyl*) |

continuous chain. The *sec*-butyl radical is the one formed when a secondary hydrogen is removed. There is branching of one methyl group at the carbon where the radical forms a bond with another atom.

There are nine 1° hydrogens and a 3° hydrogen in isobutane. Again there are two possible radicals (see #5–11). A tertiary hydrogen

**#5–11**

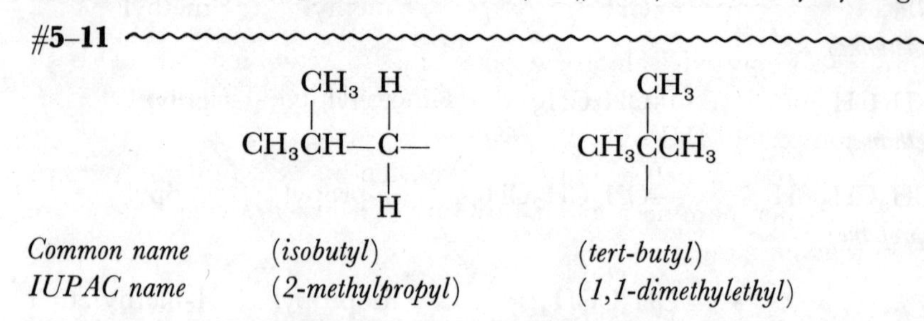


has been removed to form the *tert*-butyl group. A 1° hydrogen has been removed to form the isobutyl group. A 1° was also removed to form the *n*-butyl group. The difference between these two does not lie in the kind of hydrogen that is removed but in the nature of the carbon chain. There is an unbranched chain in the *n*-butyl group, while the chain of the isobutyl group has a methyl group on the next-to-last carbon of the chain. This branching is more clearly seen in longer chains, as indicated in #5–12.

**#5–12**

$$CH_3CH(CH_3)CH_2CH_2CH_2—$$

$$CH_3CH(CH_3)CH_2CH_2CH_2CH_2CH_2CH_2—$$

| | | |
|---|---|---|
| *Common name* | *(isohexyl)* | *(isononyl)* |
| *IUPAC name* | *(4-methylpentyl)* | *(7-methyloctyl)* |

* * * * *

Frequently used common names for a number of different classes of compounds are based on the names of the radicals containing four carbons or less. For an example, see #5–13.

**#5–13**

| | |
|---|---|
| $CH_3CH(Cl)CH_3$ | *(isopropyl chloride)* |
| $CH_3CH_2CH(OH)CH_3$ | *(sec-butyl alcohol)* |
| $CH_3CH_2OCH_2CH(CH_3)CH_3$ | *(ethylisobutyl ether)* |
| $CH_3N(H)CH_2CH_2CH_3$ | *(methyl-n-propyl amine)* |

Although one should not normally mix up the two naming systems in one name, in recent years it has become accepted practice to designate substituents with four carbons or less by their common name even though the rest of the molecule is named according to IUPAC rules. Thus, the compound in #5–14 may be named as either 2-

**#5–14**

```
          CH3
          |
    CH3—C———————————Cl
          |     /    \
          CH3   \    /
                 ————
                   |
               H—C—CH3
                   |
                   CH3
```

isobutyl-4-isopropyl-1-chlorocyclopentane (a combination of both systems) or 2(1,1-dimethylethyl)-4(1-methylethyl)-1-chlorocyclopentane (completely IUPAC).

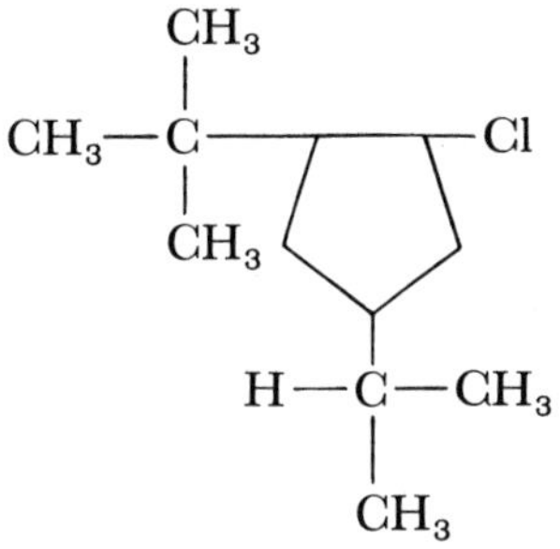

The Examples below illustrate the step-by-step thought pattern followed when naming a given structure or when drawing a structure from a given name.

*Example 1*—Name the substance represented by the two-dimensional structural formula in #5–15.

**#5–15**

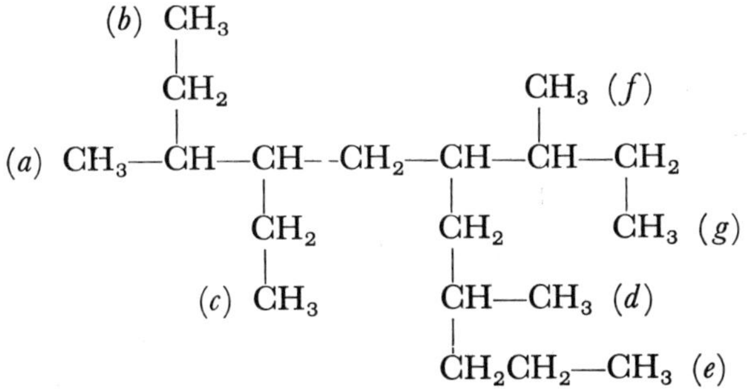

```
          (b) CH3
              |
              CH2                      CH3 (f)
              |                        |
    (a) CH3—CH—CH—-CH2—CH—CH—CH2
                  |          |         |
                  CH2        CH2       CH3 (g)
                  |          |
              (c) CH3        CH—CH3 (d)
                             |
                             CH2CH2—CH3 (e)
```

*Step 1.* Locate the longest carbon chain within the molecule.

To do so in a complicated molecule, first locate all possible chain ends. In a saturated hydrocarbon, chains end with a $—CH_3$. All possible chain ends in #5–15 are indicated by letters. There are 21 different chains in this molecule—e.g., *a–b*, *f–g*, *e–f*, etc. However, even a superficial glance shows that the three chains *a–e*, *b–e*, and *c–e* are the longest. A count reveals that they contain, respectively, 10, 11, and 10 carbons. The longest of all is *b–e*, with 11 carbons; the compound is named as a derivative of undecane.

The eleven-carbon chain has been outlined in #5–16 so that the

**#5–16**

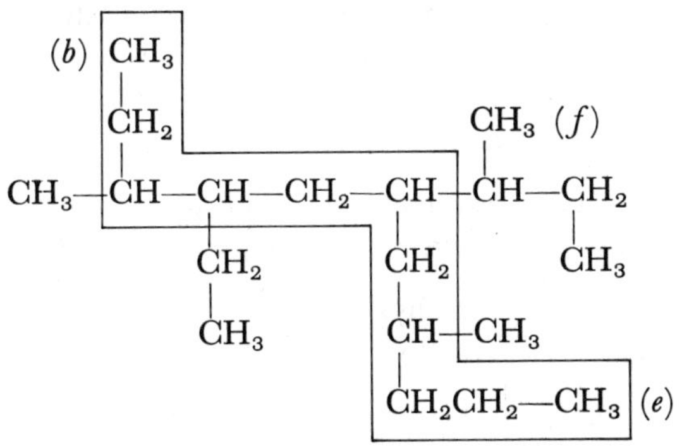

```
          (b) |CH3|
              | | |
              |CH2|___________________  CH3 (f)
              | |                     | |
          CH3—|CH—CH—CH2—CH—|CH—CH2
              |________   |   |     |  |
                  |    |  CH2 |     CH3
                  CH2  |  |   |
                  |    |  CH—|CH3
                  CH3  |  |   |_______
                       |  CH2CH2—CH3| (e)
                       |____________|
```

alkyl substituents stand out from the main chain as you look at the formula.

Note that "turning a corner" in the drawing of the carbon chain is of no significance. Bear in mind that the chain is really a three-dimensional zigzag of tetrahedrally bonded carbons. The bonding angles of all the hydrogens and of the carbons in the alkyl substituents are also tetrahedral.

*Step 2.* Number the carbon chain so that the positions of the substituents can be designated.

Draw the carbon skeleton of the undecane chain and indicate merely the position, not the nature, of the substituents as in #5–17, where the carbons with substituents are circled.

**#5–17**

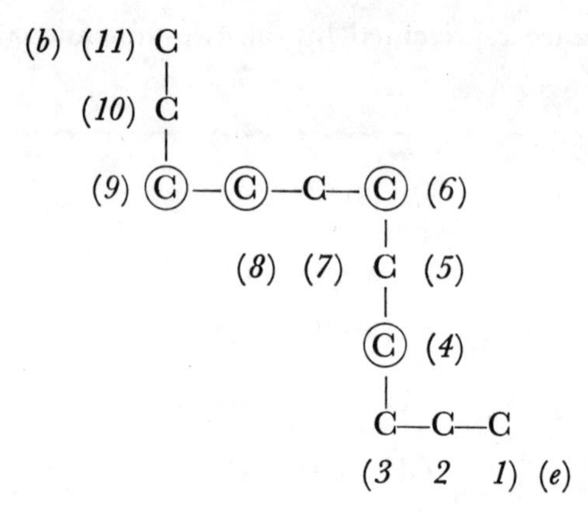


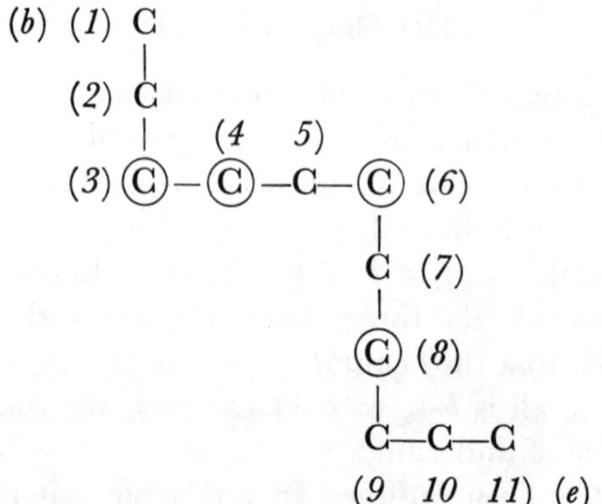


Number the chain in sequence from *b* to *e* and from *e* to *b*. For both orders of numbering, determine the sum of the digits assigned to carbons with substituents. The sum is 21 when *b* is 1 and *e* is 11, whereas the total is 27 when *e* is 1 and *b* is 11. Therefore, the correct order of numbering is from *b* to *e*.

*Step 3.* Name each substituent and assign it a number.

The substituents are shown most clearly in #5–16. They are:

| | | |
|---|---|---|
| $—CH_3$ | methyl | in position 3 |
| $—CH_3$ | methyl | in position 8 |

| | | |
|---|---|---|
| $—CH_2CH_3$ | ethyl | in position 4 |
| $—CHCH_2CH_3$ with $CH_3$ below the CH | *sec*-butyl or (1-methylpropyl) | in position 6 |

*Step 4.* Construct an IUPAC name from the above information.

- Compound is a derivative of an eleven-carbon saturated chain, and so is an *undecane.*
- The substituents in alphabetical order are: *sec*-butyl, ethyl, dimethyl.
- The position of each substituent is indicated by number, and all are assembled together with commas between consecutive numbers and hyphens between words and numbers, to read 4-ethyl-3,8-dimethyl-6-*sec*-butyl.
- Adding the substituents to the name of the hydrocarbon gives, as the IUPAC name of the compound, 4-ethyl-3,8-dimethyl-6-*sec*-butylundecane.

Note that if at any time in the naming of a compound, your result indicates that the *main* chain contains a 1-methyl, a 2-ethyl, or a 3-propyl substituent, you have made an error, and have not located the longest continuous chain. In each of these three erroneous designations, the prefix numeral corresponds to the number of carbons in the group.

*Example 2*—Draw a two-dimensional structural formula for 4,6-diethyl-2,2-dimethyl-5-isopropyl-5-propyloctane.

*Step 1.* Octane is an eight-carbon continuous chain. Draw the carbon skeleton and number it as in #5–18.

**#5–18**

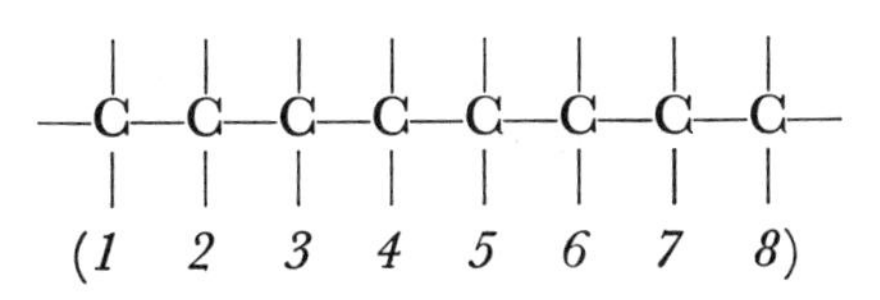

*Step 2.* List the structures of the substituents and their numbers.

| | | |
|---|---|---|
| methyl | $= —CH_3$ | position 2 and 2 |
| ethyl | $= —CH_2CH_3$ | position 4 and 6 |
| propyl | $= —CH_2CH_2CH_3$ | position 5 |
| isopropyl | $= CH_3CCH_3$ (bond up from the central C, H below it) | position 5 |

*Step 3.* Write the substituents in position on the carbon chain and fill in the hydrogens (#5–19).

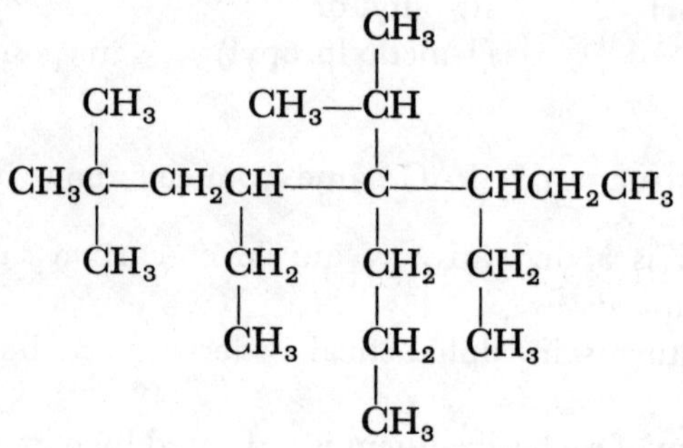

```
                          CH3
                           |
          CH3        CH3—CH
           |               |
      CH3C—CH2CH——C——CHCH2CH3
           |      |        |     |
          CH3    CH2   CH2   CH2
                  |        |     |
                 CH3   CH2   CH3
                           |
                          CH3
```

## § 6. Problems. Nomenclature of Alkanes

**1**) Draw two-dimensional structural formulas for:

*a*) octane
*b*) isobutyl chloride
*c*) 3-methylhexane
*d*) *tert*-butyl alcohol
*e*) 2,3-dimethylbutane
*f*) 2-ethylmethylcyclobutane
*g*) isobutylcyclopentane
*h*) ethylisopropyl ether
*i*) 3-ethyl-3-methylhexane
*j*) isobutyl-*sec*-butyl amine
*k*) 3-ethyl-4-propylheptane
*l*) 6,6-diethyl-4(isopropyl)-2,3,5,8-tetramethylnonane
*m*) 4-ethyl-3(3-methylbutyl)methylcyclohexane
*n*) 3,3-diethyl-2,4,7-trimethyl-6(1,1-dimethylethyl)-4-propyloctane
*o*) 5-ethyl-4(isobutyl)-2,2-dimethylheptane

**2**) Name by either IUPAC names or common names the compounds represented by these structural formulas:

*a*) $CH_3(CH_2)_7CH_3$

*b*)
```
      CH3    CH3
       |      |
   CH3CHCH2CHCH3
```

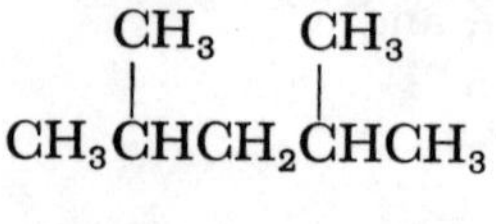

*c*) (cyclopentane ring)—$C(CH_3)_3$

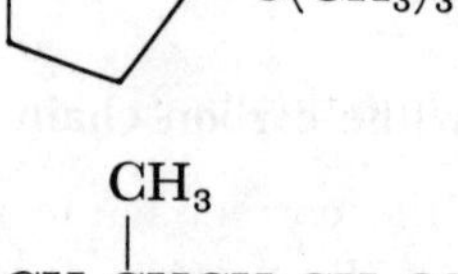

*d*)
```
      CH3
       |
   CH3CHCH2CH2OH
```

*e*) (cyclobutane ring)—$CH_2CHCH_3$ with $CH_3$ on the CH

$$\begin{array}{l} \qquad\qquad\quad CH_3 \\ \qquad\qquad\quad | \\ \square\!-\!CH_2CHCH_3 \end{array}$$

*f*)
$$\begin{array}{l} \quad\;\; CH_3 \\ \quad\;\; | \\ CH_3CHCH_2OCH_2CH_3 \end{array}$$

*g*) $CH_3CH_2CH(CH_2CH_3)CH_2CH(CH_3)_2$

*h*)
$$\begin{array}{l} \quad\;\; CH_3 \\ \quad\;\; | \\ CH_3CHBr \end{array}$$

*i*)
$$\begin{array}{l} (CH_3)_2CHCHCH(CH_3)_2 \\ \qquad\qquad | \\ \qquad\quad\;\; CH_3 \end{array}$$

*j*)
$$\begin{array}{l} \qquad\quad\;\; CH_3 \quad H \quad CH_3 \\ \qquad\qquad | \qquad\; | \qquad | \\ CH_3CH_2CH—N—CHCH_3 \end{array}$$

*k*)
$$\begin{array}{l} \qquad\quad\;\; CH_3 \;\; CH_3 \\ \qquad\qquad | \qquad\; | \\ CH_3CH_2C——CHCH(CH_3)_2 \\ \qquad\qquad | \\ \qquad\quad\;\; CH_3 \end{array}$$

*l*) (cyclopropane ring)—$CHCH_2CH_3$ with $CH_3$ on the CH

$$\begin{array}{l} \qquad\;\; CH_3 \\ \qquad\;\; | \\ \triangleright\!-\!CHCH_2CH_3 \end{array}$$

*m*) $(CH_3)_3C(CH_2)_4C(CH_3)_3$

*n*)
$$\begin{array}{l} \qquad\qquad\qquad\quad CH_3 \\ \qquad\qquad\qquad\quad | \\ CH_3CH_2CHCH_2CCH_3 \\ \qquad\qquad | \qquad\quad | \\ \qquad CH_3CH \quad HCCH_3 \\ \qquad\qquad | \qquad\quad | \\ \qquad\quad\; CH_2 \qquad CH_3 \\ \qquad\qquad | \\ \qquad\quad\; CH_3 \end{array}$$

*o*)
$$\begin{array}{l} \qquad\qquad\;\; CH_3 \\ \qquad\qquad\;\; | \\ \qquad\; CH_2CHCH_2CH_3 \\ \qquad\; | \\ CH_3CHCHCH—CHCH_2CH_3 \\ \qquad | \qquad\; | \qquad\;\; | \\ \quad\; CH_3 \quad CH_2 \;\; CH_2 \\ \qquad\qquad\quad | \qquad\;\; | \\ \qquad\qquad CH_3 \;\; CH_2CH_3 \end{array}$$

**3**) Name the structures you drew in answer to question 6, §4.

**4**) The following names are incorrect. Write the correct name and indicate what is wrong with the one given here. To do this, draw a structural formula corresponding to the name given and then name, according to the rules of the IUPAC naming system, the structure you have drawn.

*a*) 1-methyl-2-chlorobutane
*b*) 2-ethyl-4-methylpentane
*c*) 1,5-dimethylcyclohexane
*d*) 4-methyl-2-isopropylpentane
*e*) 2,4-dimethyl-3-isobutylpentane
*f*) 3-chloro-5-ethylhexane
*g*) 4-bromomethylcyclopentane
*h*) isobutylpropane
*i*) 2-ethyl-3-methyl-4-propylhexane
*j*) 2-*sec*-butyl-1-methylbutane

## § 7. Three-Dimensional Formulas and Conformation

While two-dimensional formulas of three-dimensional molecules may supply sufficient structural information for some problems, there are other problems which require a more precise designation of the spatial relationships between the atoms. Even drawings which show all three dimensions of the molecule in perspective cannot be readily understood without a great deal of practice in relating drawings to structures and vice versa with the help of models. To really understand the concepts of conformation presented in this section you will need to make models using a molecular model kit.

Construct a model of ethane, $CH_3CH_3$, with the atoms positioned as in #7–1. Hold the model in front of you so that you can sight along

#7–1

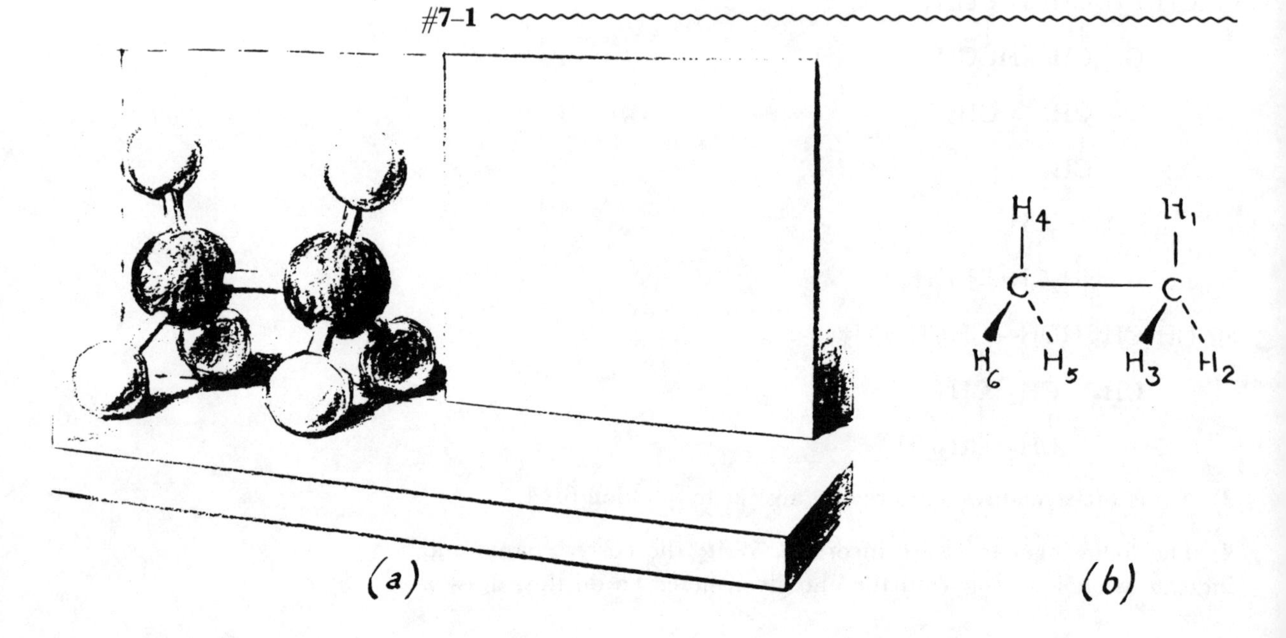

the carbon-carbon bond from C*1* to C*2*. If you have duplicated the structure in #7–1, you will find that each of the hydrogens on C*1* is aligned with a hydrogen on C*2*. You really cannot see the hydrogens on C*2* because they are eclipsed by the hydrogens on C*1*. This arrangement of the atoms is called the *eclipsed conformation* of ethane.

Let us examine the two different representations (#7–1*a* and *b*) of the eclipsed conformation so as to establish the conventions associated with each kind of drawing. Place the model on the desk in front of you with the carbon–carbon bond parallel to the front edge of the desk. Hold a 6″ × 8″ piece of cardboard upright (perpendicular to the desk top) next to the model, with the bottom edge of the cardboard also parallel to the front edge of the desk. The cardboard should be so positioned that its plane if extended would include the carbon–carbon bond and the bonds between C*1* and H*1*, and C*2* and H*4*, as shown in #7–2. Thus, you can imagine the plane of the cardboard as

**#7–2**

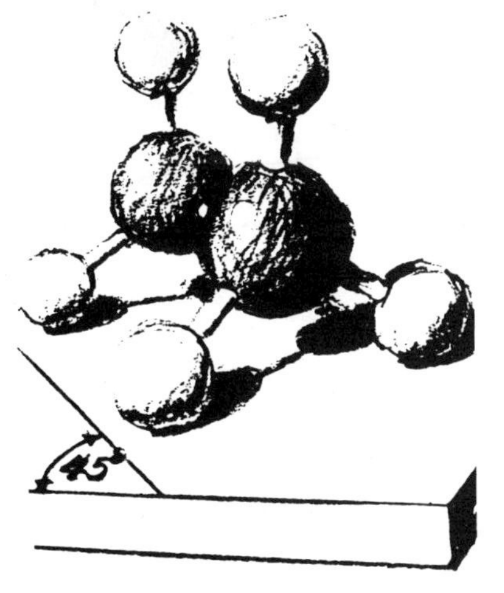


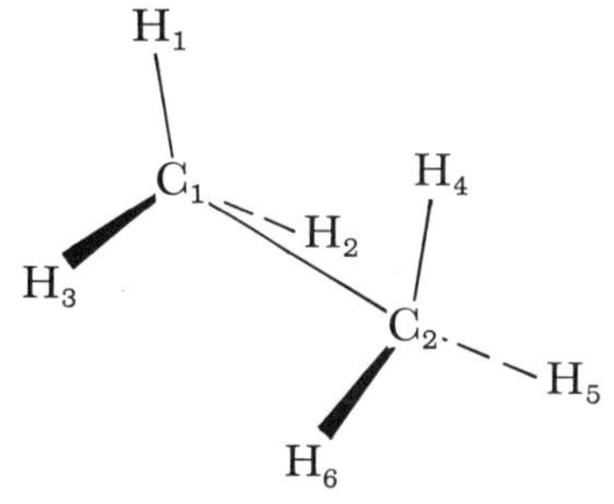


the plane of the page (plane XY). The bonds C*1*—H*2* and C*2*—H*5* project back behind the page as indicated by dotted lines, whereas the bonds C*1*—H*3* and C*2*—H*6* project out of the plane of the page toward you and are shown by wedge-shaped lines. The bonds that lie in the plane of the page, C*1*—C*2*, C*1*—H*1*, and C*2*—H*1*, are drawn with lines of even weight. This way of representing the spatial relationships between atoms is called a *projection formula*.

These drawings become cumbersome at times and the shorthand conventions of the *sawhorse* figure in #7–1*b* have been developed to make it easier to show molecular conformations. Both #7–1*a* and #7–1*b* represent the same eclipsed conformation of ethane, but the molecule is drawn in a different position relative to the page.

To visualize the position shown in #7–1*b* refer back to the model. Move C*1* so that the carbon–carbon bond, instead of being parallel to the edge of the desk, forms a 45° angle with it. The atom C*1* is now closer to you while C*2* remains in the plane of the upright cardboard representing the plane of the page. The carbon–carbon bond is to be visualized as projecting out of the page toward you at an angle of about 45°.

The two tetrahedral —$CH_3$ clusters are each represented by **λ**. The carbon atom is at the point of intersection of the three prongs. The closer —$CH_3$ group is drawn lower on the page and to the left of the more remote —$CH_3$ group. The two are joined by a diagonal line representing the carbon–carbon bond. No attempt is made to indicate that the prongs of the front cluster project toward the viewer in reference to C*1* while the prongs of the rear cluster project away from the viewer in reference to C*2*. You just have to remember the relationship as you have seen it in the model.

* * * * *

Now change the relative position of the atoms in your model, i.e., change its conformation. Rotate C*1* in the model so that the hydrogens on C*1* are no longer aligned with the hydrogens on C*2* but alternate with them. This arrangement is called the *staggered conformation.* If you sight along the carbon–carbon bond from C*1* to C*2* you can now see all six hydrogens. Before you look at #7–3 draw both a projection formula and a sawhorse diagram for the staggered conformation of ethane.

**#7–3**

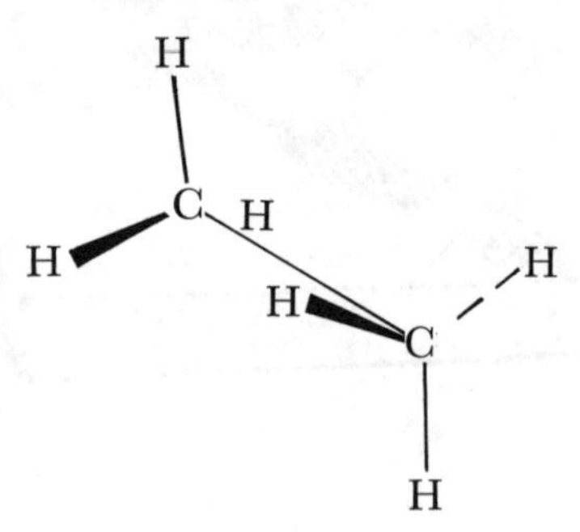


The structure in #7–3 is a *conformer* of the structure in #7–1. Conformers are different forms of the same substance attributable to variations in the relative positions of the atoms in the molecule. Hold your right hand in front of you, palm down with the fingers and thumb spread out. Then close your right hand into a fist. The difference between your open hand and your fist is the same kind of difference as that between conformers; the relative positions of the components differ. Your right hand is your right hand whether it is open or closed; ethane is ethane whether the hydrogens are aligned or staggered.

No bonds are made or broken during the transition from one conformer to another. The two conformers in #7–1 and #7–3 are possible because *atoms joined by a single bond are free to rotate.* The eclipsed and the staggered conformers represent only two of the infinite number of arrangements that are possible as the carbon rotates through 360° at an extremely rapid rate. Rotate C*1* in your model so that one specific hydrogen atom travels a complete circle of 360° and note the varying relationship between that hydrogen and those on C*2*.

Consider next the conformation of *n*-pentane at the bond between C*2* and C*3* ($CH_3CH_2$—$CH_2CH_2CH_3$). The substituents on C*2* are —$CH_3$, —H, and —H; those on C*3* are —$CH_2CH_3$, —H, and —H. Figure #7–4 presents sawhorse structures for some of the conformers that arise during the course of a single complete rotation of C*2* or C*3*. To help visualize these structures, make a model of *n*-pentane using different colors to represent different kinds of substituents and rotate C*2* through 360°. As you form one conformer after the other with your model, match each one with the appropriate drawing. Make some drawings of your own.

A parallel series of conformers drawn according to the *Newman convention* has been included in #7–4 to acquaint you with this second method of depicting conformations. In these drawings, the circle represents the front carbon C*2* and the lines extending from the center of the circle are the bonds from C*2*. The lines originating at the edge of the circle represent the bonds of the back carbon (C*3*) which is assumed to be directly behind C*2* and hidden by it.

As you rotate the carbons at the C*2*—C*3* bond, note that of the three eclipsed conformers there is one in which the two alkyl groups are aligned. This conformer is the *cisoid* form. Similarly, of the staggered conformers there is one in which the two alkyl groups are on opposite sides of the carbon–carbon bond. This conformer is the *transoid* or the *anti* staggered form.

* * * * *

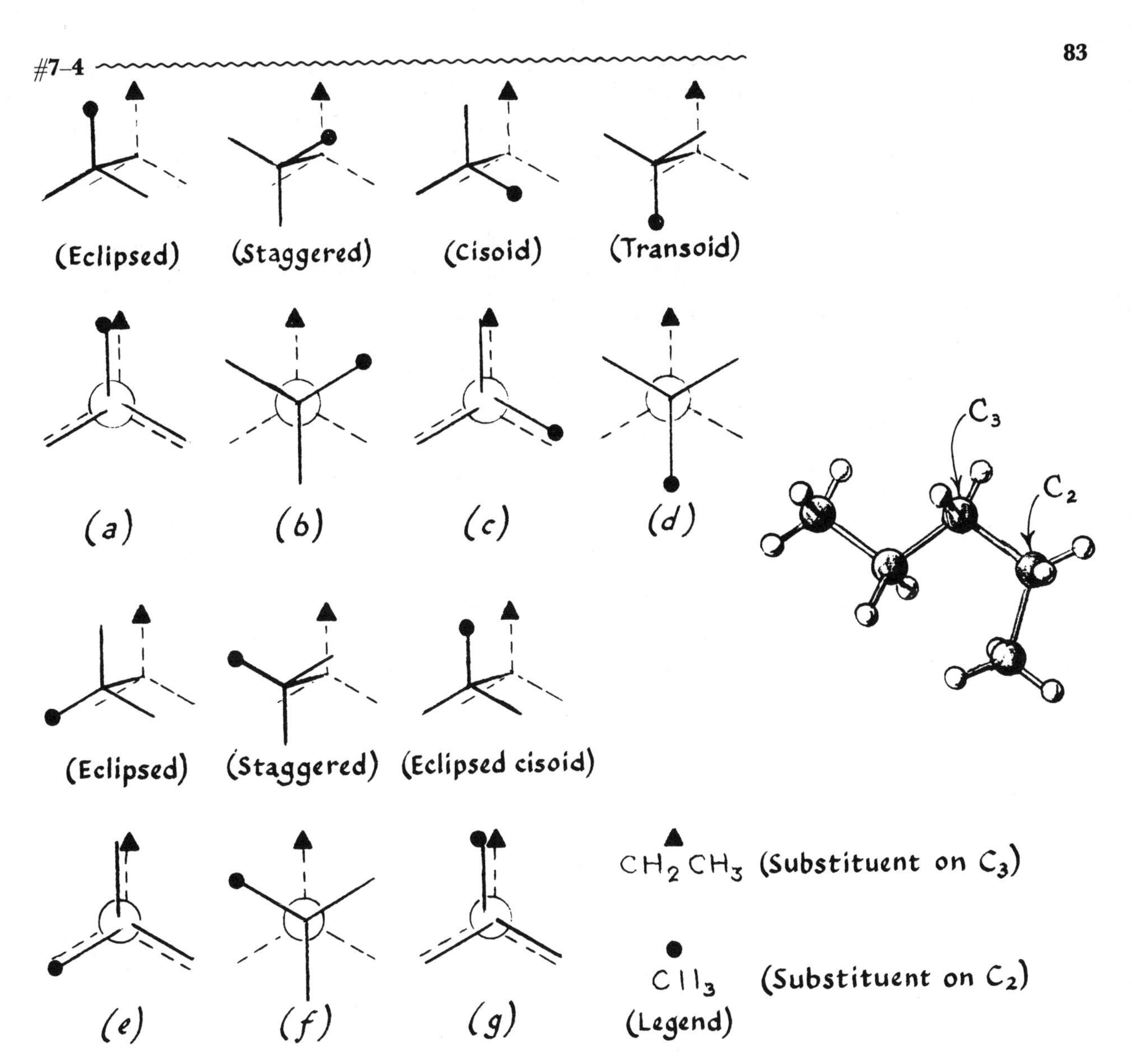


Free rotation at an unhindered single bond leads to a continually changing spatial relationship between the substituents on the two carbons. So one might expect a given sample of a particular substance to contain a random distribution of all possible conformers. However, there are stability factors that favor one conformer over another. The torsional strain is greater in an eclipsed conformer than in a staggered one, making the latter more stable. When substituent groups approach each other, closer than the van der Waal radii, repulsion occurs. Therefore, staggered conformers that place the largest group on one carbon between the smaller groups on the second carbon are more stable than staggered conformers that place the larger groups on each carbon close to each other. At any given moment the majority of the molecules in a sample of *n*-pentane are in the anti staggered conformation.

Conformers are rarely isolated or even isolatable. However, the excess of one particular conformer over another plays an important role in determining the products of many reactions.

Example 1 below has been included to give you some idea as to the actual mechanics of drawing sawhorse structures. This is the notation that we will use most frequently.

*Example 1*—Draw a sawhorse structure to illustrate the most stable conformation at the bond between C*3* and C*4* in *n*-hexane.

*Step 1.* Identify the substituents on C*3* and C*4* by writing the formula in such a way as to emphasize the bond under consideration.

As shown in #7–5, the substituents on C*3* are $CH_3CH_2$—, H—, and H—. Those on C*4* are the same.

#7–5

$$\begin{array}{c} \phantom{CH_3CH_2}\mathrm{H}\;\;\mathrm{H}\phantom{CH_2CH_3} \\ CH_3CH_2\overset{|}{\underset{|}{C}}\text{—}\overset{|}{\underset{|}{C}}CH_2CH_3 \\ \phantom{CH_3CH_2}\mathrm{H}\;\;\mathrm{H}\phantom{CH_2CH_3} \\ (3\;\;\;4) \end{array}$$

*Step 2.* Draw the sawhorse framework.

First draw the ⅄ for the rear unit. Then draw the front unit to the left and below the rear unit. The front ⅄ unit may point either up or down depending upon whether the structure is eclipsed or staggered. Connect the two units. These steps are shown in #7–6.

#7–6

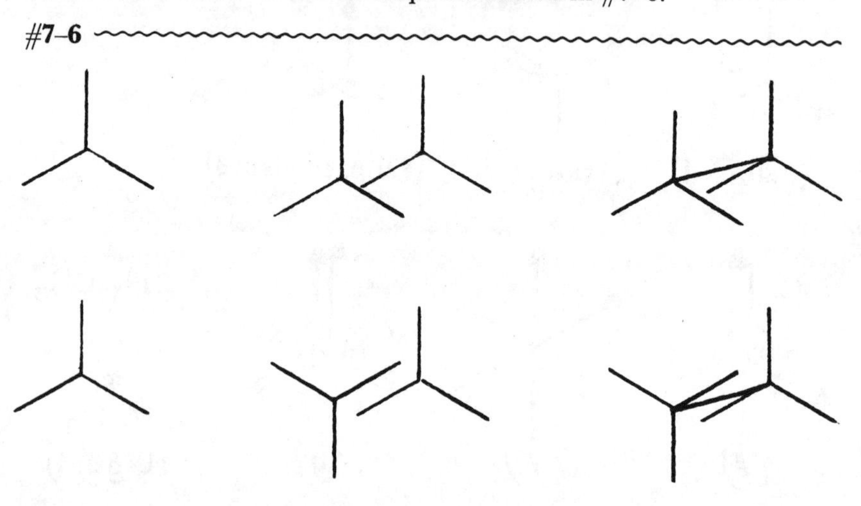

*Step 3.* Add the substituents to the sawhorse framework.

Position the substituents on the rear unit first and then draw all possible arrangements for the substituents on the front unit. Do this by moving each substituent around the ⅄ for each succeeding structure, as in #7–7 and #7–8.

#7–7

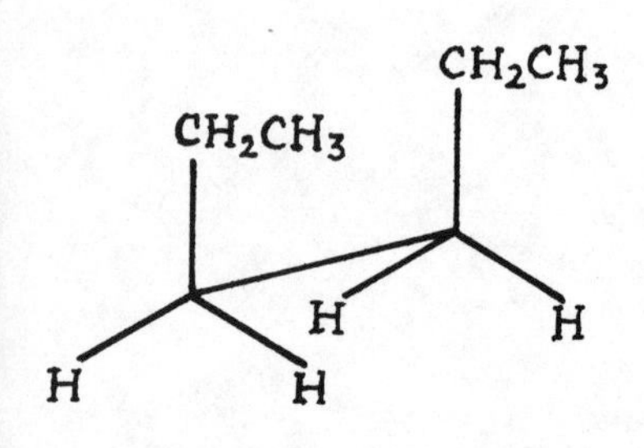


CH₂CH₃
H
H
H
H
CH₂CH₃

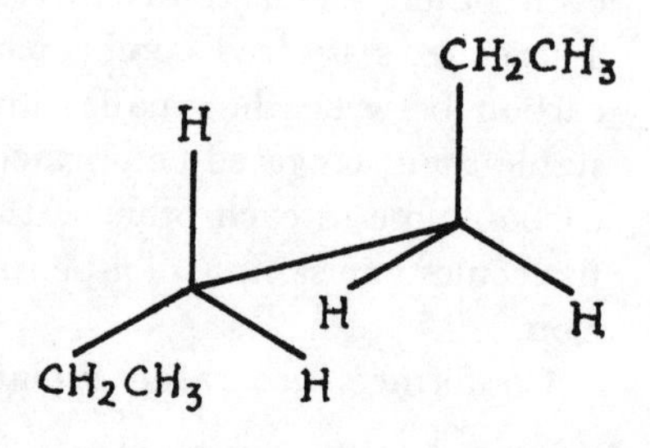

#7-8

*Step 4.* Decide which conformer is the most stable.

Eclipsed conformers are less stable than staggered conformers. Therefore, the structures in #7–7 can be eliminated.

The $CH_3CH_2$— groups are farthest apart in the anti or transoid conformation of the staggered conformers in #7–8. This, therefore, is the most stable conformer.

* * * * *

Free rotation cannot occur at the carbon–carbon bonds in a cyclic hydrocarbon. Yet there are numerous conformers for each cyclic structure. These arise from variations in the relative positions of the atoms in the carbon skeleton of the ring, variations that change the relative positions of the substituents on the ring.

Using components from your model kit, try to make the four-carbon cyclobutane in the form of a flat ring. Unless the material used to represent bonds is flexible, you cannot do so. As indicated in #7–9*a*, the angles at the corners of a square are 90°, considerably less than the

#7–9

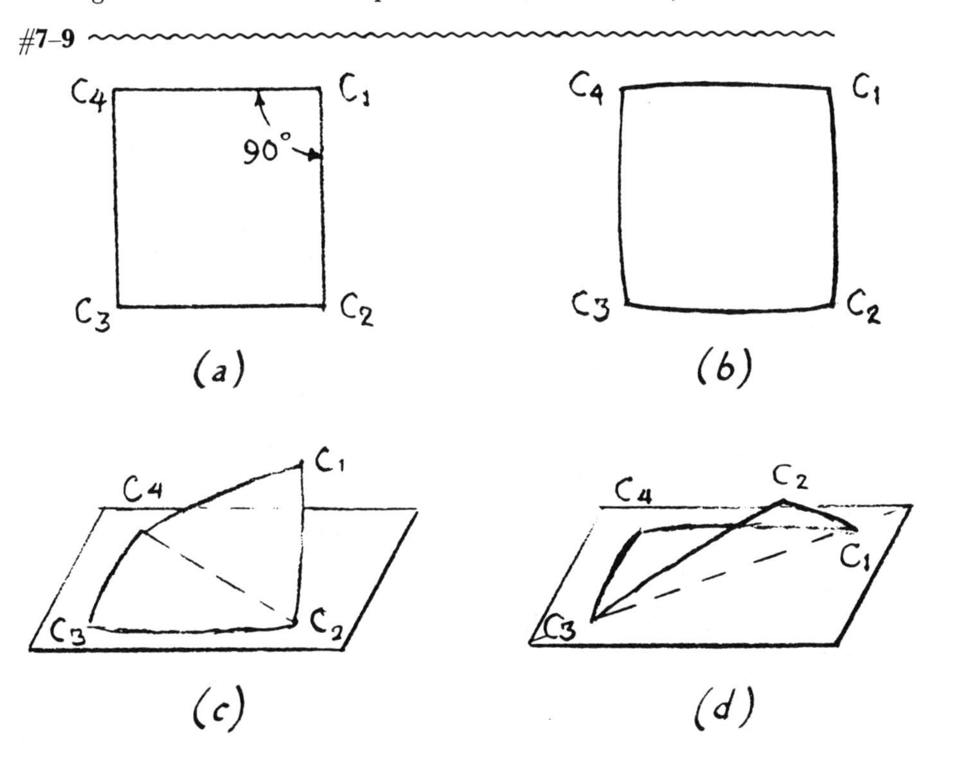

109.5° of tetrahedral bonding. If a flexible material is available for representing the bonds, it will curve outward and the model will assume the shape shown in #7–9*b*. This flat conformer represents a strained, unstable situation.

The strain can be reduced somewhat, as follows: Place your model on a table top with all four carbons touching the flat surface. Move C*1* up out of the plane which includes the other three carbons. Keep the other three carbons on the table, but allow them to rotate as you move C*1* so that the model does not separate. We do not want to "make" or "break" any bonds. We are merely changing the relative positions of the four carbons.

You now have the conformer shown in #7–9*c*. When the ring is "bent" rather than flat, the angles of bonding can more readily approximate 109.5° and there is less strain. Return C*1* to its original position on the table and raise C*2* out of the plane of the flat conformer. The ring is now "bent" in a different manner. Thus, the structures in #7–9*c* and *d* represent two different conformers. Raising C*1* produces the conformer in #7–9*c* while raising C*2* produces the conformer in #7–9*d*.

All the conformations between the extremes of *c* and *d* exist, including *b*. However, in any given sample of cyclobutane a greater number of molecules have the more stable conformations of *c* and *d* than have the less stable conformation of *b*. Despite the "bending" of the ring, the angles of bonding do not actually become tetrahedral; so cyclobutane is a less stable molecule than cyclohexane because of bond strain.

* * * * *

The six-carbon ring, cyclohexane, is a more stable molecule than cyclobutane because there are conformers in which the angles of bonding for all atoms are tetrahedral.

Try to make a flat ring using six carbon atoms from your model kit. As with cyclobutane you will have trouble unless you have flexible material to use for bonds. The strain in this unit causes the bonds to bow toward the center of the ring rather than outward, as they did in the model of cyclobutane. Compare #7–9*a* and #7–10*a*.

**#7–10**

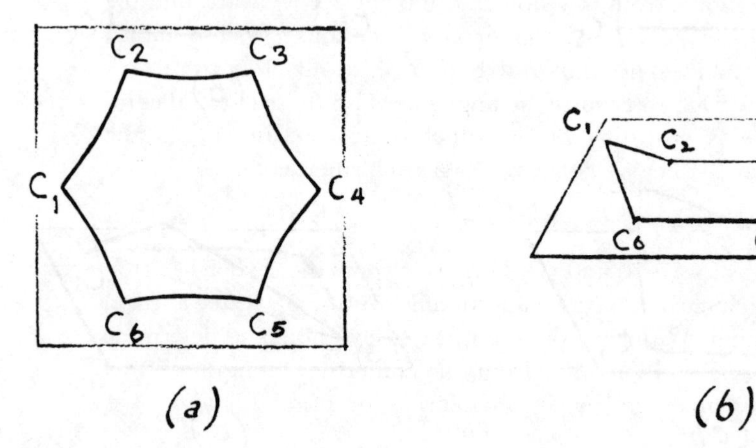

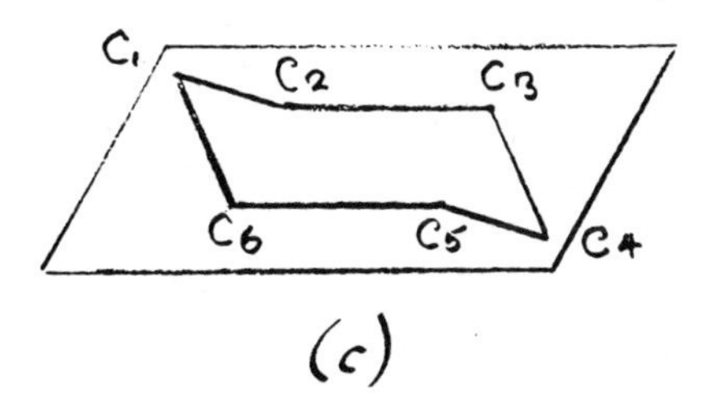


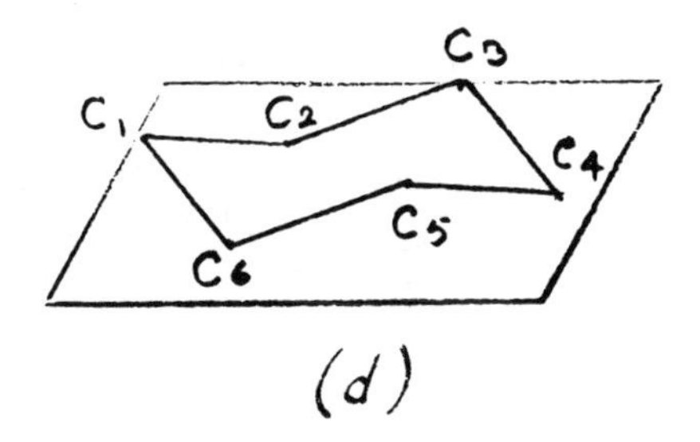


Place your model on a flat surface and move C*1* upward, as in #7–10*b*, from that surface until there is no longer any strain in the bonding at C*1*, C*2*, and C*6*. The angles of bonding approach 109.5° and you can now use rigid bars to hold these three carbons together. There is no longer any tendency for the bond to bow in.

Lift the model, keeping the bonds C*2*—C*3* and C*5*—C*6* parallel with the table top. Move C*4* downward so that the bond angles at C*4*, C*3*, and C*5* also become tetrahedral. A little imagination should lead you to see why this conformation is referred to as "the chair" (#7–10*c*). If you place this chair conformer on the table you find that C*2*, C*4*, and C*6* rest on the flat surface while C*1*, C*3*, and C*5* are all in a higher parallel plane, as shown in #7–10*d*.

There are two chair conformers and we shall now convert the model from one into the other. Move C*4* upward until the model assumes the shape shown in #7–11*b*. This figure represents the boat

**#7–11**

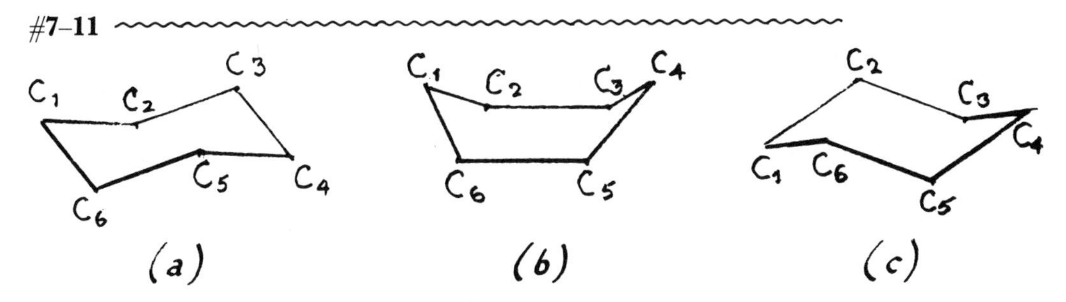


conformer, formed during the transition from one chair conformer to the other. In the boat conformer C*1* and C*4* lie in one plane, while all four remaining carbons are in a second parallel plane.

Remove the model from the table and move C*1* downward into the position shown in #7–11*c*. The chair conformations (#7–11*a* and *c*) are the most stable. The angles of bonding are all close to 109.5°. Although in the boat conformer the angles are also those of tetrahedral bonding, there is crowding of the other atoms substituent on the carbons, so that the boat is not as stable a conformer as the chair.

* * * * *

The diagrams in #7–10 and #7–11 merely show the carbon skeleton of the ring. Each carbon has two additional substituents at tetrahedral angles of bonding. We next examine these twelve bonds to determine how their relative positions vary from one conformer to another.

Arrange your model in the chair conformation of #7–12*a* and place

**#7–12**

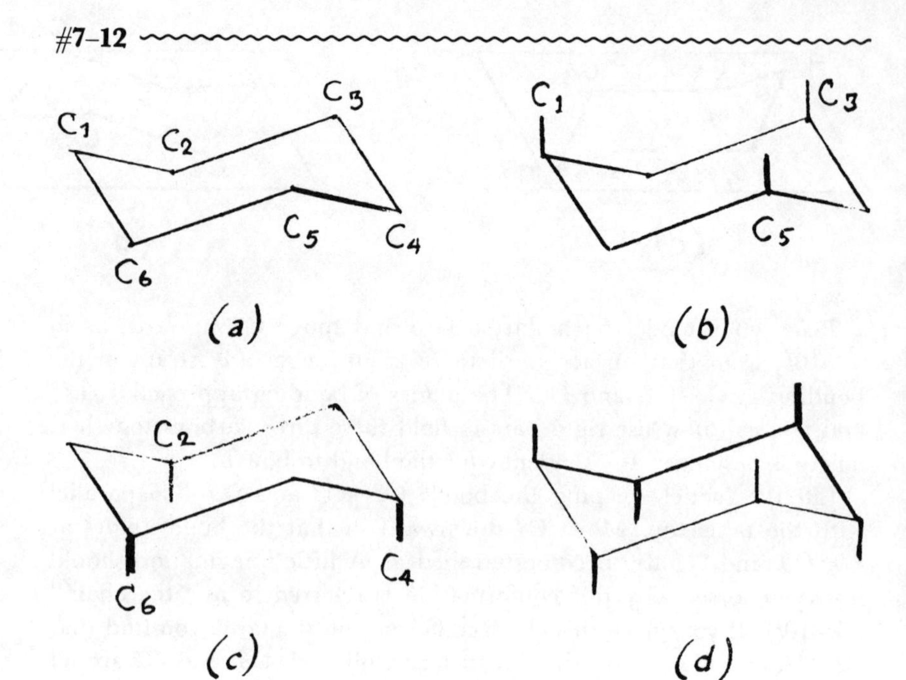


it on a table. There are three bonds, one each on C*1*, C*3*, and C*5*, which point upwards (#7–12*b*) in a direction that is perpendicular to the table top, and three bonds, C*2*, C*4*, and C*6*, which point downward (#7–12*c*), also in a direction perpendicular to the table top. These upward and downward bonds alternate with each other (#7–12*d*). They are called *axial bonds* because, if you imagine the ring as a wheel with an axle, these bonds all would be parallel to the axle, as illustrated in #7–13.

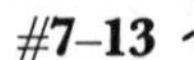

**#7–13**

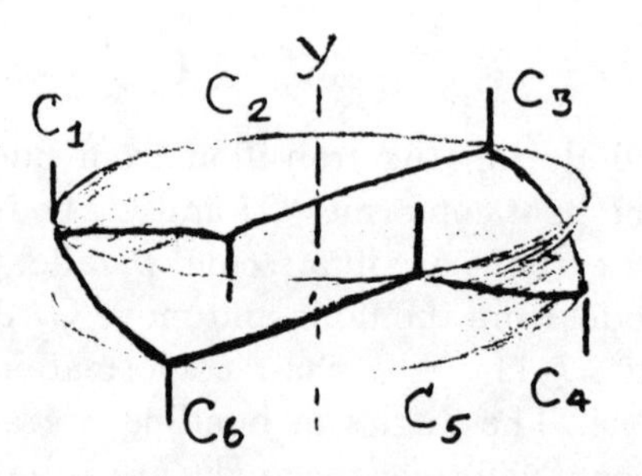


The remaining six bonds project outward from what would be the circumference of the wheel in #7–13 and are called *equatorial bonds*. Each equatorial bond is at an angle of 109.5° with the axial bond on the same carbon. When the axial bond is directed upward, as on C*1*, C*3*, and C*5* in #7–12*b*, the corresponding equatorial bond is directed downward, and when the axial bond is directed downward (C*2*, C*4*, and C*6*), the equatorial bond is directed upward.

The easiest way to position the equatorial bonds on a skeleton drawing of a chair conformer of cyclohexane is to keep in mind (*a*) that the direction of the equatorial bonds alternates up and down, and (*b*) that each equatorial bond is parallel to two of the carbon–carbon bonds in the ring skeleton as summarized below:

- Equatorial bonds on C*1* and C*4* are parallel to the bonds C*2*—C*3* and C*5*—C*6*, as shown in #7–14*a*.
- Equatorial bonds on C*2* and C*5* are parallel to the bonds C*1*—C*6* and C*3*—C*4*, as shown in #7–14*b*.
- Equatorial bonds on C*3* and C*6* are parallel to the bonds C*4*—C*5* and C*1*—C*2*, as shown in #7–14*c*.

**#7–14**

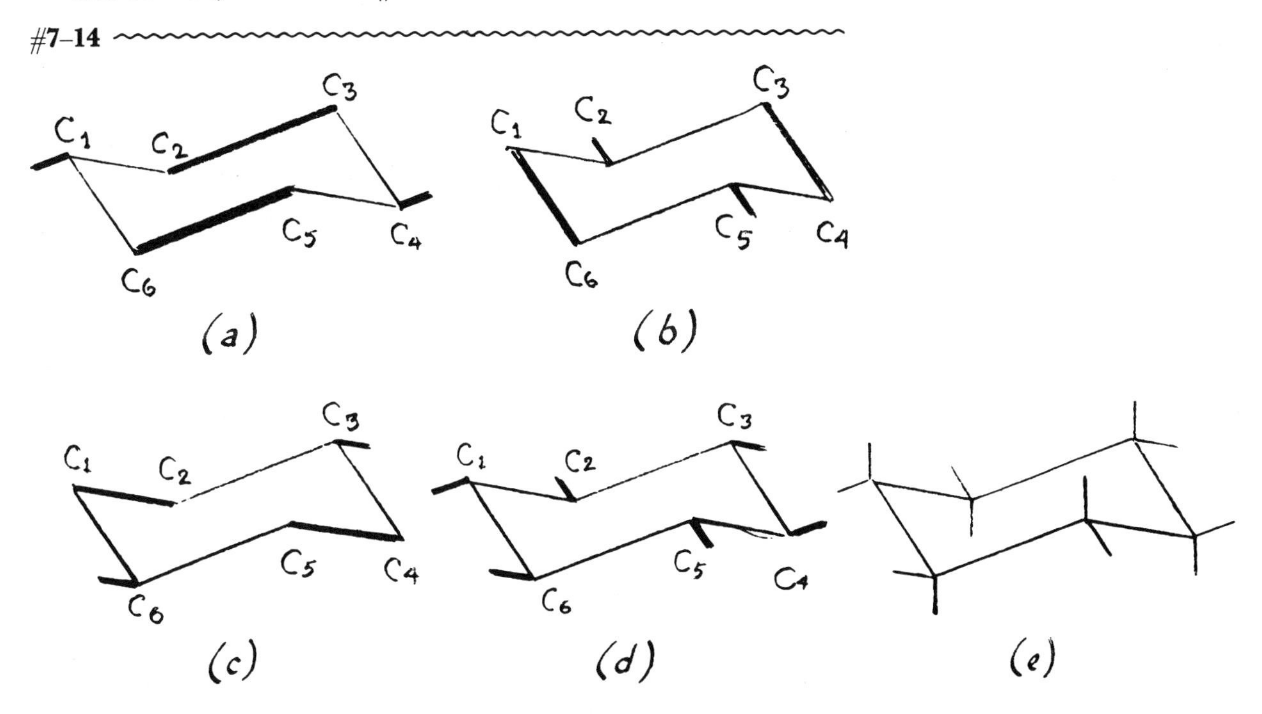


Figure #7–14*d* shows all six equatorial bonds for this chair conformer, and #7–14*e* shows both the axial and the equatorial bonds.

* * * * *

Make a model of the chair conformer shown in #7–15*a* and include both axial and equatorial bonds. Mark the axial bonds in some way that is distinctive. In the drawing they are represented by a broader

**#7–15**

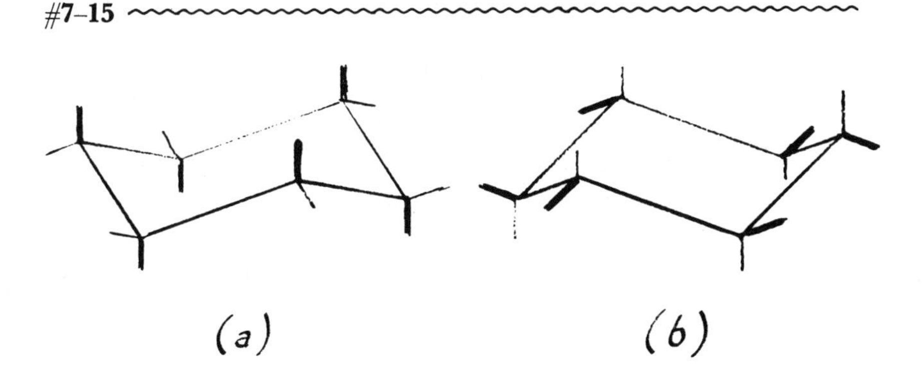

line. Convert the model of the conformer in #7–15*a* into a model of the conformer in #7–15*b*. You have already carried out this maneuver when working with #7–11. However, your model consisted of only the carbon skeleton without the equatorial and axial bonds. To accomplish the conversion, move C*1* downward and C*4* upward as you did previously, allowing the rest of the atoms to rotate so as to maintain the tetrahedral angle of bonding.

When you have completed the conversion you will find that all the axial bonds in conformer (*a*) are in the equatorial position in conformer (*b*). No bonds have been made or broken—the conformers merely represent two of the many different positions that can be assumed by the atoms in cyclohexane.

The conformers in #7–15*a* and *b* are of equal stability in unsubstituted cyclohexane. However, if there are substituents present, the more stable conformer is the one that places the bulkiest groups in the equatorial position, because this reduces crowding. Compare *a* and *b* in #7–16.

**#7–16**

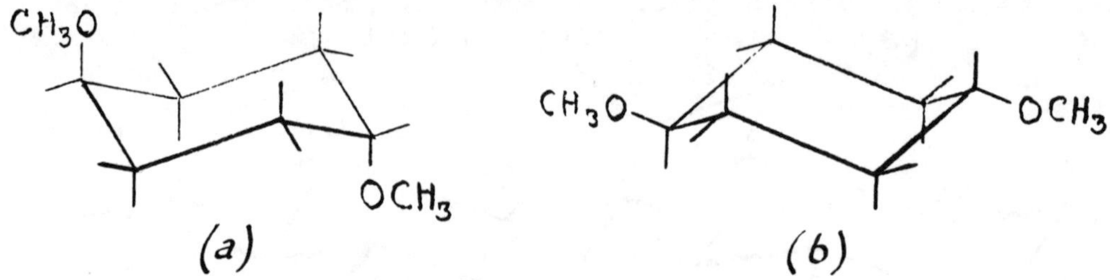

(a) (b)

*trans* – 1,4 dimethylcyclohexane

As an exercise, work out the bond positions in the boat conformer of cyclohexane and of the compound from #7–16. Relate bond directions of the non-ring bonds to those in the ring, determining which sets are parallel as we did in #7–14.

* * * * *

You have learned that both axial and equatorial bonds may be directed either upward or downward relative to the carbon skeleton of a cyclic structure. Two substituents on different carbons of such a ring, both attached by bonds that are directed upward (or downward), are said to be *cis* to each other (on the same side of the molecule). If one substituent is attached by a bond that is directed upward and the other is attached by a bond that is directed downward, the two substituents are said to be *trans* to (across from) each other. Thus, there are both *cis*- and *trans*-1,2-dichlorocyclohexanes. These are isomers. You cannot convert a cis into a trans isomer merely by shifting the relative positions of the atoms within the molecular framework. The order of bonding in two isomers is different. Each isomer is a unique substance, whereas conformers are different forms of the same substance.

Each isomer has numerous conformers, the most stable of which for a cyclohexane derivative is the chair conformation that places the largest number of bulky groups in the less crowded equatorial position. In Example 2 we apply these concepts and learn how to show cis and trans isomerism in a three-dimensional drawing of chair conformers.

*Example 1*—Draw the most stable conformation for *cis*-1,2-dichlorocyclohexane and for *trans*-1,2-dichlorocyclohexane.

*Step 1.* Draw the outline of the two chair conformers as shown in #7–17.

**#7–17**

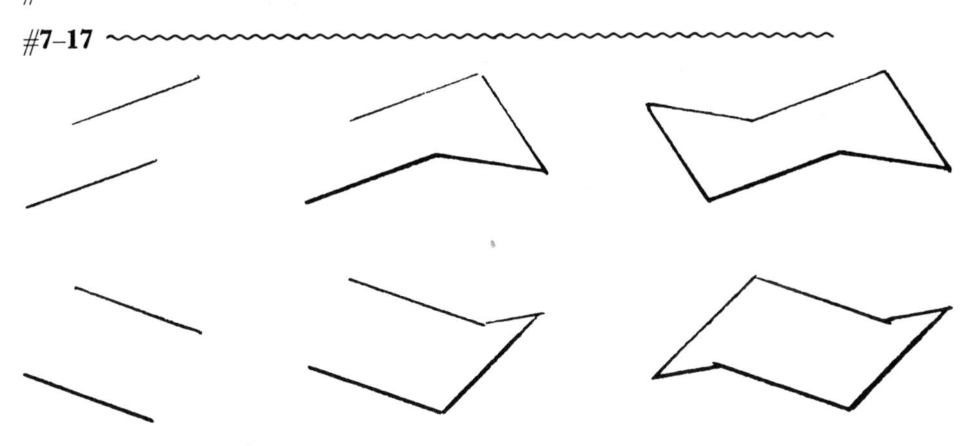

*Step 2.* Fill in the axial bonds.

**#7–18**

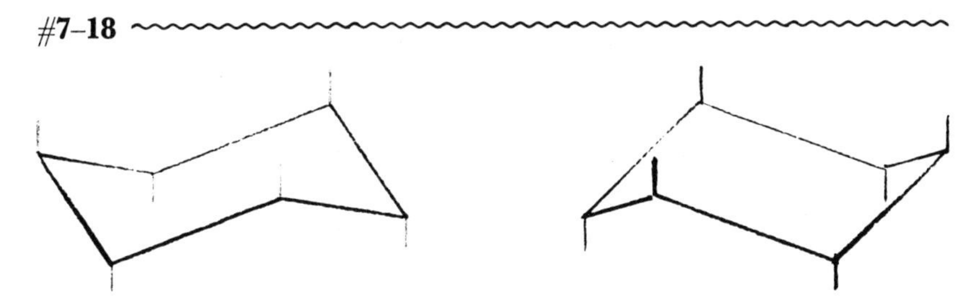

*Step 3.* Fill in the equatorial bonds keeping in mind the various sets of parallel bonds in the structure (see #7–14).

**#7–19**

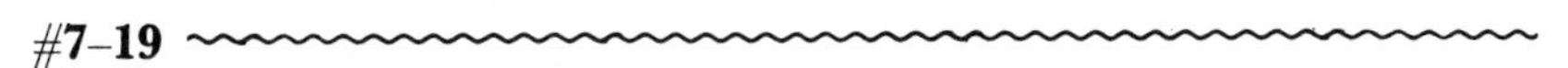

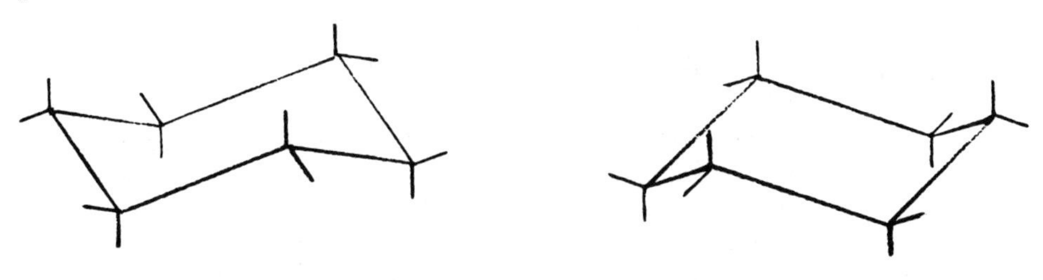

*Step 4.* Fill in the halogens to show the trans isomer.

In a trans isomer the bonds to the trans substituents must alternate, one being directed upward and the other downward. The axial bonds alternate in this manner on adjacent carbons. Thus, a trans isomer could have both halogens in the axial position, as shown in #7–20*a*.

#7–20

Cl

Cl

(a)

Axial-axial

Cl

Cl

(b)

Equatorial-equatorial

The direction of equatorial bonds on adjacent carbons also alternates, so that a trans isomer could also have both halogens in the equatorial position as indicated in #7–20*b*. The more stable of the two conformers is the one in *b* which places the bulky halogens in the equatorial position.

*Step 5.* Fill in the halogens to show the cis isomer.

In a cis isomer, both bonds to the cis substituents must be in the same direction relative to the ring skeletons. Thus, there could be one halogen on an upward axial bond and a second halogen on an adjacent upward equatorial bond, as shown for two conformers in #7–21. These two conformers of the cis isomer are of equal stability. Each has one

#7–21

Cl

Cl

(a)

Axial-equatorial

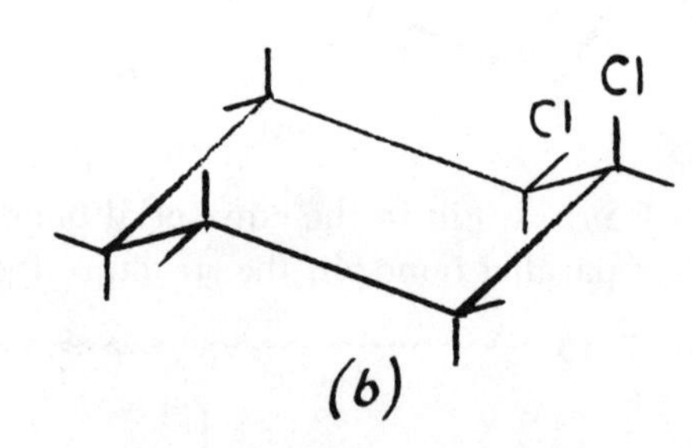


(b)

Equatonial-axial

substituent in an equatorial position and one in an axial position, so that there is no difference between the two in terms of crowding.

The positions of the substituents in the two chair conformations for the cis and trans isomers of 1,2-, 1,3-, and 1,4-disubstituted cyclohexanes are summarized in Table 7–1. When you can easily draw all the structures referred to in this table, you understand the concept of ring conformations. The problems in §8 will provide the necessary practice.

**Table 7–1**

**Disubstituted Cyclohexanes—Relationships of Substituents in the Chair Conformations**

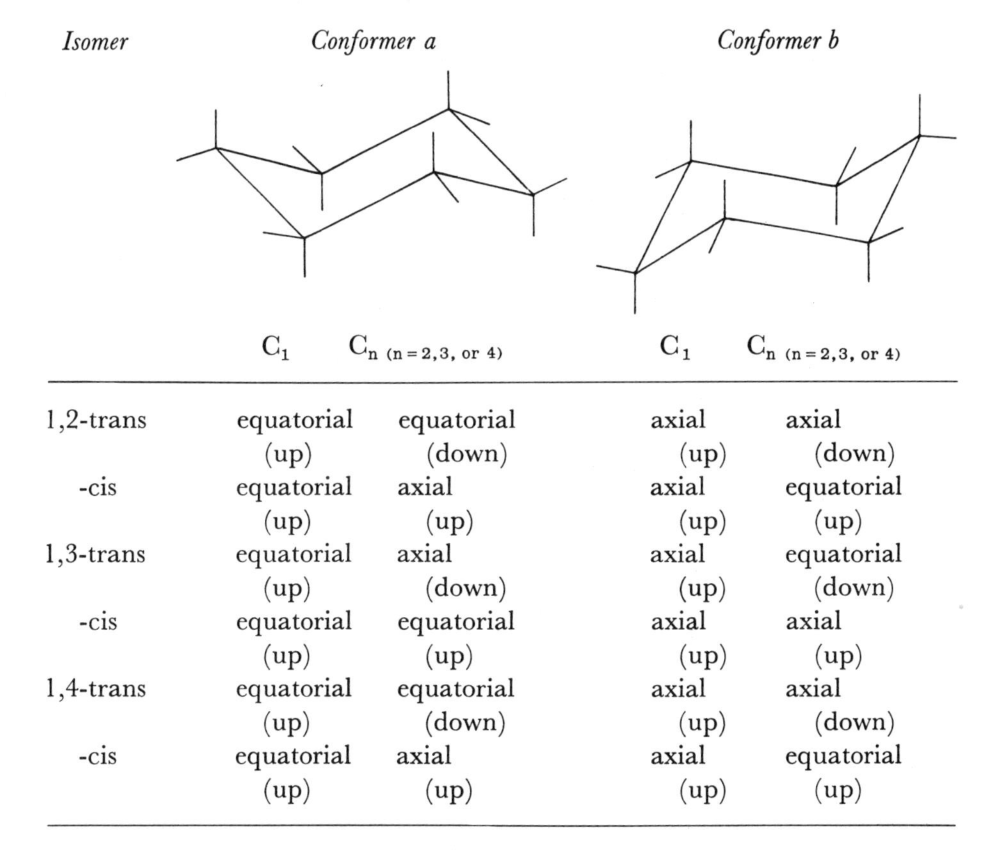

| *Isomer* | *Conformer a* | | *Conformer b* | |
|---|---|---|---|---|
| | $C_1$ | $C_n$ (n = 2, 3, or 4) | $C_1$ | $C_n$ (n = 2, 3, or 4) |
| 1,2-trans | equatorial (up) | equatorial (down) | axial (up) | axial (down) |
| -cis | equatorial (up) | axial (up) | axial (up) | equatorial (up) |
| 1,3-trans | equatorial (up) | axial (down) | axial (up) | equatorial (down) |
| -cis | equatorial (up) | equatorial (up) | axial (up) | axial (up) |
| 1,4-trans | equatorial (up) | equatorial (down) | axial (up) | axial (down) |
| -cis | equatorial (up) | axial (up) | axial (up) | equatorial (up) |

## § 8. Problems. Structural Formulas. Conformation

As you work on these problems, make a model of each structure from your drawing.

**1**) Draw both sawhorse and Newman diagrams for the *cisoid* (eclipsed) and the *transoid* (staggered) conformers of 1,2-dichloroethane. Which conformer is more stable?

**2**) Draw sawhorse and side-view projection diagrams to show conformational changes that arise from free rotation at the bond between C*2* and C*3* in butane. Include three eclipsed and three staggered conformers and indicate which would be most stable.

**3**) Draw sawhorse and Newman diagrams for the cisoid and the transoid conformers at the bond between C*2* and C*3* in heptane. Repeat for the bond between C*3* and C*4*. In each case indicate the more stable conformation.

**4**) Draw sawhorse diagrams showing three eclipsed and three staggered conformations for the bond between C*2* and C*3* in 3-ethyl-2-methylpentane. Indicate which conformer would be the most stable. Repeat for the bond between C*3* and C*4*.

**5**) Draw three conformations, two of equal stability and one of lesser stability, for these derivatives of cyclobutane:

*a*) *trans*-1,2-dichloro
*b*) *cis*-1-chloro-2-bromo
*c*) *trans*-1-bromo-3-hydroxy
*d*) *cis*-1,3-dihydroxy

**6**) Draw three conformations, one flat and two bent, for these derivatives of cyclopentane:

*a*) *trans*-1,2-dichloro
*b*) *cis*-1,2-dichloro
*c*) *trans*-1-amino-3-bromo
*d*) *cis*-1-amino-3-bromo

**7**) Draw three conformations, two chair and one boat, for these derivatives of cyclohexane:

*a*) *trans*-1,2-dibromo
*b*) *cis*-1,2-dihydroxy
*c*) *trans*-1-chloro-3-hydroxy
*d*) *cis*-1,3-dicyano
*e*) *trans*-1,4-dichloro
*f*) *cis*-1-bromo-4-hydroxy

## § 9. Cis-Trans Isomerism

The double bond associated with $sp^2$ hybridization results in a molecule with two less hydrogens than the corresponding alkane. Triple bond formation with *sp* hybridization reduces the number of hydrogens by four as compared to the alkane. A molecule with fewer hydrogens than the maximum is referred to as *unsaturated*. An unsaturated alk*ene* has double bonds, while an unsaturated alk*yne* has triple bonds. Position isomers are formed when the center of unsaturation is located between different carbons. These facts are summarized by the four-carbon structures in #9–1.

**#9–1**

| | | |
|---|---|---|
| $CH_3CH_2CH_2CH_3$ | $CH_2{=}CHCH_2CH_3$ | $HC{\equiv}CCH_2CH_3$ |
| | $CH_3CH{=}CHCH_3$ | $CH_3C{\equiv}CCH_3$ |
| $C_4H_{10}$ | $C_4H_8$ | $C_4H_6$ |
| $C_nH_{2n+2}$ | $C_nH_{2n}$ | $C_nH_{2n-2}$ |
| alkane | alkene | alkyne |
| butane | 1-butene | 1-butyne |
| | 2-butene | 2-butyne |

The structural relationships imposed by $sp^2$ hybridization were presented in §1. The molecule is flat at the site of a double bond and the angles of bonding at the two unsaturated carbons are approximately 120°. Rotation is restricted at a double bond. In order for rotation to occur, the $\pi$ bond has to be broken, an act which requires 40 kcal per mole, and which therefore occurs only with the addition of more energy than is normally available in an organic reaction.

Restricted rotation at the double bond leads to cis-trans isomerism. We learned in §1 that a pair of doubly bonded carbons and their substituent atoms all lie in one plane. The plane of the molecule is shown as the XY plane in #9–2. Each pair of substituents either R or T can be described as being located on the "same side" of the double bond in the structure in #9–2*a* or on "opposite sides" of the double bond in #9–2*b*. What is meant by these terms is that the substituents are either on the same side or on opposite sides of the XY plane that passes through the $\pi$ bond perpendicular to the plane of the molecule. The isomer in which the two substituents are across from each other is the *trans isomer*; the *cis isomer* is the one in which the substituents are on the same side of the XZ plane.

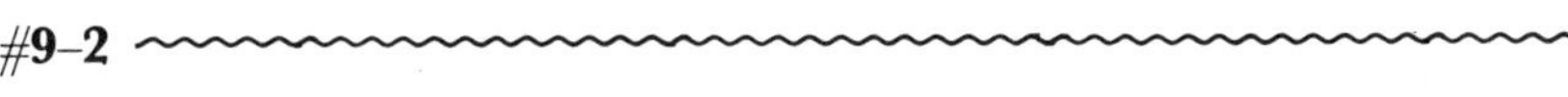


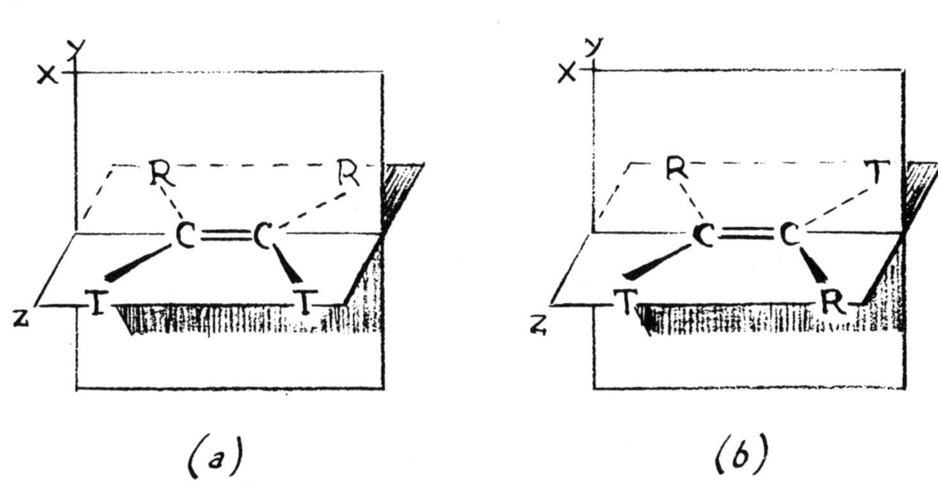


Cis-trans isomerism can occur only when the two substituents on each carbon differ from each other. There is only one structural form for 1-butene because, although there are two different substituents (an ethyl group and a hydrogen) on one of the unsaturated carbons, the other unsaturated carbon has two like substituents (two hydrogens). The substituents on C*1* must differ from each other and so must the substituents on C*2*. There is also only one isobutene, whereas there are two isomers of 2-butene, as shown in #9–3.

Cis-trans isomerism in cyclic structures was discussed in §7, starting with #7–16. The basic factor contributing to the effect is the same for the cyclic structures as for the olefinic structures. In both instances there is restricted rotation at carbon–carbon bonds. Cis-trans isomerism

```
H       CH3         H       CH3
 \     /             \     /
  C==C        ≡       C==C          (isobutene)
 /     \             /     \
H       CH3         H       CH3

H       H           H       CH2CH3
 \     /             \     /
  C==C        ≡       C==C          (1-butene)
 /     \             /     \
H       CH2CH3      H       H

CH3       CH3         H         CH3
   \     /             \       /
    C==C        ≢       C==C          (2-butene)
   /     \             /     \
  H (cis) H         CH3 (trans) H
```

is an example of *geometric isomerism*, isomerism that arises because of differences in the order in which the same atoms are bonded together in different molecules. The various types of isomerism have been summarized in Table 15–1.

Cis-trans isomerism does not occur at a triple bond. The geometry of *sp* hybridization arises from the linear relationship of *sp* orbitals. Both substituents on the two acetylenic carbons lie in a straight line. Therefore, there is no opportunity for cis-trans isomerism even though there is restricted rotation at a triple bond (see #1–14).

Cyclobutane, $C_4H_8$, a saturated hydrocarbon, has two fewer hydrogens than the noncyclic four-carbon alkane, $C_4H_{10}$. This reduction in the number of hydrogens arises because there are no terminal —$CH_3$ groups in the cyclic structure. Thus, the generic formula $C_nH_{2n}$ may represent an alkene with one double bond or a cyclic alkane, because each of these structural features reduces the hydrogen count by two. A triple bond reduces the count by four, giving a generic formula of $C_nH_{2n-2}$ for alkynes.

The relationship between the number of carbons and the number of hydrogens in a molecule provides a useful guide when you undertake to draw all the possible structures for a given molecular formula. For example, a deficiency of two hydrogens below saturation indicates either a double bond or a ring; a deficiency of four hydrogens means that the molecule contains a triple bond, or two double bonds, or two rings, or a double bond and a ring. These relationships are summarized down to $C_nH_{n-4}$ in Table 9–1. Example 1 shows you how to use this information. A halogen can be counted as a hydrogen for this purpose. It represents only a single bond to carbon. Thus, $C_5H_{11}Cl$ is a saturated noncyclic compound, just as $C_5H_{12}$ ($C_nH_{2n+2}$) is a saturated noncyclic compound.

**Table 9–1**

**Structural Features as Related to the Carbon–Hydrogen Ratio**

| *Generic Formula* | *Possible structural features for this carbon–hydrogen ratio* |
|---|---|
| $C_nH_{2n+2}$ (e.g., $C_6H_{14}$) | saturated hydrocarbon |
| $C_nH_{2n}$ (e.g., $C_6H_{12}$) | 1 double bond |
| | 1 ring |
| $C_nH_{2n-2}$ (e.g., $C_6H_{10}$) | 1 triple bond |
| | 2 double bonds |
| | 1 double bond and 1 ring |
| | 2 rings |
| $C_nH_{2n-4}$ (e.g., $C_6H_8$) | 1 triple bond and 1 double bond |
| | 1 triple bond and 1 ring |
| | 3 double bonds |
| | 2 double bonds and 1 ring |
| | 1 double bond and two rings |
| | 3 rings |

*Example 1*—Draw all the isomeric structures that can be represented by the molecular formula $C_5H_8$, including the cis-trans isomers.

*Step 1.* Summarize the information that can be deduced from the formula itself based on the relationships given in Table 9–1.

The relationship between carbon and hydrogen in this compound is $C_nH_{2n-2}$, which indicates a deficiency of four hydrogens below saturation. This means that the following structures are possible:

- 1 triple bond, *or*
- 2 double bonds, *or*
- 1 double bond and 1 ring, *or*
- 2 rings.

*Step 2.* Write all the possible noncyclic five-carbon skeletons.

*a*) continuous chain of five:

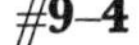

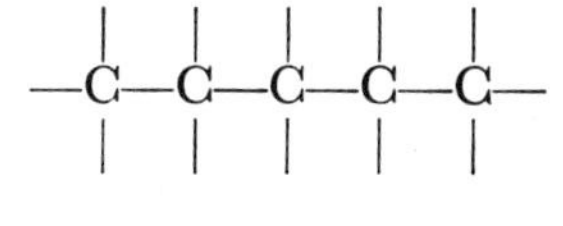

*b*) chain of four with one substituent carbon:

#9–5

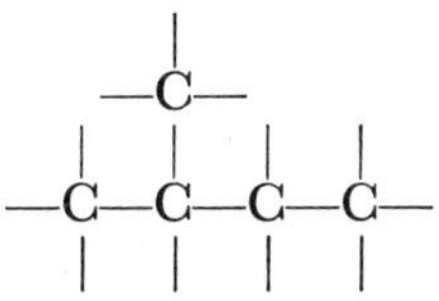

*c*) chain of three with two substituent carbons:

#9–6

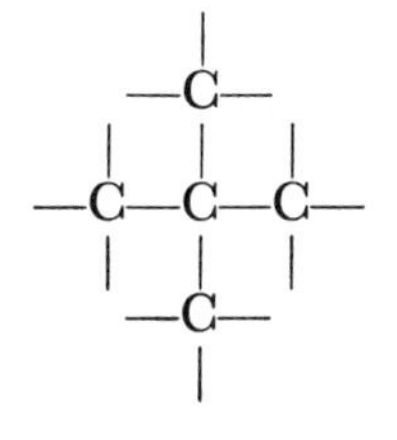

*Step 3.* Determine all the possible positions for a triple bond in each of the above skeletons. Fill in the hydrogens to give the structure of the compound. Count the hydrogens in each as a check on yourself.

*a*) continuous five-carbon chain:

#9–7

$HC{\equiv}CCH_2CH_2CH_3$ $CH_3C{\equiv}CCH_3$

*b*) chain of four with one substituent carbon:

#9–8

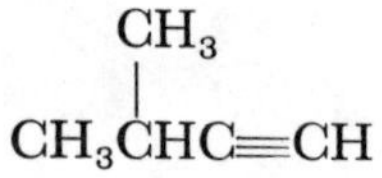

$CH_3CH(CH_3)C{\equiv}CH$

*c*) chain of three with two substituent carbons:
none.

*Step 4.* Determine all the possible positions for two double bonds in the skeletons outlined in Step 2 and fill in the hydrogens. Count the hydrogens as a check on yourself.

*a*) continuous five-carbon chain:

Position one double bond between C*1* and C*2* and the other double bond successively between each of the other pairs of carbons.

#9–9

$CH_2{=}C{=}CHCH_2CH_3$ $CH_2{=}CHCH{=}CHCH_3$ $CH_2{=}CHCH_2CH{=}CH_2$

Position one double bond between C*2* and C*3* and the other double bond successively between each of the other pairs of carbons. Eliminate any structures that duplicate those above.

#9–10

$CH_3CH{=}C{=}CHCH_3$ ~~$CH_3CH{=}CHCH{=}CH_2$~~

*b*) chain of four with one substituent carbon:

#9–11

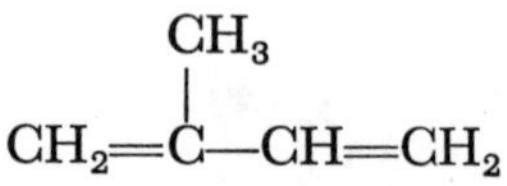

$CH_2{=}C(CH_3){-}CH{=}CH_2$

*c*) chain of three with two substituent carbons:
none.

*Step 5.* Go back and examine the structures in Step 4 to determine which of them can exist as cis-trans isomers.

For cis-trans isomerism each of the double-bonded carbons must have two different substituents.

Only one of the double bonds is so constituted. Therefore, only one of these structures can exist as a pair of cis-trans isomers.

#9–12

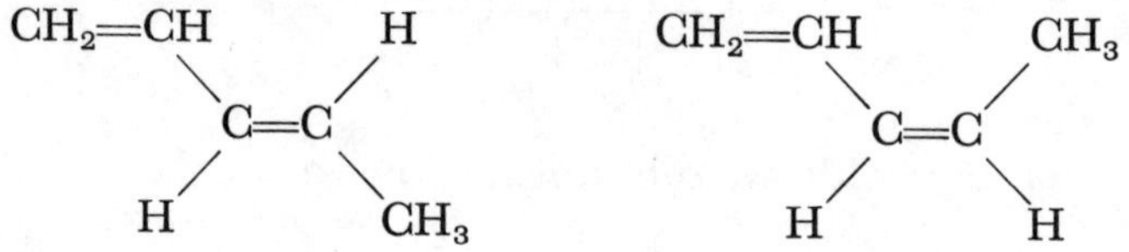

(trans) $CH_2{=}CH$ and H on one carbon, H and $CH_3$ on the other: $CH_2{=}CH(H)C{=}C(H)CH_3$, with $CH_2{=}CH$ and $CH_3$ on opposite sides

(cis) $CH_2{=}CH$ and H on one carbon, $CH_3$ and H on the other: $CH_2{=}CH$ and $CH_3$ on the same side

*Step 6.* Draw the carbon skeletons containing five carbons and one ring.

Start with the largest ring and work down to cyclopropane, systematically positioning the substituent carbons.

*a*) five-carbon ring:

**#9–13**

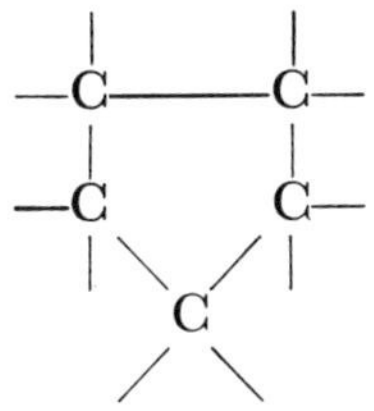

```
     |       |
    —C———————C—
     |       |
    —C       C—
     | \   / |
        C
       / \
```

*b*) four-carbon ring with one substituent carbon:

**#9–14**

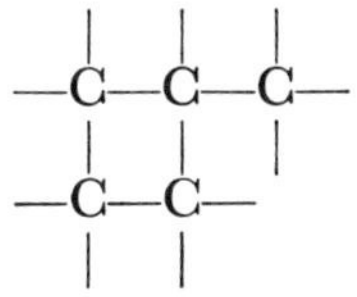

```
     |  |  |
    —C—C—C—
     |  |  |
    —C—C—
     |  |
```

*c*) three-carbon ring with two substituent carbons:

**#9–15**

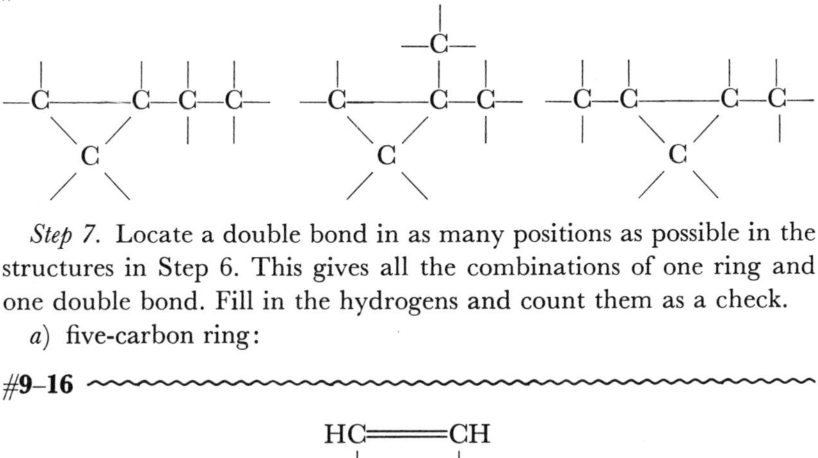

```
                                        |
                                       —C—
  |        |  |  |        |             |  |       |  |        |  |
 —C————————C—C—C—        —C—————————————C—C—     —C—C————————C—C—
   \      /   |  |         \           /   |      |    \    /    |
      C                         C                         C
     / \                       / \                       / \
```

*Step 7.* Locate a double bond in as many positions as possible in the structures in Step 6. This gives all the combinations of one ring and one double bond. Fill in the hydrogens and count them as a check.

*a*) five-carbon ring:

**#9–16**

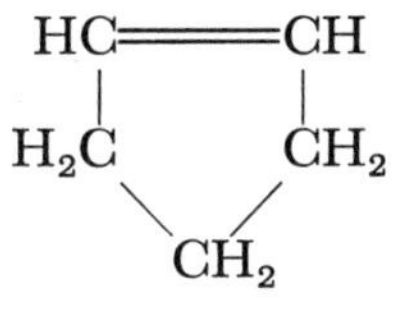

```
   HC=====CH
    |      |
  H2C      CH2
     \    /
      CH2
```

*b*) four-carbon ring with one substituent:

**#9–17**

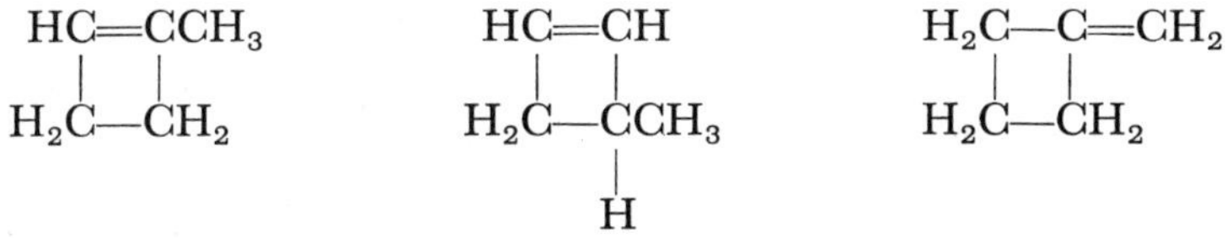

```
  HC=CCH3          HC=CH            H2C—C=CH2
   |  |             |  |              |  |
 H2C—CH2          H2C—CCH3          H2C—CH2
                       |
                       H
```

*c*) three-carbon ring with two substituents:

**#9–18**

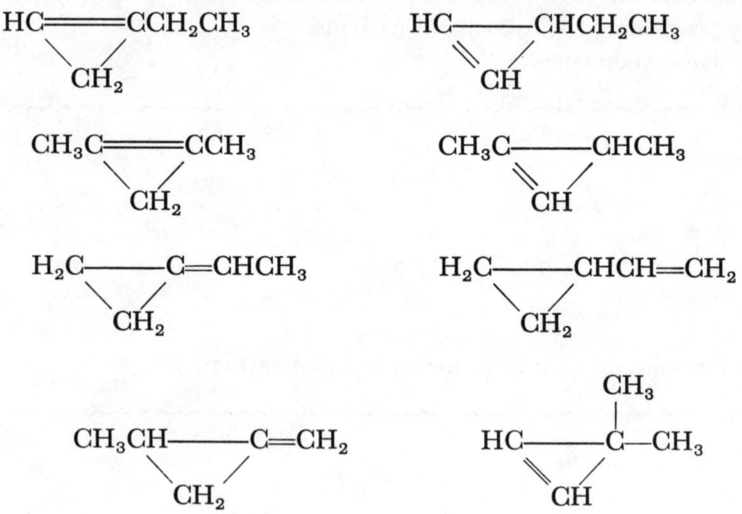

$HC{=}CCH_2CH_3$ (ring closed by $CH_2$) $\qquad$ $HC{-}CHCH_2CH_3$ (ring closed by $CH$, double bond $HC{=}CH$)

$CH_3C{=}CCH_3$ (ring closed by $CH_2$) $\qquad$ $CH_3C{-}CHCH_3$ (ring closed by $CH$, double bond $C{=}CH$)

$H_2C{-}C{=}CHCH_3$ (ring closed by $CH_2$) $\qquad$ $H_2C{-}CHCH{=}CH_2$ (ring closed by $CH_2$)

$CH_3CH{-}C{=}CH_2$ (ring closed by $CH_2$) $\qquad$ $HC{-}C(CH_3){-}CH_3$ (ring closed by $CH$, double bond $HC{=}CH$)

*Step 8.* Go back and check the structures in Step 7 for possible cis-trans isomers, both at a double bond and on a ring.

There are none.

*Step 9.* Draw the carbon skeleton for all possible fused ring systems with five carbon atoms.

Two rings are said to be *fused* when they share two or more carbon atoms and the bond between the carbons. To determine how many different fused ring systems can be formed, proceed as follows:

*a*) Draw the largest possible ring. Choose one particular carbon in the ring and successively draw a bond between that carbon and each of the other carbons on the ring, as illustrated in #9–19. Only two such structures can be drawn starting with cyclopentane and these do

**#9–19**

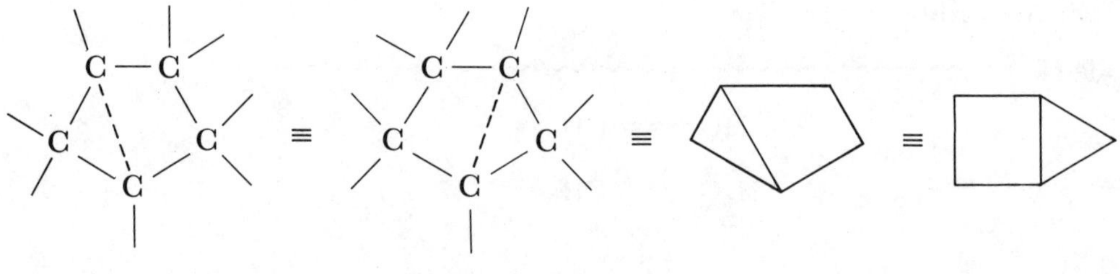

≡ ≡ ≡

not represent two different compounds. There is only one fused ring system based on a five-carbon ring. The shorthand representations of this compound, bicyclo[1.1.0] pentane, are also given in #9–19.

*b*) Draw the next smaller ring, i.e., cyclobutane, positioning the leftover carbon as a substituent on the ring. Determine how many fused ring systems can be formed with this ring as a base. This process is shown in #9–20. The three-membered ring is the smallest ring, so

#9–20

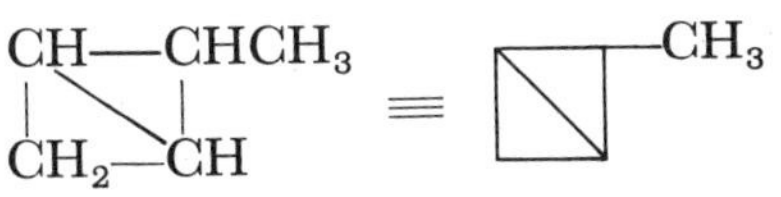

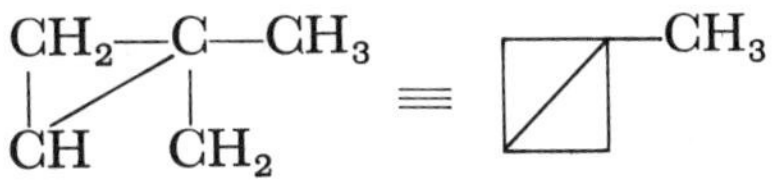

$$\begin{array}{ll} CH\text{—}CHCH_3 & \\ | \quad\ \ | & \\ CH_2\text{—}CH & \end{array} \equiv \square\text{—}CH_3 \qquad \begin{array}{l} CH_2\text{—}C\text{—}CH_3 \\ | \qquad\ \ | \\ CH \quad\ CH_2 \end{array} \equiv \square\text{—}CH_3$$

we have completed our analysis of possible fused rings.

*Step 10.* Draw the carbon skeleton for all possible spirane ring systems with five carbons.

When two rings share a single carbon atom, the resulting compound is called a *spirane*. To determine how many different spirane ring compounds can be formed, proceed as follows:

*a*) Draw a three-membered ring and form a second ring from the remaining atoms, using one carbon from the three-membered ring as the starting point for the second ring, as shown in #9–21.

#9–21

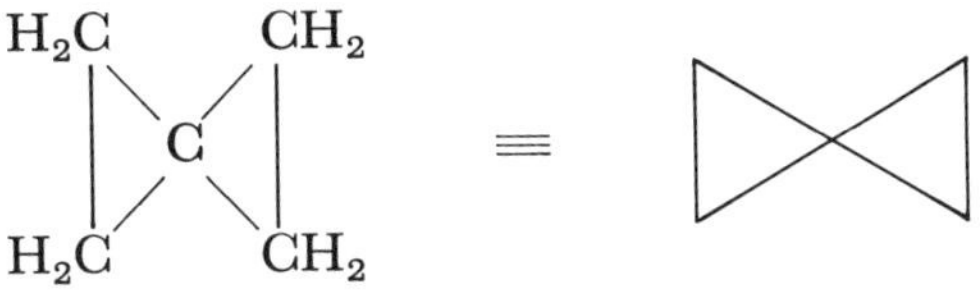

$$\begin{array}{ccc} H_2C & & CH_2 \\ | & C & | \\ H_2C & & CH_2 \end{array} \equiv \bowtie$$

*b*) Draw a four-membered ring and form a second ring with the remaining carbons. This second operation is not possible with only five carbons, so there is only one spirane structure. With more than five carbons the operation is repeated in order of increasing ring size until no new structures are formed. With more carbons it would also be necessary to consider alkyl substituents on the ring system. For example, there could be a methyl substituent on spiro[2.2]pentane, the structure in #9–21, if the problem had been based on a molecular formula of $C_6H_{10}$.

Neither this spirane nor these fused ring systems offer the opportunity for cis-trans isomerism; so, except for considering the possibility of optical isomers (discussed in §15), we have completed the problem. There are 25 isomers.

## § 10. Problems. Cis-Trans Isomerism

**1**) Some of these compounds can exist as both cis and trans isomers; others cannot. Draw the cis and the trans structure for each compound that can form the two geometric isomers.

*a*) $CH_3CH_2\overset{\overset{\large CH_3}{|}}{C}{=}CHCH_3$

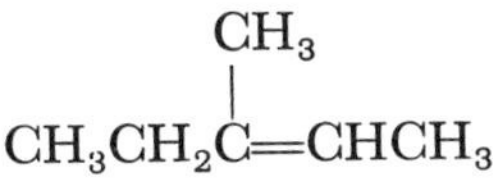

*b*) $CH_3\overset{\overset{\large CH_3}{|}}{C}{=}\overset{\overset{\large CH_3}{|}}{C}CH_3$

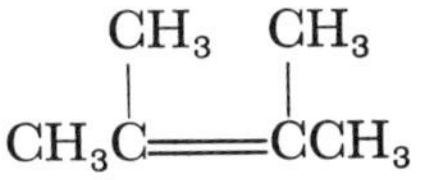

*c*) $\mathrm{CH_2{=}CHCH{=}CH_2}$
*d*) $\mathrm{CH_2{=}CHCH{=}CHCH_3}$
*e*) $\mathrm{CH_3\overset{Cl}{\overset{|}{C}}HCH{=}\overset{CH_3}{\overset{|}{C}}CH_3}$
*f*) $\mathrm{CH_3\overset{Cl}{\overset{|}{C}}HCH{=}CHCH_2CH_3}$
*g*) $\mathrm{CH_3CH{=}CHCH{=}CHCH_3}$
*h*) 1,3-dichlorocyclobutane
*i*) 1,3-dibromocyclopentane
*j*) 1,2,3-trihydroxycyclopentane
*k*) $\mathrm{CH_3\overset{Br}{\overset{|}{C}}{=}CHCH_3}$
*l*) $\mathrm{C_6H_5CH{=}CHCH_2CH_3}$
*m*) 1,4-di(chloromethyl)cyclopentane
*n*) 1,1-dibromo-3-methylcyclohexane
*o*) $\mathrm{CH_3\overset{OH}{\overset{|}{C}}HCH{=}CHC_6H_5}$

**2**) Draw all the isomeric structures that can be represented by these molecular formulas. Include both position and cis-trans isomers. Keep your answers and, after you have finished studying §11, name by the IUPAC system all the structures you have drawn.
*a*) $C_4H_8$ *b*) $C_5H_{10}$ *c*) $C_4H_7Cl$ *d*) $C_5H_8$ *e*) $C_6H_6$

**3**) Draw all the structures with the formula $C_7H_{12}$ that have a four-carbon ring and one double bond.

**4**) Draw all the unsubstituted two-ring systems, both fused and spirane, with the molecular formulas *a*) $C_8H_{14}$, *b*) $C_9H_{16}$.

**5**) Draw all the structures composed of three fused rings with the molecular formula $C_9H_{14}$ and no alkyl substituents.

**6**) Draw all the structures with the molecular formula $C_7H_{12}$ that have two rings and one methyl substituent.

## § 11. Alkenes and Alkynes. Nomenclature

The names for the alkenes and the alkynes are built up logically using the basic rules you learned for the alkanes plus some special rules for unsaturated compounds. Examples of the application of each rule are given in Table 11–1.

*a*) The extent of unsaturation is specified in the name ending:

- *ane* = saturated.
- *ene* = double bond.
- *yne* = triple bond.

*b*) A multiplicity of double or triple bonds is indicated by the

**Table 11–1**

**Alkenes and Alkynes. Nomenclature**

| *Formula* | *IUPAC Name* |
|---|---|
| $CH_3CH_2CH_2CH_3$ | butane |
| $CH_2{=}CHCH_2CH_3$ | 1-butene |
| 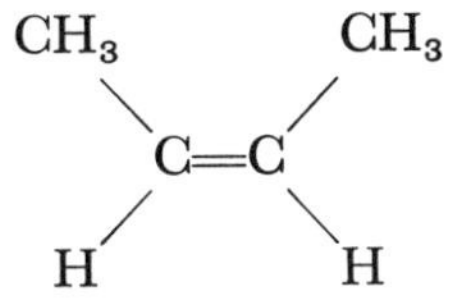 $(CH_3)HC{=}CH(CH_3)$ (both $CH_3$ on same side) | *cis*-2-butene |
| 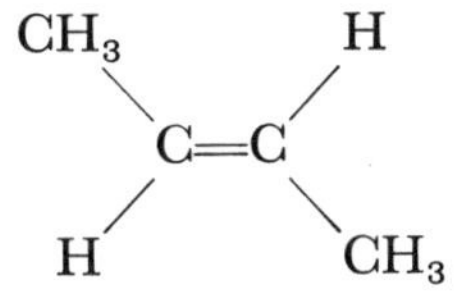 $(CH_3)HC{=}CH(CH_3)$ ($CH_3$ on opposite sides) | *trans*-2-butene |
|  □ | cyclobutene |
| $HC{\equiv}CCH_2CH_3$ | 1-butyne |
| $CH_3C{\equiv}CCH_3$ | 2-butyne |
| $CH_2{=}CHCH{=}CH_2$ | 1,3-butadiene |
| $HC{\equiv}CC{\equiv}CH$ | 1,3-butadiyne |
| $HC{\equiv}CCH{=}CH_2$ | 1-buten-3-yne |
| 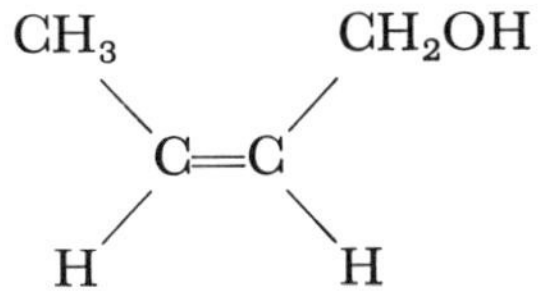 $(CH_3)HC{=}CH(CH_2OH)$ ($CH_3$ and $CH_2OH$ on same side) | *cis*-2-butenol |
| $HC{\equiv}CCH_2CH_2OH$ | 3-butynol |
| 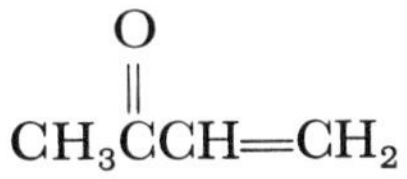 $CH_3\overset{O}{\overset{\Vert}{C}}CH{=}CH_2$ | 3-buten-2-one |

appropriate prefix placed before the ending and after the word root indicating the number of carbons:

- *di* = 2,
- *tri* = 3,
- *tetra* = 4,
- *penta* = 5, etc.

For the sake of euphony, the *a* from the *ane* ending is kept in front of the consonant of the prefix; e.g., hexene, but hex*a*-diene, not hexdiene; pentyne, but pent*a*diyne, not pentdiyne.

*c*) The position of the multiple bond is designated by number. The longest continuous chain which includes the multiple bond is numbered in the order which yields the lowest value for the multiple bond. A single number designates the bond. It is the lower of the two

**#11–1**

*(6 5 4 3 2 1)*

$CH_3CHCH_2CH{=}CHCH_3$ (*is* 5-methyl-2-hexene, *is not* 2-methyl-4-hexene)

$\quad\;\; |$

$\quad\; CH_3$

*(1 2 3 4 5 6)*

values assigned to the carbons forming the bond by whichever order of numbering is chosen.

*d*) If there is a multiplicity of double or triple bonds, the sum of the numbers used to indicate their position must be at a minimum as in #11–2.

**#11–2**

*(7 6 5 4 3 2 1)*

$CH_3CH{=}CHCH{=}CHCH{=}CH_2$ (*is* 1,3,5-heptatriene, *is not* 2,4,6-heptatriene)

*(1 2 3 4 5 6 7)*

*e*) A compound containing both double and triple bonds is named as an *en-yne.* The number referring to the position of the double bond goes in front of the Greek root, and the numbers referring to the triple bond go between the two endings, as in 1-buten-3-yne

$$(CH_2{=}CH{-}C{\equiv}CH).$$

The order of numbering must result in a sum that is the minimum. If there are two ways to number which give the same sum, then the lower number is assigned to a double bond. For example, see #11–3, #11–4.

**#11–3**

*(4 3 2 1)*

$CH_2{=}CH{-}C{\equiv}CH$ (*is* 1-buten-3-yne, *is not* 3-buten-1-yne)

*(1 2 3 4)*

**#11–4**

*(8 7 6 5 4 3 2 1)*

$CH_2{=}CHCH_2CH_2CH{=}CHC{\equiv}CH$ (*is* 3,7-octadiene-1-yne, *is not* 1,5-octadiene-7-yne)

*(1 2 3 4 5 6 7 8)*

*f*) When cis-trans isomerism is possible, it is necessary to designate which isomer is present. When there is only one double bond and each of the unsaturated carbons contains like substituents, the naming is obvious.

**#11–5**

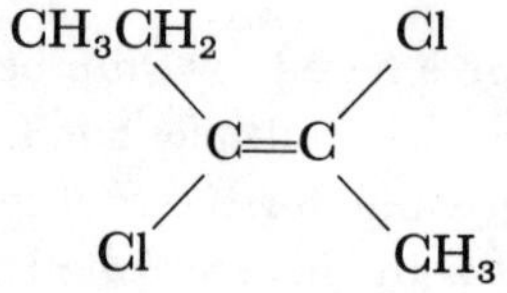

$CH_3CH_2 \qquad\quad Cl$

$\qquad\quad \diagdown \quad \diagup$

$\qquad\qquad C{=}C$ (*trans*-2,3-dichloro-2-butene)

$\qquad\quad \diagup \quad \diagdown$

$\qquad Cl \qquad\quad CH_3$

However, all four substituents may be different. In this instance, locate the longest chain and use the prefix to indicate the relative positions of the groups comprising this longest chain.

**#11–6**

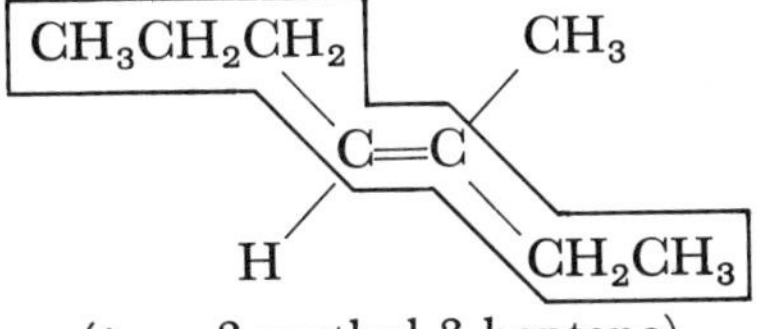


(*trans*-3-methyl-3-heptene)

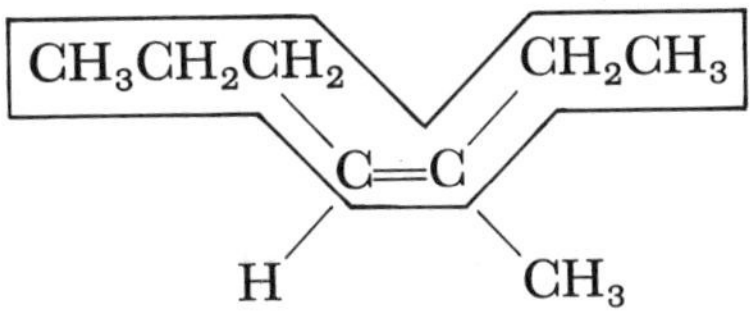

$CH_3CH_2CH_2$ $CH_2CH_3$

C=C

H $CH_3$

(*cis*-3-methyl-3-heptene)

If more than one center of cis-trans isomerism is present, the arrangement of the atoms at each needs to be indicated. There are four combinations of cis-trans arrangements for 2,4-hexadiene.

**#11–7**

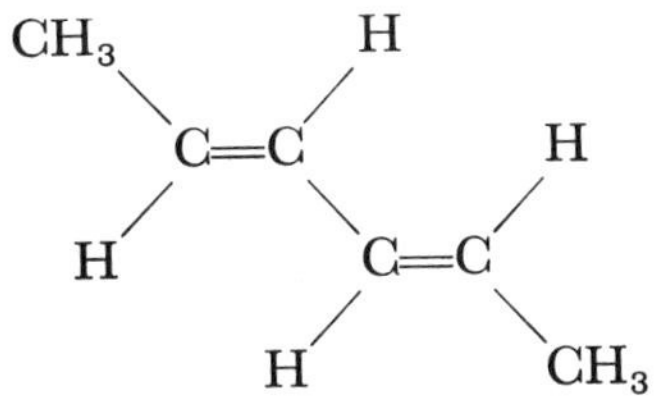

$CH_3$ H

C=C H

H C=C

H $CH_3$

(trans, trans)

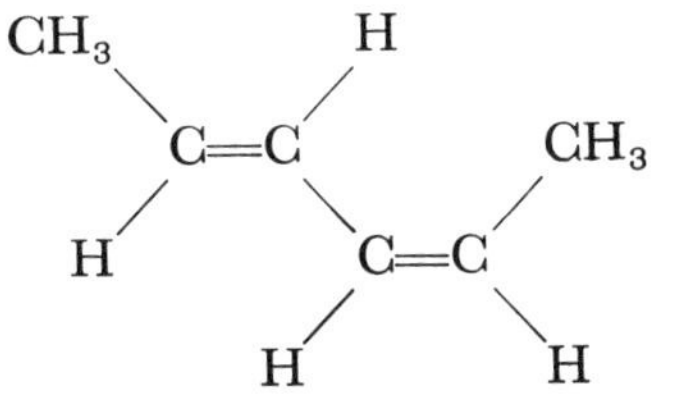

$CH_3$ H

C=C $CH_3$

H C=C

H H

(trans, cis)

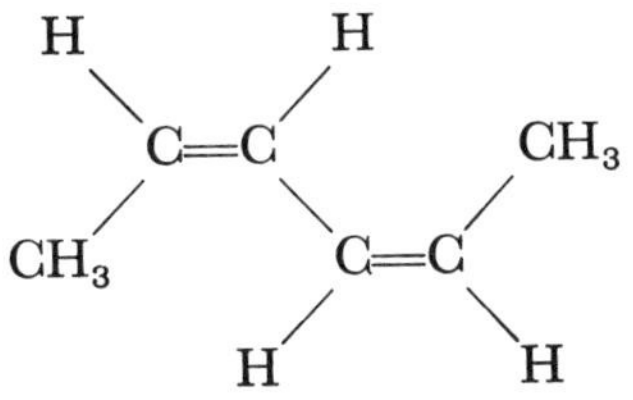

H H

C=C $CH_3$

$CH_3$ C=C

H H

(cis, cis)

H H

C=C H

$CH_3$ C=C

H $CH_3$

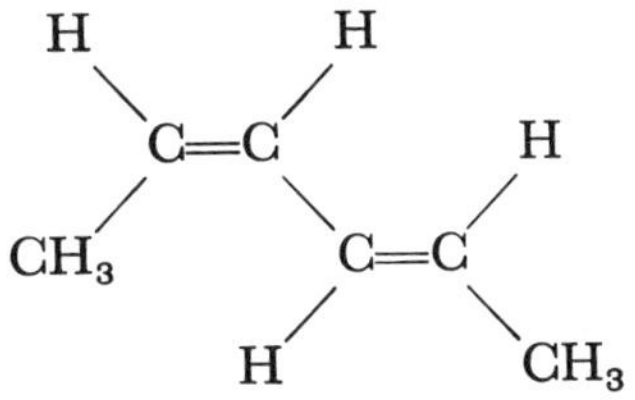

(cis, trans)

The logical application of these rules is illustrated by the following examples.

*Example 1*—Name the substance $CH_3CHCH_2CH_3$.
with C≡CH attached to the CH (second carbon)

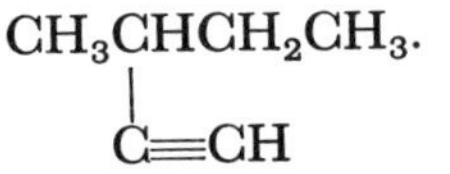

*Step 1.* Determine the number of carbons in the longest chain containing the multiple bond.

**#11–8**

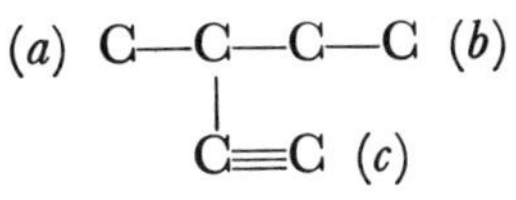

(*a*) C—C—C—C (*b*)
with C≡C (*c*) attached to the second carbon

(Chain *a* to *c* = 4 carbons)
(Chain *b* to *c* = 5 carbons)
There are five carbons in the longest chain which includes the triple bond.

*Step 2.* Determine the numbering order of the main chain.

**#11–9**

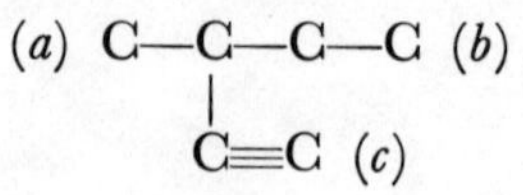

```
(a) C—C—C—C (b)
      |
      C≡C (c)
```

Numbering from *b* to *c* gives 4 as the designation for the triple bond; numbering from *c* to *b* gives a value of 1.

*Step 3.* Name and number the substituents on the chain.
There is a —$CH_3$ (methyl group) in position 3.

*Step 4.* Put all this information together and write the name:
*a*) 5 carbons = *pent.*
*b*) triple bond = *yne.*
*c*) *yne* in position 1 = *1-yne* or number may be omitted.
*d*) methyl in position 3 = 3-methyl.
Name = 3-methylpentyne.

*Example 2*—Draw a two-dimensional structural formula for *trans*-1,4-hexadiene.

*Step 1.* Draw a skeleton of carbon chain and number it (#11–10). *Hex* means six.

**#11–10**

C—C—C—C—C—C
(1 *2 3 4 5 6*)

*Step 2.* Locate the multiple bonds. *Diene* means two double bonds. 1,4 locates them between C*1* and C*2*, and between C*4* and C*5*.

**#11–11**

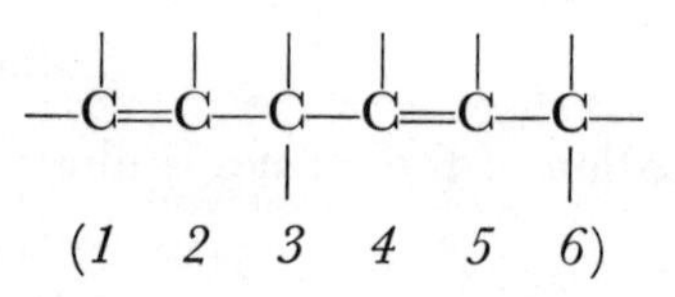

*Step 3.* Fill in the hydrogens and examine the structure to determine whether or not cis-trans isomerism can occur at either of the double bonds.

**#11–12**

$CH_2{=}CH{-}CH_2CH{=}CHCH_3$

C*1* has two like substituents, 2 hydrogens; so cis-trans isomerism is not possible at the bond between C*1* and C*2*.

There are two different substituents on C*4* and two different substituents on *C5*. The *trans* in the name must, therefore, specify the structure at this bond.

*Step 4.* Draw the compound.

First write $\diagdown C{=}C \diagup$ for the *C4*—*C5* bond. Then fill in the substituents, placing the alkyl groups trans to each other.

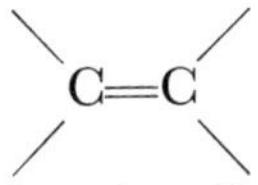

**#11–13**

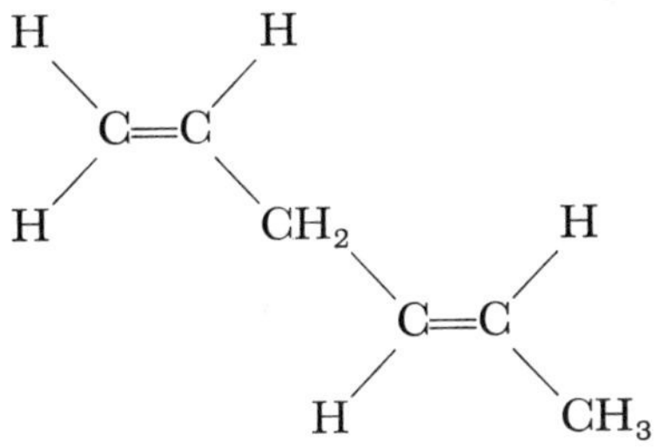

```
H         H
 \       /
  C═══C
 /       \
H         CH2           H
             \         /
              C═══C
             /        \
            H          CH3
```

(*trans*-1,4-hexadiene)

*Example 3*—Draw a structural formula for trans-2(isopropyl)-5-(2-hydroxyethyl)-3-cyclohexenone.

*Step 1.* Draw the carbon skeleton of the longest continuous chain. *Cyclohex* means a cyclic six-carbon ring.

**#11–14**

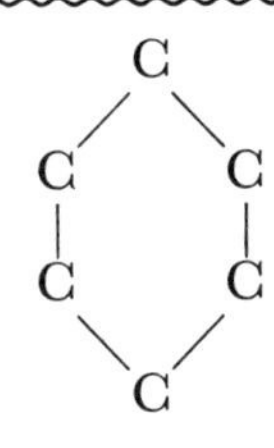


*Step 2.* Put in the class-determining functional group and number the main carbon chain.

The compound is a ketone. The 3 in front of the cyclohex*en*one locates the position of the double bond. The lack of a number in front of the syllable *one* implies a 1. Thus we have the structure shown in #11–15.

**#11–15**

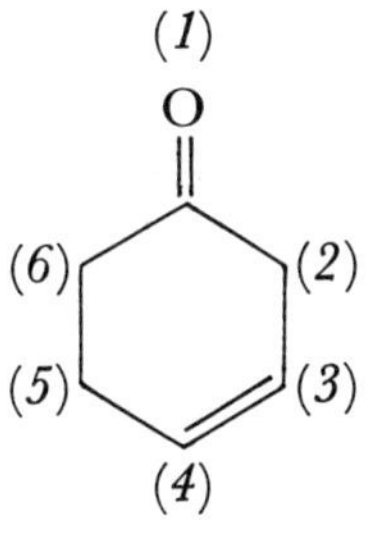


*Step 3.* Determine the structure of the substituents and position them on the ring.

| | | |
|---|---|---|
| isopropyl | $= CH_3\overset{|}{C}HCH_3$ | position = 2 |
| 2-hydroxyethyl | $= —CH_2CH_2OH$ | position = 5 |

**#11–16**

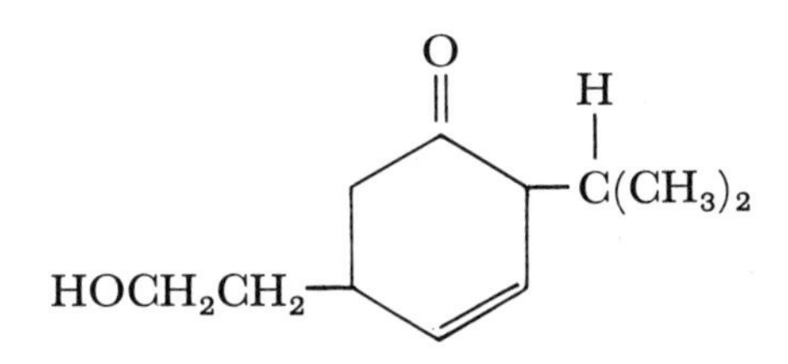

*Step 4.* Redraw the structure so as to show that the two substituents are trans to each other.

#11–17

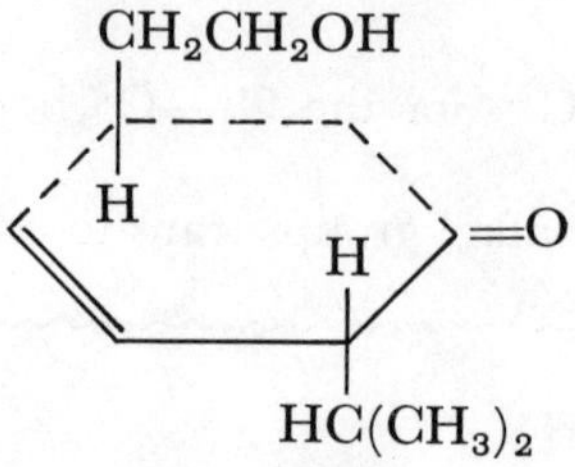

$CH_2CH_2OH$

H

H =O

$HC(CH_3)_2$

*trans*-2(isopropyl)-5-(2-hydroxyethyl)-3-cyclohexenone

## § 12. Problems. Alkenes, Alkynes. Nomenclature

**1**) Draw two-dimensional structural formulas for:

*a*) *trans*-2-pentene
*b*) *cis*-2-pentene
*c*) *cis*-1,3-pentadiene
*d*) 2,3-dimethyl-2-butene
*e*) 1,4-cyclooctadiene
*f*) 1-hexen-4-yne
*g*) 2,4-heptadiyne
*h*) 2,6-dimethyl-1,5-heptadiene-3-yne
*i*) 4-(2-methylpropyl)cyclohexene
*j*) 3(1-methylethyl)-5-methyl-*cis*-3-hexen-1-yne
*k*) *cis,cis,trans*-2,4,6-nonatriene

**2**) Name by the IUPAC system:

*a*) H H
C=C
H $CH_2CH_2CH_3$

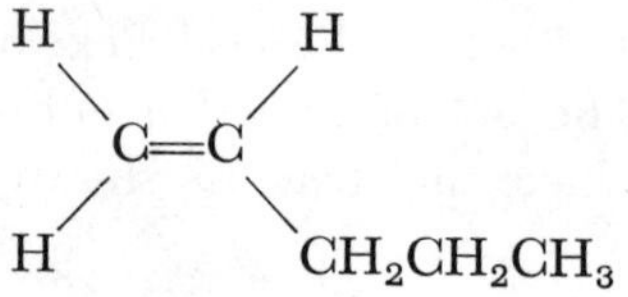

*b*) H H
C=C H
H C=C
H H

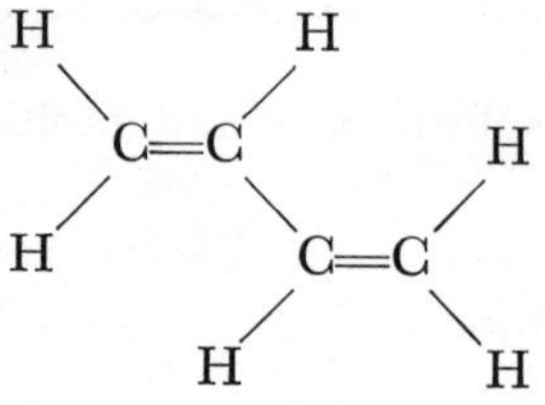

*c*) $CH_3C{\equiv}CH$
*d*) $CH_3C{\equiv}CCH_2C{\equiv}CH$

*e*)

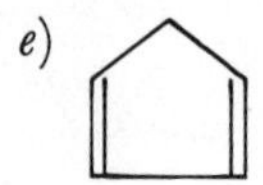

*f*)

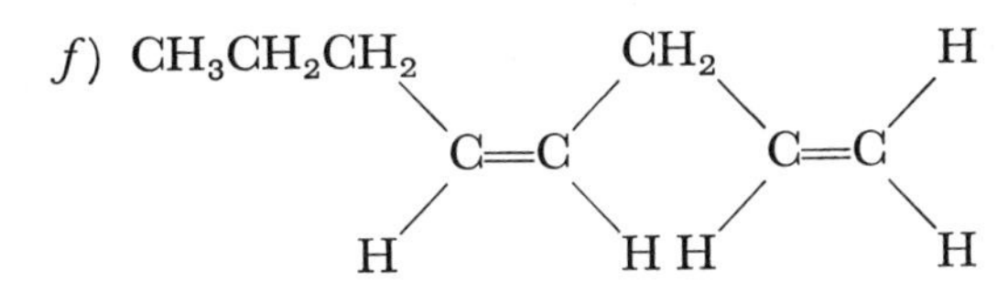


*g*) H CH$_3$

C=C

H CH$_2$C≡CCH$_3$

*h*) CH$_3$ CH$_3$ CH$_3$ H

C=C C=C

CH$_3$ CH$_2$ CH$_3$

*i*)

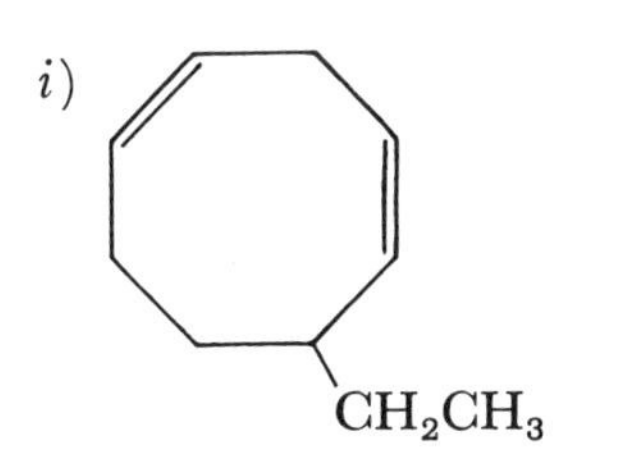


*j*)

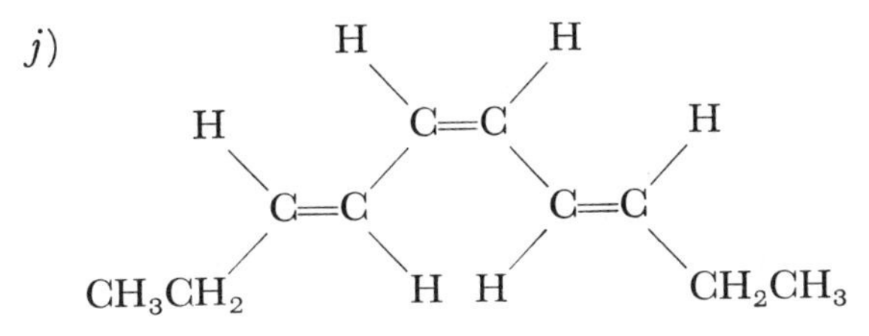


**3**) Name the structures for which formulas are given in problem 1, §10. Use the IUPAC naming system and include the designation of cis or trans for the particular geometric isomers you drew in answer to this question.

**4**) Name the structures you drew in answer to question 2*c*, *d*, and *e*, §10 (omit the condensed rings and the spirane rings).

**5**) The following names are incorrect. Draw the structure indicated by the wrong name and then give it the correct IUPAC designation.

*a*) *trans*-1-butene
*b*) 3-pentyne
*c*) 2-chloro-3-butene
*d*) 1-methyl-3-cyclopentene
*e*) 5-methylcyclohexene
*f*) hex-2-en-5-yne
*g*) pent-4-en-1-yne
*h*) 2-methyl-3-pentene
*i*) 2,4-pentadiene
*j*) 3,5,7-nonatriene
*k*) *trans*-1,1-dibromocyclohexane
*l*) *cis*-3-methylcyclopentene

## § 13. Hydrocarbon Derivatives. Nomenclature

Compounds such as $CH_3CH_2CH_2OH$, $CH_3\overset{\overset{O}{\|}}{C}CH_3$, $CH_3CH_2CH_2Cl$, and $CH_3CH_2COOH$ can be considered as hydrocarbon derivatives. The chemical reactivity of these substances is directly attributable to the functional group. All substances that can be represented as ROH where R = alkyl can be expected to have similar chemical properties in varying degrees. On the basis of chemical similarity all ROH compounds are classified as alcohols. All $R\overset{\overset{O}{\|}}{C}H$ substances are aldehydes and $R\overset{\overset{O}{\|}}{C}OH$ compounds are carboxylic acids. Table 13–1 lists the common functional groups along with the corresponding class name.

The class to which a substance belongs is indicated in the IUPAC system by replacing the terminal *e* in the ending of names like butan*e*, buten*e*, butyn*e* with the suffix assigned to that particular class. The common suffixes are given in Table 13–2, which also illustrates how they are incorporated into the basic hydrocarbon name. Examples of named compounds from each main class are given in Table 13–3. Many substances known for a long time have acquired common names which frequently replace the IUPAC name in daily usage. A number of these are also included in Table 13–3 so that you may become familiar with them. These require more memorization than the IUPAC names because they do not follow a pattern.

The functional group which determines the classification of a substance also determines the order of numbering. The longest continuous carbon chain associated with the functional group is numbered so that the lowest possible value is assigned either to the carbon atom to which the group is attached, as in $-\overset{|}{\underset{|}{C}}-OH$ or to the carbon atom which itself is a part of the group, as in $-\overset{\overset{O}{\|}}{C}OH$. For example, see #13–1.

**#13–1**

$$CH_3CH_2\overset{\overset{OH}{|}}{C}HCH_3 = \text{(2-butanol, } \textit{not} \text{ 3-butanol)}$$

(*1* *2* *3* *4*)
(*4* *3* *2* *1*)

$$CH_3CH_2CH_2\overset{\overset{O}{\|}}{C}H = \text{(1-butanal, } \textit{not} \text{ 4-butanal)}$$

(*1* *2* *3* *4*)
(*4* *3* *2* *1*)

**Table 13–1**
**Common Functional Groups**

| *Empirical formula* | *Functional group* | *Class name* | *IUPAC designation* |
|---|---|---|---|
| RCOOH | —C(=O)OH | acid | -oic acid |
| RC(=O)$NH_2$ | —C(=O)$NH_2$ | amide | -amide |
| RC(=O)R | —C(=O)— | ketone | -one |
| RCHO | —C(=O)H | aldehyde | -al |
| ROH | —OH | alcohol | -ol |
| RSH | —SH | mercaptan | -thiol |
| RCN | —CN | nitrile | -nitrile |
| $RNH_2$ | —$NH_2$ | amine | -amine |
| ROR | —OR | ether | alkoxy- |

**Table 13–2**
**IUPAC Endings Corresponding to Class Designations**

| *Hydro-carbon* | *Alcohol* **ol** | *Aldehyde* **al** | *Ketone* **one** | *Acid* **oic** | *Amine* **amine** | *Amide* **amide** |
|---|---|---|---|---|---|---|
| -ane | -anol | -anal | -anone | -anoic | -anamine | -anamide |
| -ene | -enol | -enal | -enone | -enoic | -enamine | -enamide |
| -yne | -ynol | -ynal | -ynone | -ynoic | -ynamine | -ynamide |

If the class-determining functional group appears more than once, the appropriate prefix—*di* (2), *tri* (3), *tetra* (4), *penta* (5), etc.—is placed in front of the syllable which characterizes the functional group. The numbering of these multiple functional groups must be such that the sum of their values is as low as possible (#13–2).

**#13–2**

$$CH_3\underset{}{\overset{OH}{\overset{|}{C}}}HCH_2CH_2OH$$ = (butan-1,3-diol, *not* butan-2,4-diol)

(*1* *2* *3* *4*)
(*4* *3* *2* *1*)

**Table 13–3**

**The Various Classes of Compounds**

| *Formula* | *IUPAC name* | *Common name* |
|---|---|---|
| $CH_4$ | methane | |
| $CH_3OH$ | methanol | methyl alcohol |
| $H\overset{O}{\overset{\|}{C}}H$ | methanal | formaldehyde |
| $H\overset{O}{\overset{\|}{C}}OH$ | methanoic acid | formic acid |
| $CH_3NH_2$ | methanamine | methylamine |
| $CH_3CH_3$ | ethane | |
| $CH_3CH_2OH$ | ethanol | ethyl alcohol |
| $CH_3\overset{O}{\overset{\|}{C}}H$ | ethanal | acetaldehyde |
| $CH_3\overset{O}{\overset{\|}{C}}OH$ | ethanoic acid | acetic acid |
| $CH_3CH_2NH_2$ | ethanamine | ethyl amine |
| $CH_3CH_2CH_3$ | propane | |
| $CH_3CH_2CH_2OH$ | 1-propanol | *n*-propyl alcohol |
| $CH_3CHOHCH_3$ | 2-propanol | isopropyl alcohol |
| $CH_3CH_2\overset{O}{\overset{\|}{C}}H$ | propanal | propionaldehyde |
| $CH_3\overset{O}{\overset{\|}{C}}CH_3$ | propanone | acetone |
| $CH_3CH_2\overset{O}{\overset{\|}{C}}OH$ | propanoic acid | propionic acid |
| $CH_3CH_2CH_2NH_2$ | 1-propanamine | *n*-propyl amine |
| $CH_3\overset{NH_2}{\overset{\vert}{C}}HCH_3$ | 2-propanamine | isopropyl amine |

The classification of a substance is obvious if the molecule contains only one functional group. However, when there is more than one, a decision has to be made as to which functional group is the ruling factor. For example, should $CH_3\overset{O}{\overset{\|}{C}}CH_2CH_2OH$ be named as an alcohol or a ketone? Common usage and a concerted effort by chemical abstracts for consistency has led to a generally accepted order of group preference. Table 13–4 lists the various common classes of compounds in this priority order. The class which is highest on the list takes precedence over those below it. A ketone is above an alcohol, so

$$CH_3\overset{\overset{\displaystyle O}{\|}}{C}CH_2CH_2OH$$

is classified as a ketone rather than an alcohol. The carbon chain must be numbered so that the keto rather than the hydroxy group has the lowest value even if the sum of the numbers is lower the other way (#13–3).

**#13–3**

$$\underset{\substack{(1\\(4}}{CH_3}\underset{\substack{2\\3}}{\overset{\overset{\displaystyle O}{\|}}{C}}\underset{\substack{3\\2}}{CH_2}\underset{\substack{4)\\1)}}{CH_2OH}$$

= (4-hydroxy-2-butanone, *not* 3-oxo-1-butanol)

When more than one functional group is present, those functional groups of lower priority which do not determine the classification are named as substituent radicals. As illustrated in #13–3, an —OH is referred to as an *hydroxy* group and the oxygen of a ketone is an *oxo* (see Table 13–4). The numbers assigned to such substituents are determined by the number originally assigned to the class-determining functional group.

A halogen is not a class-determining functional group. There are no "halides" in the IUPAC system, merely hydrocarbons with halogen substituents. These are dealt with exactly like alkyl groups, cyano groups, hydroxy groups, etc. If such substituents are the only ones present, the numbers are assigned so as to give the lowest sum. However, if a class-determining functional group is also present, it will determine the numbering of the chain.

**#13–4**

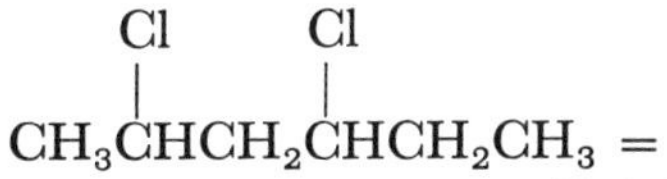

$$CH_3\overset{\overset{\displaystyle Cl}{|}}{C}HCH_2\overset{\overset{\displaystyle Cl}{|}}{C}HCH_2CH_3 =$$

(2,4-dichlorohexane, *not* 3,5-dichlorohexane)

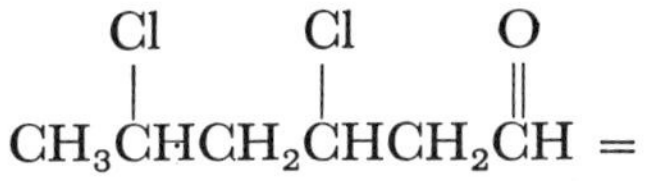

$$CH_3\overset{\overset{\displaystyle Cl}{|}}{C}HCH_2\overset{\overset{\displaystyle Cl}{|}}{C}HCH_2\overset{\overset{\displaystyle O}{\|}}{C}H =$$

(3,5-dichlorohexanal, *not* 2,4-dichloro-6-hexanal)

*Example 1*—Name by the IUPAC system

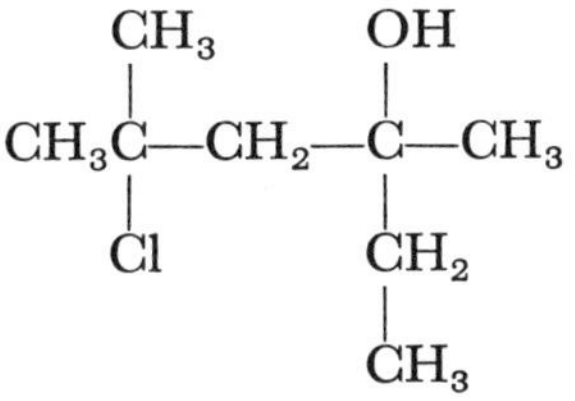

$$CH_3\underset{\underset{\displaystyle Cl}{|}}{\overset{\overset{\displaystyle CH_3}{|}}{C}}—CH_2—\underset{\underset{\underset{\underset{\displaystyle CH_3}{|}}{\displaystyle CH_2}}{|}}{\overset{\overset{\displaystyle OH}{|}}{C}}—CH_3$$

*Step 1.* Identify the class-determining group.

**Table 13–4**

**Preference Order of Functional Groups in Naming Compounds**

| *Substituent group* | *Name when group is class determining* | *Name when group is a substituent* |
|---|---|---|
| *1)* —COOH | -oic acid | carboxy |
| *2)* $—SO_3H$ | -sulfonic acid | sulfo- |
| *3)* $—\overset{\overset{O}{\|}}{C}H$ | -al | formyl |
| *4)* —CN | -nitrile | cyano |
| *5)* $—\overset{\overset{O}{\|}}{C}—$ | -one | oxo |
| *6)* —OH | -ol | hydroxy |
| *7)* —SH | -thiol | mercapto |
| *8)* $—NH_2$ | amine | amino |
| *9)* —O— | ether | alkoxy (oxa) |

Note: F, Cl, Br, I and $—NO_2$ are not class-determining and are named only as substituents: fluoro, chloro, bromo, iodo, and nitro.

A halogen is not a class-determining group; so the compound is an alcohol (Table 13–1). The name ends in *ol*.

*Step 2.* Locate the longest continuous carbon chain to which the —OH is attached.

There are five chains involving the —OH group. Each chain end is labeled with a letter in #13–5. The chains from *a* to *c* and from *b* to *c*

**#13–5**

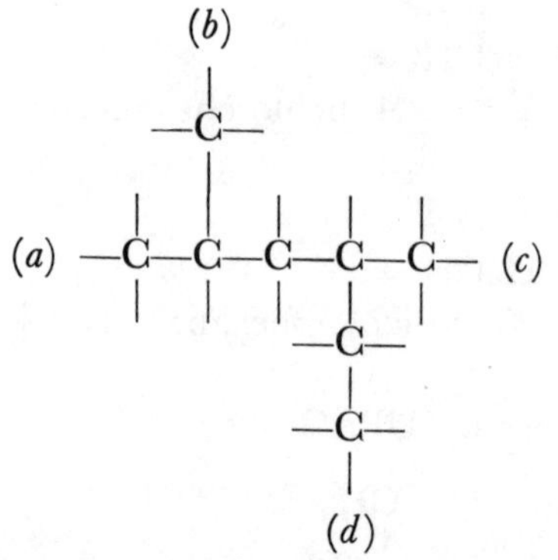

```
              (b)
               |
              —C—
               |
         |     |    |    |    |
(a)     —C—————C————C————C————C— (c)
         |     |    |    |    |
                        —C—
                         |
                        —C—
                         |
                        (d)
```

are five-carbon chains. There is a four-carbon chain from *c* to *d*. The two six-carbon chains between *a* and *d* and between *b* and *d* are identical. A six-carbon chain means that the name is to be built around *hex*.

*Step 3.* Determine the order in which the longest chain is to be numbered.

Numbering from *a* to *d* gives —OH the value of 4, whereas numbering from *d* to *a* gives —OH a value of 3. The compound is a 3-ol.

*Step 4.* Name and assign numbers to the other substituents.

*a*) chloro position 5
*b*) methyl position 3
*c*) methyl position 5

*Step 5.* Organize available information and combine into a name.

*a*) 6 carbons = hex- (Table 1–5–1)
*b*) saturated carbon chain = ane
*c*) alcohol = -ol in place of e in -ane
*d*) —OH on carbon 3 = -3-ol
*e*) —Cl on carbon 5 = 5-chloro-
*f*) two methyl groups = dimethyl-
*g*) methyl groups on carbon 3 and 5 = 3,5-dimethyl-
*h*) alphabetize substituents according to name

The name of the compound is: 5-chloro-3,5-dimethyl-3-hexanol.

*Example 2*—Name $CH_3\overset{\overset{O}{\|}}{C}CH_2\overset{\overset{O}{\|}}{C}OH$.

*Step 1.* Identify the class-determining group.

The compound could be either a ketone or an acid. Acids are higher on the list in Table 13–4; therefore, the compound is classified as an acid.

*Step 2.* Assign numbers to main chain.

The class-determining functional group gets the lowest number.

#13–6

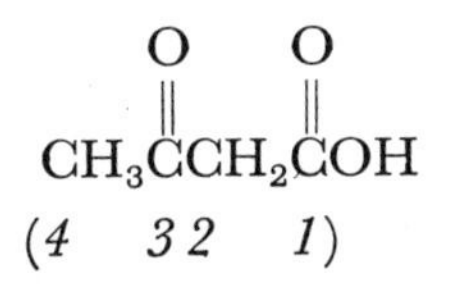

*Step 3.* Organize available information and build into a name.

*a*) 4 carbons = but- (Table 13–1)
*b*) saturated chain = -ane
*c*) carboxylic acid = -oic in place of e of -ane
*d*) carboxyl group is in position 1. Number goes in front of hydrocarbon stem: 1- -anoic acid
*e*) =0 as substituent = -oxo- (Table 13–4)
*f*) =0 on carbon 3 = 3-oxo

The name of the compound is: 3-oxo-1-butanoic acid.

## § 14. Problems. Hydrocarbon Derivatives. Nomenclature

**1**) Name both by the IUPAC system and by their common names.

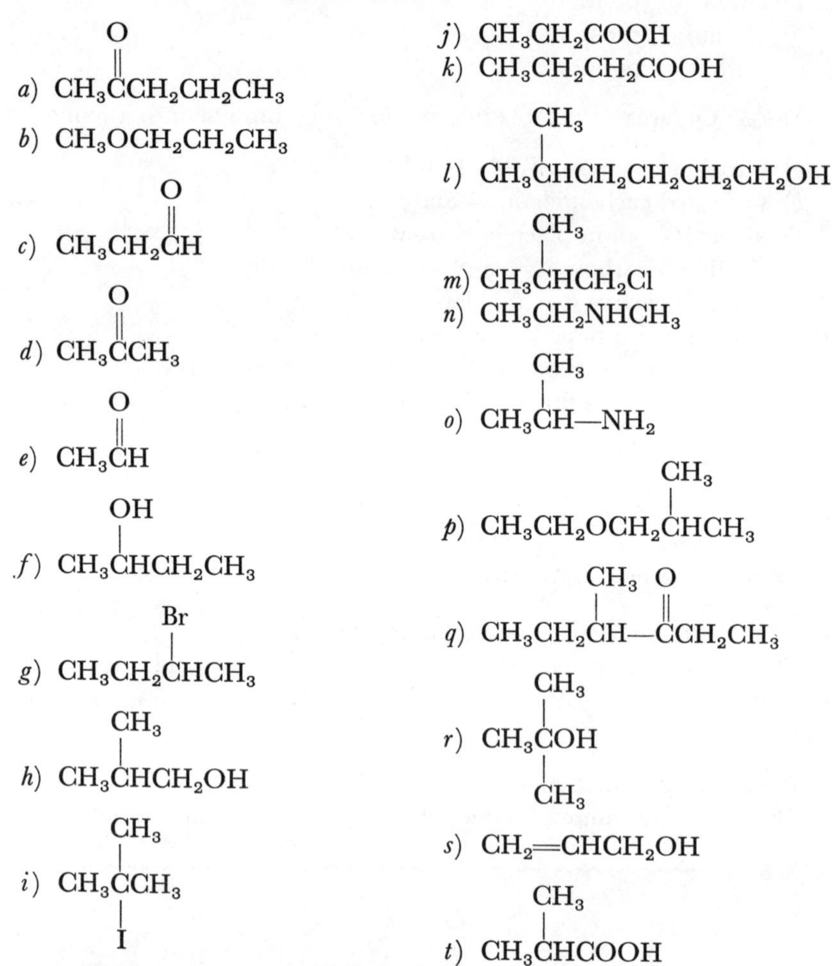


**2**) After you have finished naming the compounds in problem 1 allow some time to lapse and then work from the answers to write out the structures. This will provide additional drill.

**3**) Write formulas for the following compounds:

*a*) pentanal
*b*) *cis*-3-pentenal
*c*) 3-pentynal
*d*) 2-butanone
*e*) 3-buten-2-one
*f*) 3-butyn-2-one
*g*) 3-methylhexanoic acid
*h*) *trans*-3-methyl-2-hexenoic acid
*i*) 3-pentynoic acid
*j*) 3-hydroxybutanoic acid
*k*) 4-oxo-pentanoic acid
*l*) *trans*-4-oxo-2-pentenoic acid
*m*) 3-methoxypropanol
*n*) 3-methoxy-2-propanol
*o*) 4-hydroxy-5-methoxy-2-pentanol
*p*) ethanamine
*q*) isobutyl amine
*r*) *cis*-2-butenamine

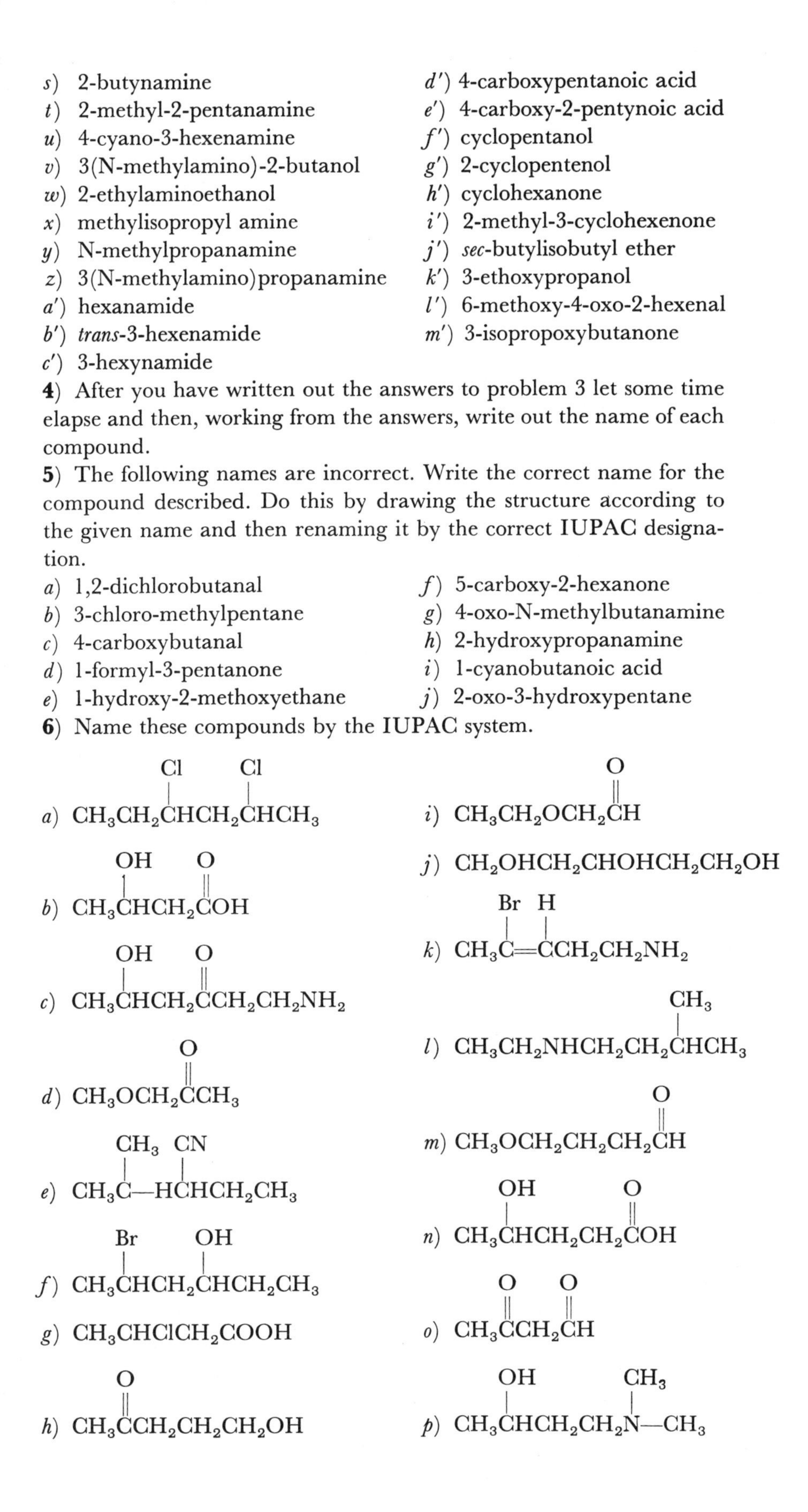

*s*) 2-butynamine
*t*) 2-methyl-2-pentanamine
*u*) 4-cyano-3-hexenamine
*v*) 3(N-methylamino)-2-butanol
*w*) 2-ethylaminoethanol
*x*) methylisopropyl amine
*y*) N-methylpropanamine
*z*) 3(N-methylamino)propanamine
*a′*) hexanamide
*b′*) *trans*-3-hexenamide
*c′*) 3-hexynamide
*d′*) 4-carboxypentanoic acid
*e′*) 4-carboxy-2-pentynoic acid
*f′*) cyclopentanol
*g′*) 2-cyclopentenol
*h′*) cyclohexanone
*i′*) 2-methyl-3-cyclohexenone
*j′*) *sec*-butylisobutyl ether
*k′*) 3-ethoxypropanol
*l′*) 6-methoxy-4-oxo-2-hexenal
*m′*) 3-isopropoxybutanone

**4**) After you have written out the answers to problem 3 let some time elapse and then, working from the answers, write out the name of each compound.

**5**) The following names are incorrect. Write the correct name for the compound described. Do this by drawing the structure according to the given name and then renaming it by the correct IUPAC designation.

*a*) 1,2-dichlorobutanal
*b*) 3-chloro-methylpentane
*c*) 4-carboxybutanal
*d*) 1-formyl-3-pentanone
*e*) 1-hydroxy-2-methoxyethane
*f*) 5-carboxy-2-hexanone
*g*) 4-oxo-N-methylbutanamine
*h*) 2-hydroxypropanamine
*i*) 1-cyanobutanoic acid
*j*) 2-oxo-3-hydroxypentane

**6**) Name these compounds by the IUPAC system.

*a*) $CH_3CH_2\overset{Cl}{\overset{|}{C}}HCH_2\overset{Cl}{\overset{|}{C}}HCH_3$

*b*) $CH_3\overset{OH}{\overset{|}{C}}HCH_2\overset{O}{\overset{\|}{C}}OH$

*c*) $CH_3\overset{OH}{\overset{|}{C}}HCH_2\overset{O}{\overset{\|}{C}}CH_2CH_2NH_2$

*d*) $CH_3OCH_2\overset{O}{\overset{\|}{C}}CH_3$

*e*) $CH_3\overset{CH_3}{\overset{|}{C}}{-}H\overset{CN}{\overset{|}{C}}HCH_2CH_3$

*f*) $CH_3\overset{Br}{\overset{|}{C}}HCH_2\overset{OH}{\overset{|}{C}}HCH_2CH_3$

*g*) $CH_3CHClCH_2COOH$

*h*) $CH_3\overset{O}{\overset{\|}{C}}CH_2CH_2CH_2OH$

*i*) $CH_3CH_2OCH_2\overset{O}{\overset{\|}{C}}H$

*j*) $CH_2OHCH_2CHOHCH_2CH_2OH$

*k*) $CH_3\overset{Br}{\overset{|}{C}}{=}\overset{H}{\overset{|}{C}}CH_2CH_2NH_2$

*l*) $CH_3CH_2NHCH_2CH_2\overset{CH_3}{\overset{|}{C}}HCH_3$

*m*) $CH_3OCH_2CH_2CH_2\overset{O}{\overset{\|}{C}}H$

*n*) $CH_3\overset{OH}{\overset{|}{C}}HCH_2CH_2\overset{O}{\overset{\|}{C}}OH$

*o*) $CH_3\overset{O}{\overset{\|}{C}}CH_2\overset{O}{\overset{\|}{C}}H$

*p*) $CH_3\overset{OH}{\overset{|}{C}}HCH_2CH_2\overset{CH_3}{\overset{|}{N}}{-}CH_3$

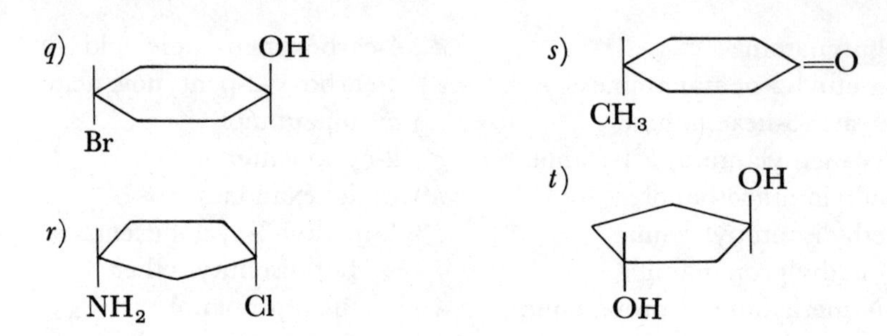


## § 15. Optical Isomerism

As discussed in §7, the designation of *conformer* is applied to the various spatial relationships that can be assumed by the atoms in a particular molecule without breaking any bonds. The term *isomer* is used when two molecules contain the same numbers and kinds of atoms but these atoms are bonded together differently. Thus, conformers are merely interchangeable forms of the same compound, whereas isomers are different compounds with the same molecular formula. The various types of isomerism are summarized in Table 15–1. This section develops the concepts of optical isomerism.

It is possible for the same atoms to be bonded together in two different molecules with only the order of bonding being different. A difference in bond order is responsible for stereoisomerism, which may appear as optical isomerism (#15–1*a*) or as geometric isomerism (#15–1*b*).

**#15–1**

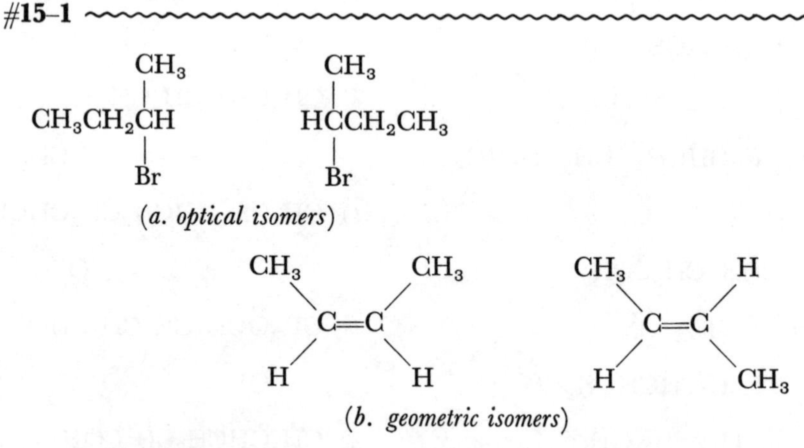


(*a. optical isomers*)

(*b. geometric isomers*)

A carbon bonded to four different substituents is referred to as an asymmetric carbon. "Different substituents" means $—CH_3$, $—CH_2CH_3$, $—CH_2D$, —H, —D, —Cl, —Br, —OH, $—OCH_3$, $—OCH_2CH_3$, etc.—any substituents that are not identical. There are two ways in which four different substituents can be arranged around a tetrahedral carbon. Molecules that are identical except for the order of bonding at an asymmetric carbon are optical isomers. The structural difference leading to optical isomerism can best be visualized through models. The description here can be adapted to any model set you are using. You can even use your cardboard model of tetrahedral bonding (from #1–29) with four different-colored arms. You will need two tetrahedral units.

Attach a black and a white plastic tube to each of two tetrahedral clusters. Hold each unit in the position shown in #15–2*a*, with the black

#15–2

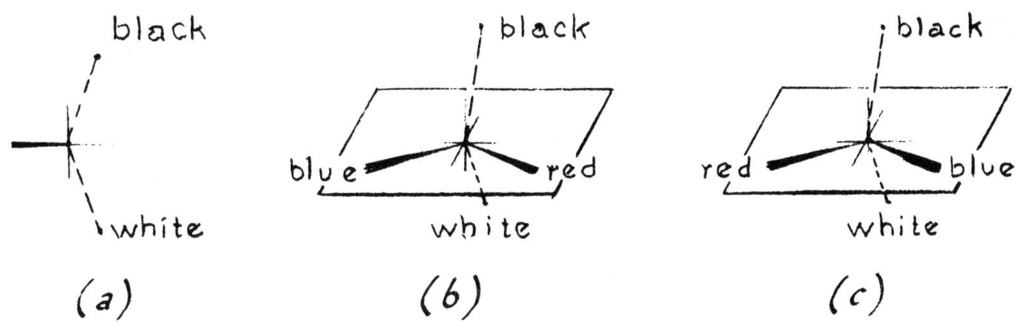


tube projecting up and away from you in the YZ reference plane. The empty prongs will be toward you. Attach a red tube to the left prong of the cluster and a blue one to the right prong. The blue and red tubes should project toward you in the horizontal position in the XZ plane. Reverse the order with the second cluster, placing the blue tube on the left prong and the red one on the right prong. You now have the two structures shown in #15–2*b* and *c*. Try to superimpose them, i.e., try to position them so that each of the four colors from one structure projects in the same direction as each of the four colors in the other structure. As you move the models around, you will find that they just cannot be placed in identical positions. Although any two pairs of colors can be matched, the remaining two will always be reversed as illustrated in #15–3. In #15–3, the tetrahedral structure is

#15–3

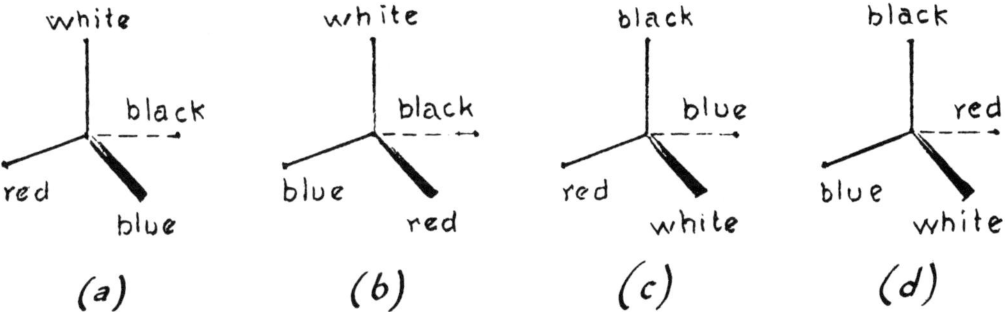


positioned so that three of the bonds form the base of a tripod, with the fourth bond extending upward from the base. In these drawings the upward and the left-hand prongs are in the plane of the paper (XY reference plane), while one of the remaining prongs extends behind the plane of the paper and the other projects in front of it.

Structure *b* is the mirror image of structure *c*. If you hold *c* up in front of a mirror, the image that you see is *b*. Similarly, if you hold *b*

**Table 15–1**

**Types of Isomerism**

| *The type and its characteristics* | Example | |
|---|---|---|
| *1) Position isomerism*—Different atoms bonded together | $CH_3CH_2CH_2Br$ (*1-bromopropane*) | $CH_3CH(Br)CH_3$ (*2-bromopropane*) |
| | $CH_3OCH_3$ (*dimethyl ether*) | $CH_3CH_2OH$ (*ethanol*) |
| | $HC{\equiv}CCH_2CH_3$ (*1-butyne*) | $CH_3C{\equiv}CCH_3$ (*2-butyne*) |
| *2) Stereoisomerism*—same atoms bonded together in different relative positions | H(CH₃)C=C(H)CH₃ (*cis-2-butene*) | CH₃(H)C=C(H)CH₃ (*trans-2-butene*) |
| *Geometrical isomerism*—due to restricted rotation<br>substituents on opposite sides of a double bond | CH₃ CH₃ / H H (*cis-1,2-dimethyl-cyclobutane* | |

up in front of a mirror, the image that you see is *c*. The structures *b* and *c* bear the same relationship to each other as your two hands do; that which is on the right in one is on the left in the other; they are nonsuperimposable mirror images.

The model represents an *asymmetric carbon*—a tetrahedral carbon with four different substituents. The second carbon in the chain of 2-chlorobutane has four different substituents, as shown in #15–4, and is an asymmetric carbon. There are two ways in which these four

**#15–4**

$CH_3CH(Cl)CH_2CH_3$
(2-chlorobutane)

substituents can be bonded to C2, just as there are two possible arrangements for the four colored tubes in your model. Thus, there are two 2-chlorobutanes with structures that are nonsuperimposable mirror images. These two structures are identical except for the order of bonding and are called *enantiomers*.

The most readily measurable difference in the physical and chemical

substituents on opposite sides of molecular plane

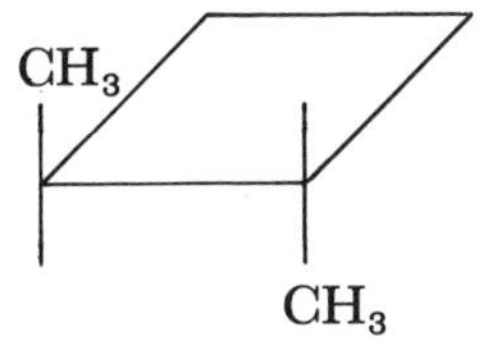

$CH_3$ / $CH_3$

(*trans-1,2-dimethylcyclobutane*)

*Optical isomerism—*
order of bonding at one carbon is different
four different substituents on one carbon

enantiomers
mirror images

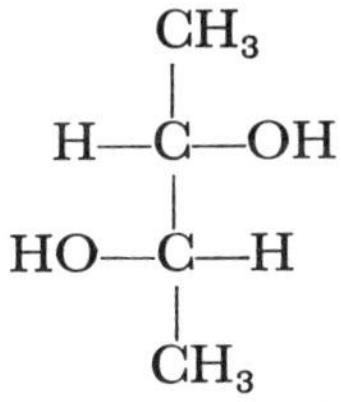

$$\begin{array}{c} CH_3 \\ | \\ H-C-OH \\ | \\ HO-C-H \\ | \\ CH_3 \end{array} \qquad \begin{array}{c} CH_3 \\ | \\ HO-C-H \\ | \\ H-C-OH \\ | \\ CH_3 \end{array}$$

(**R,R**)-*2,3-dihydroxybutane* (**S,S**)-*2,3-dihydroxybutane*

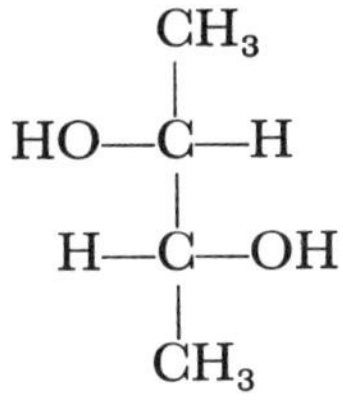

diastereoisomers
meso compound

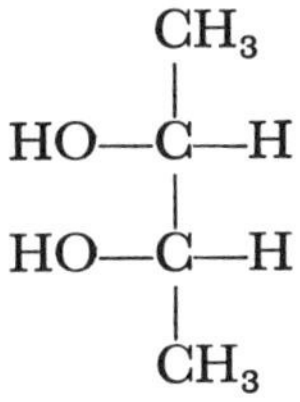

$$\begin{array}{c} CH_3 \\ | \\ HO-C-H \\ | \\ HO-C-H \\ | \\ CH_3 \end{array}$$

(**R,S**)-*2,3-dihydroxybutane*

---

properties of a pair of enantiomers is to be found in their effect upon plane polarized light. While one isomer rotates the plane of polarized light to the left, the other rotates it to the right. A mixture which contains equal numbers of two enantiomers is a *racemic modification.* Light emerges from a racemic modification with the same polarization that it had when it entered, because the effect of one enantiomer is offset by its mirror image and both enantiomers are present in equal quantity.

In general, laboratory preparations produce racemic modifications except under specially controlled circumstances. However, each of nature's reactions usually produces a single optical isomer. Sugars, proteins, and drugs of natural origin are optically active substances. Frequently one optical isomer will have a marked physiological effect while the other isomer is relatively inactive. Lysergic acid is such a compound. These differences are attributed to the way in which the molecule "fits" into the biological substrate.

We can show the two different configurations of the substituents around an asymmetric carbon by drawings such as those in #15–3 and we will use this notation extensively. However, certain conventions have been set up which make it possible also to show these relationships in two-dimensional formulas. These conventions are illustrated

**#15–5**

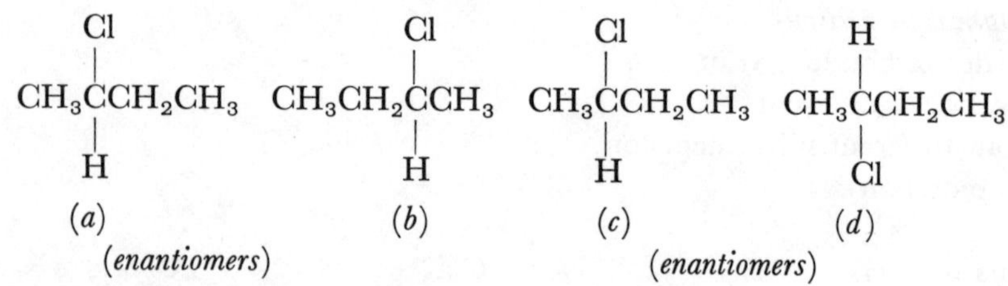


in #15–5. Structural formulas of mirror images may be drawn with the two vertical substituents in the same relative positions but with the two horizontal ones reversed from left to right, as illustrated in #15–5*a* and *b*. Or they may be drawn with the horizontal substituents in the same relative positions and the vertical substituents reversed from top to bottom, as illustrated by *c* and *d*.

You concluded that your models were mirror images because you were unable to superimpose one upon the other. We need also to be able to determine whether two written formulas represent the same structure or mirror images. We cannot pick up one formula and place it over the other to discover whether the two are superimposable. We can, however, imagine doing so. In order to compare two structural formulas, you may mentally rotate them within the plane of the paper. You may not turn them over or do any maneuver that would take them out of the plane of the paper.

Compare *a* and *b* of #15–5. You can rotate *b* so that the —$CH_3$ and —$CH_2CH_3$ groups coincide in both structures, but the —H and the —Cl cannot also be made to coincide. These two formulas represent mirror images. Compare *b* and *d*. If you rotate *d* in the plane of the paper by 180° you can superimpose it on *b*. These two formulas represent the same compound. You need to be able to recognize mirror images from flat drawings. Problems 1 and 2, §16, have you do some of this. Whenever there is reversal (left and right or up and down) of the position of two substituents while the relative positions of the other two substituents remain the same, the formulas represent enantiomers (mirror images).

The name of a compound with an asymmetric carbon should differentiate between the optical isomers by indicating the order in which substituents are bonded to the asymmetric carbon. A notation system which effectively accomplishes this has been developed by R. S. Cahn, C. Ingold, and V. Prelog. It is based on two concepts: the assignment of a rating order from 1 to 4 for the substituents on the asymmetric carbon; the utilization of the atom with the lowest priority to establish a point of reference in space. The simpler rules of this system and their application follow:

*a*) The rating system—

Atoms are rated relative to each other on the basis of their atomic numbers. The higher the atomic number, the higher the priority assigned to that atom. Atoms with high priority come first and are assigned the lower of the numbers from 1 to 4. Some of the more commonly encountered atoms are listed in order from the highest priority to the lowest.

Br (*35*) Cl (*17*) S (*16*) F (*9*) O (*8*) N (*7*) C (*6*) H (*1*)

Isotopes are rated according to atomic weight with the higher atomic weight being given higher priority.

T (3) D (2) H (1)

For example, the substituents on the asymmetric carbon in 1-amino-ethanol would be numbered as indicated in figure #15–6, which

**#15–6**

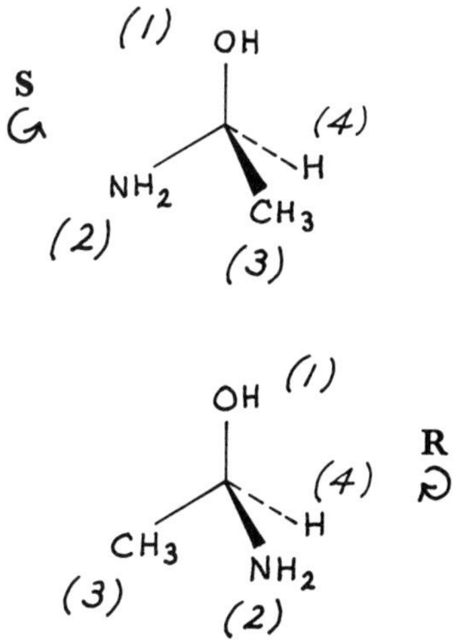

S: (1) OH, (2) $NH_2$, (3) $CH_3$, (4) H

R: (1) OH, (2) $NH_2$, (3) $CH_3$, (4) H

represents the enantiomeric pair of optical isomers. Note the reversal of the positions of the methyl and the amino groups in the same structures. An O has first priority, an N second priority, a C third priority, and an H has fourth priority, in order of decreasing atomic number.

When substituent groups are so constituted that they present the same bonding atom to the asymmetric carbon, the rating of the whole group is determined by considering the ratings of the substituents on the second atom in the group. For example, in 1-chloro-3-methyl-pentane the initial bonding of each alkyl group to the asymmetric carbon occurs through the carbons marked *a* in #15–7. All carbons are rated equally, so the next set of atoms bonded to each of these carbons must be rated to differentiate between the three groups. The substituents on C*a* in the methyl group are H,H,H; in the ethyl group

**#15–7**

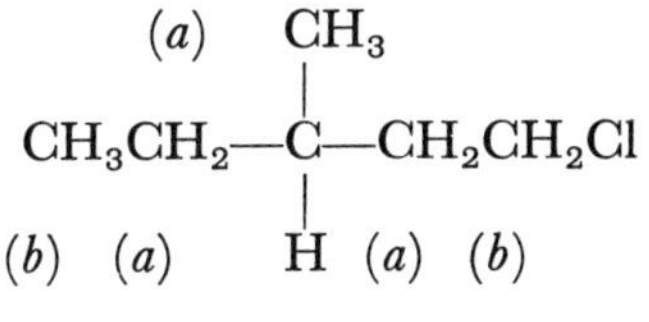

$$\begin{array}{c} (a)\ CH_3 \\ | \\ CH_3CH_2—C—CH_2CH_2Cl \\ | \\ (b)\ (a)\quad H\ (a)\ (b) \end{array}$$

they are C,H,H, and in the chlorethyl group they are also C,H,H. Three H's are considered of lower priority than 2 H's plus a C. Thus, evaluation of the atoms bonded to C*a* in each group reveals that the methyl group should be assigned a lower priority than either of the other groups, but it does not provide a means for distinguishing between the other two groups. The final rating must come from a consideration of the atoms bonded to C*b* in each group. These are H,H,H for the ethyl group and H,H,Cl for the chlorethyl group. The higher atomic number of a Cl as compared to an H places the chlorethyl group ahead of the ethyl group in the priority order. This gives the numbering shown in #15–8 for the substituents on the asymmetric carbon in the enantiomers of 1-chloro-3-methylpentane.

A double bond between carbon and an atom Y is rated as if there

#15–8

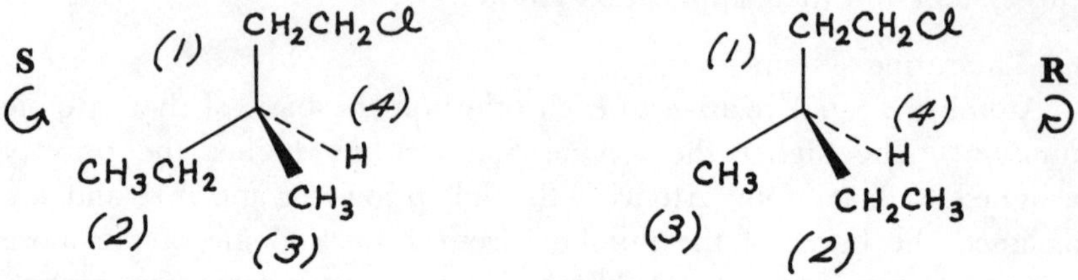

S (1) $CH_2CH_2Cl$, (4) H, $CH_3CH_2$ (2), $CH_3$ (3)

(1) $CH_2CH_2Cl$, (4) H, $CH_3$ (3), $CH_2CH_3$ (2) R

(Note reversal of methyl and ethyl groups)

were two Y's singly bonded to the carbon. Similarly, a triple bond is treated as though there were three single bonds each to an atom of Y. If in another group there really are two Y's singly bonded to a carbon, this group is rated higher than the group with $>C{=}Y$. This point is illustrated by the rating sequence shown in #15–9. Two O's have priority over a double-bond O, which has priority over an O, which, in turn, has priority over an N.

#15–9

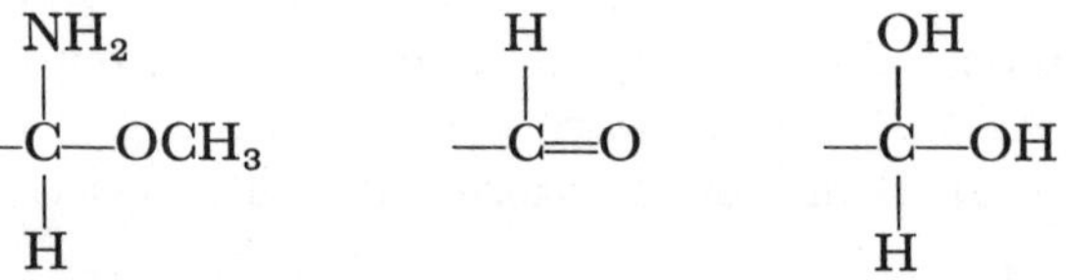

| $-CH(NH_2)-OCH_3$ | $-CH{=}O$ | $-CH(OH)-OH$ |
|---|---|---|
| (3) | (2) | (1) |

The determining factor is the comparative ratings of the atoms on each successive carbon in the chain of each group substituent on the asymmetric carbon. The comparative ratings of the total number of substituent atoms does not come under consideration. Figure out for yourself why the groups in #15–10 were given the indicated ratings in relations to each other (answer given in appendix).

#15–10

| $-CH_2CH_2CH_2Cl$ | $-CH_2CH_2CHCl_2$ |
|---|---|
| (4) | (3) |
| $-CH_2CHClCH_3$ | $-CH_2CHClCH_2Cl$ |
| (2) | (1) |

*b*) The spatial orientation of the molecule—

Each optical isomer has to be designated in terms of the relative positions of four different substituents around a tetrahedral carbon. To assure consistency of nomenclature it has been necessary to specify a standard position from which the tetrahedral unit is viewed while determining the order of bonding.

In the beginning this process can be understood better from an actual model than from a mental image. Make models of the enantiomeric structures #15–2*b* and *c*. Label each color with a number as follows: black = 1, red = 2, blue = 3, white = 4. Place the two models on the desk in front of you with the white (4) tube projecting away from you behind the other three and with the black (1) tube in the vertical position, as illustrated in #15–11*a*.

**#15–11**

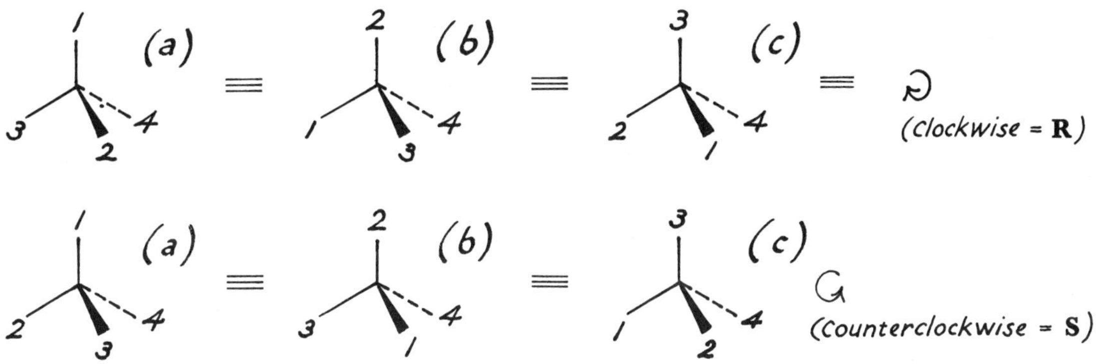

(a) ≡ (b) ≡ (c) ≡ (Clockwise = **R**)

(a) ≡ (b) ≡ (c) (Counterclockwise = **S**)

In one model, if you look at the number on the vertical tube and then at the number on the tube to the right and then at the number on the tube to the left, while the white (4) tube remains in back, you will find that the numbers are in ascending order, 1, 2, 3. You have looked at the numbers in a clockwise direction.

- The configuration at an asymmetric carbon is described as **R** if the priority numbers assigned to the substituents ascend in a clockwise direction when the tetrahedral unit is viewed from a point in space that places the substituent with the lowest priority on the opposite side of the carbon atom from the viewer. **R** *stands for rectus, or right.*

In the other model, you have to look at the tubes in a counterclockwise order: vertical, left, right, if the numbers are to progress 1, 2, 3.

- The configuration at an asymmetric carbon is described as **S** when the priority numbers assigned to the substituents increase in a counterclockwise direction when the tetrahedral unit is viewed from a point in space that places the substituent with the lowest priority on the side of the carbon atom away from the viewer. **S** *stands for sinister, or left.*

The actual positions of the numbers in the model are unimportant; what matters is the direction in which the numbers increase: 1, 2, 3. To emphasize this point, pick up one model by the white tube, keeping it in the same position as it had on the desk. Twirl it like a pinwheel between your thumb and fingers. The **R** model will have to be revolved clockwise for the numbers to move 1, 2, 3, while the **S** model will have to be revolved counterclockwise. Other possible positions for each model are shown in #15–11*b* and *c*. As an example of how the system is applied to naming a formula, determine the correct designation for the enantiomers of 2-chlorobutane as written in #15–12.

**#15–12**

Cl<br>
H—C—$CH_2CH_3$<br>
$CH_3$<br>
(*a*)

Cl<br>
$CH_3CH_2$—C—H<br>
$CH_3$<br>
(*b*)

*Example 1*—Give a complete name to each of the structural formulas in #15–12.

*Step 1.* Assign ratings to the four substituents.

An —H, if present, always is of lowest priority and is assigned the number 4.

The remaining three atoms bonded to the asymmetric carbon are C, C, and Cl. The Cl has the highest atomic number so it is assigned number 1.

The carbon in the methyl group has 3 H's substituent on it. The carbon in the ethyl group has 2 H's and 1 C. A C has priority over an H. Therefore, the ethyl group has a lower number than the methyl (a higher priority).

The entire molecule can now be numbered as in #15–13.

**#15–13**

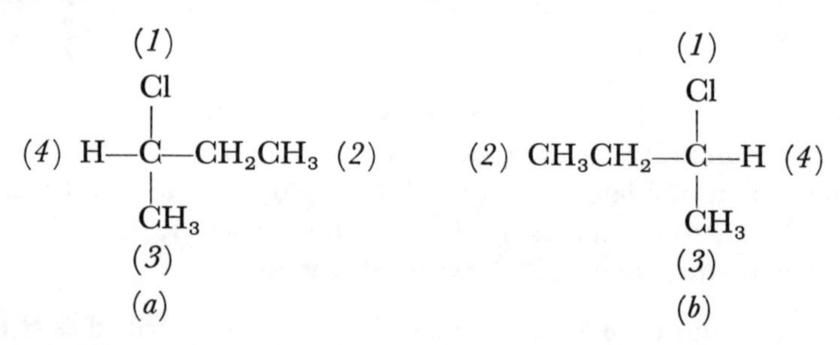


*Step 2.* Draw the molecule using the three-dimensional notation and indicating the substituent with the lowest rating projecting away from the viewer.

A flat drawing such as that in #15–13 has spatial implications that were established by Fischer during his early work on optically active sugars. Horizontal bonds are assumed to project toward the viewer in reference plane XY, while vertical bonds project away from the viewer, in reference plane YZ. In order to visualize the actual relationships between the groups, make a three-dimensional drawing of each structure (#15–14) in the position shown by the formulas, keeping in mind the *Fischer convention.*

**#15–14**

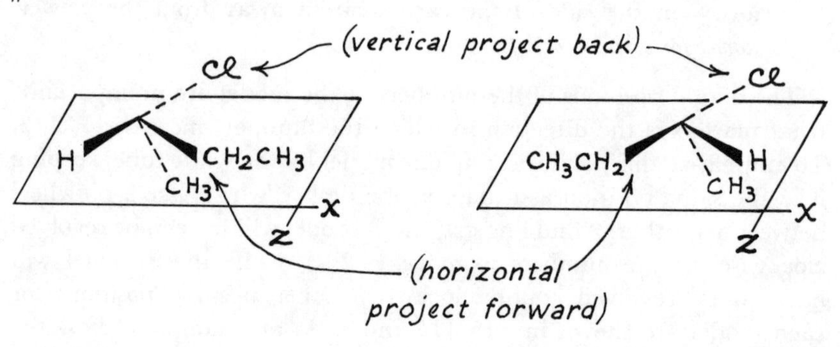

Now tilt the molecule forward, placing the two horizontal projections on a flat surface along with the lower of the two vertical projections. Thus three of the substituents become the feet of a tripod and the Cl projects upward in the plane of the paper. For structure *a*, this places the H on the left in the plane of the paper, the —$CH_3$ behind the paper, and the —$CH_2CH_3$ in front of the paper. In the mirror image *b*, the positions of —$CH_2CH_3$ and —H are reversed, as shown in #15–15.

**#15–15**

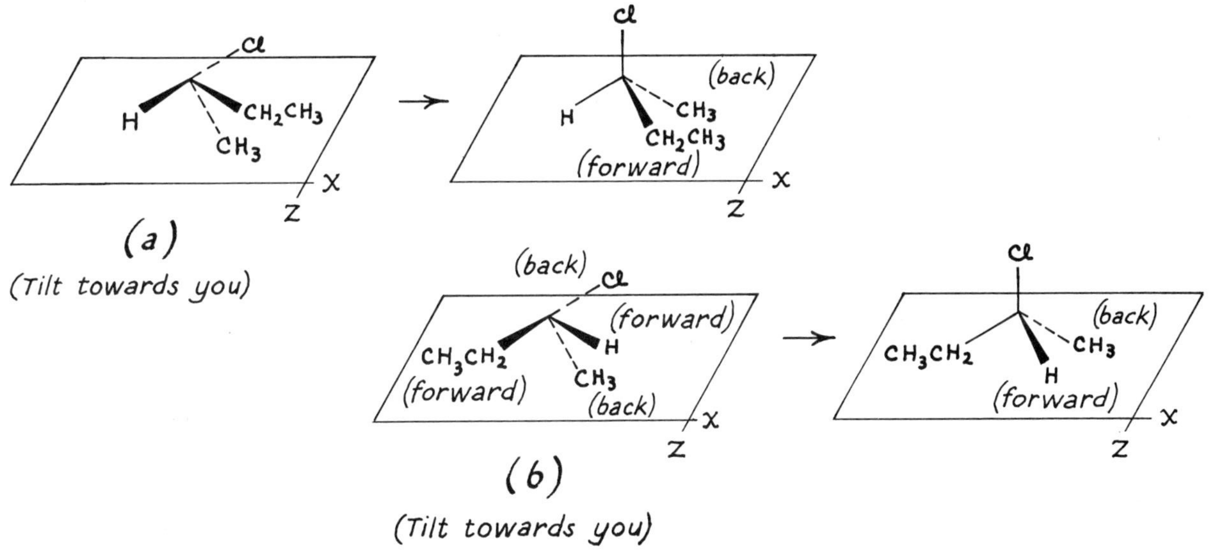

(a)
(Tilt towards you)

(b)
(Tilt towards you)

Imagine that you can hold the Cl and rotate the tripod until the H (the lowest priority substituent) projects behind the plane of the paper. As the H in structure #15–15 is moved back and around, the ethyl group moves forward and around, and the methyl group moves toward the left, giving the structure shown in #15–16*a*. Similarly, the *b*

**#15–16**

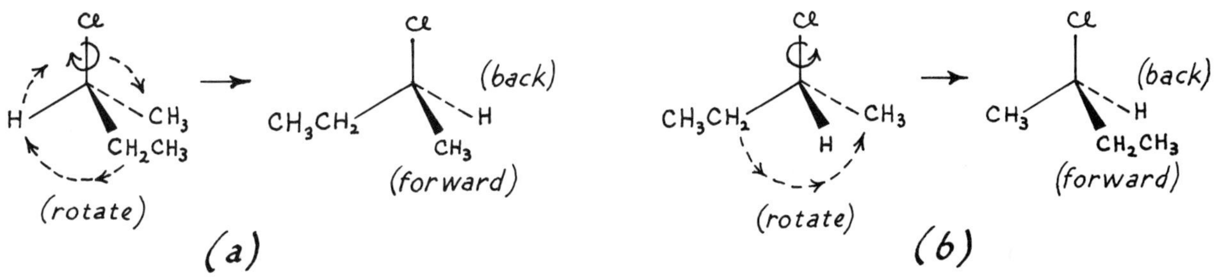

(a)

(b)

configuration can be rotated to place the H at the rear, projecting behind the paper. Visualization of this process will be easier the first few times if you follow along with a model.

*Step 3.* Number the substituents according to the ratings established in Step 1 and determine whether the increasing numerical order proceeds clockwise (**R**) or counterclockwise (**S**) when you view the molecule from the side opposite substituent 4 (#15–17).

#15–17

Cl (1)　　CH$_2$CH$_3$ (2)　　H (4)　　CH$_3$ (3)　　R

Cl (1)　　CH$_3$ (3)　　H (4)　　CH$_2$CH$_3$ (2)　　S

*Step 4.* Name the compounds.

Formula *a* in figure #15–12 is **S**-2-chlorobutane.

Formula *b* in figure #15–12 is **R**-2-chlorbutane.

The designation of **R** or **S** does not give any information about the direction or the extent of optical rotation produced by a given structure. It merely enables you to determine how the substituents are bonded to a particular asymmetric carbon. The mere examination of a molecular structure does not yield information as to its effect on plane polarized light. This information has to be obtained experimentally. If one wishes to indicate that a compound is detrorotatory, a (+) or a (*d*) is placed in front of the name. A (−) or an (*l*) indicates a compound that is levorotatory, while a (±), a (*dl*), or an **RS** designates a racemic modification.

When a molecule contains more than one asymmetric carbon, the number of possible isomers is $2^n$, where $n$ = the number of asymmetric carbons present. The compound 2,3-dichlorobutane has two asymmetric carbons, C*2* and C*3*, and would be expected to form the four

#15–18

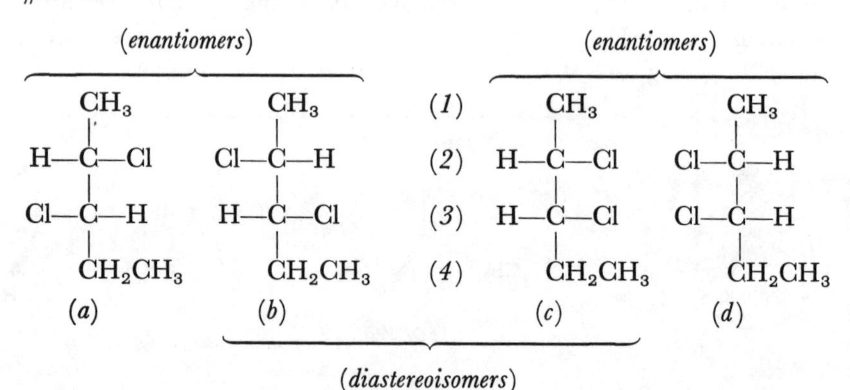


optical isomers indicated in #15–18. Drawing these structures as vertical chains makes it easier to show the enantiomers. The mirror image is obtained by merely interchanging the right and left hand substituents, as in #15–18*a* and *b* and in #15–18*c* and *d*. The four optical isomers consist of two enantiomeric pairs. Structures which are optical isomers but are not enantiomers, such as *b* and *c* or *a* and *d*, are called *diastereoisomers*. A molecule must have more than one asym-

THE tetrahedral geometry of $sp^3$ hybridized carbon is a basic concept of organic chemistry. This card will enable you to visualize that geometry. The card is for use with . . .

# ORGANIC CHEMISTRY—*How to Solve It,*

by Professor Ruth A. Walker, Herbert H. Lehman College, C.U.N.Y.


*Instructions* (see text for illustrations of procedures, fig. 1-28): Below are 2 forms. Each makes a 3-D model of an $sp^3$ hybridized carbon atom. To assemble: (*a*) Note that a form has *4 parts*, each of these consisting of 3 small triangles. The 4 parts are *not* to be separated. Cut out the top (1 part shaded) form . . . (*b*) Punch or cut out the small central black spot in each of the 4 parts . . . (*c*) Bend *up* 2 triangles of a part along the 2 dotted lines in the part; bend *only* until the 2 separated points of a part meet. Affix a bit of tape to the back of each part so that those 2 points are held together where they meet . . . (*d*) Bend *down* along the solid line between each pair of parts until 2 small adjacent triangles (1 in each part) are back to back . . . (*e*) Check your folding: all 4 punched out centers should meet, and each of the 4 sides is recessed . . . (*f*) Then, tape or glue the structure together permanently.

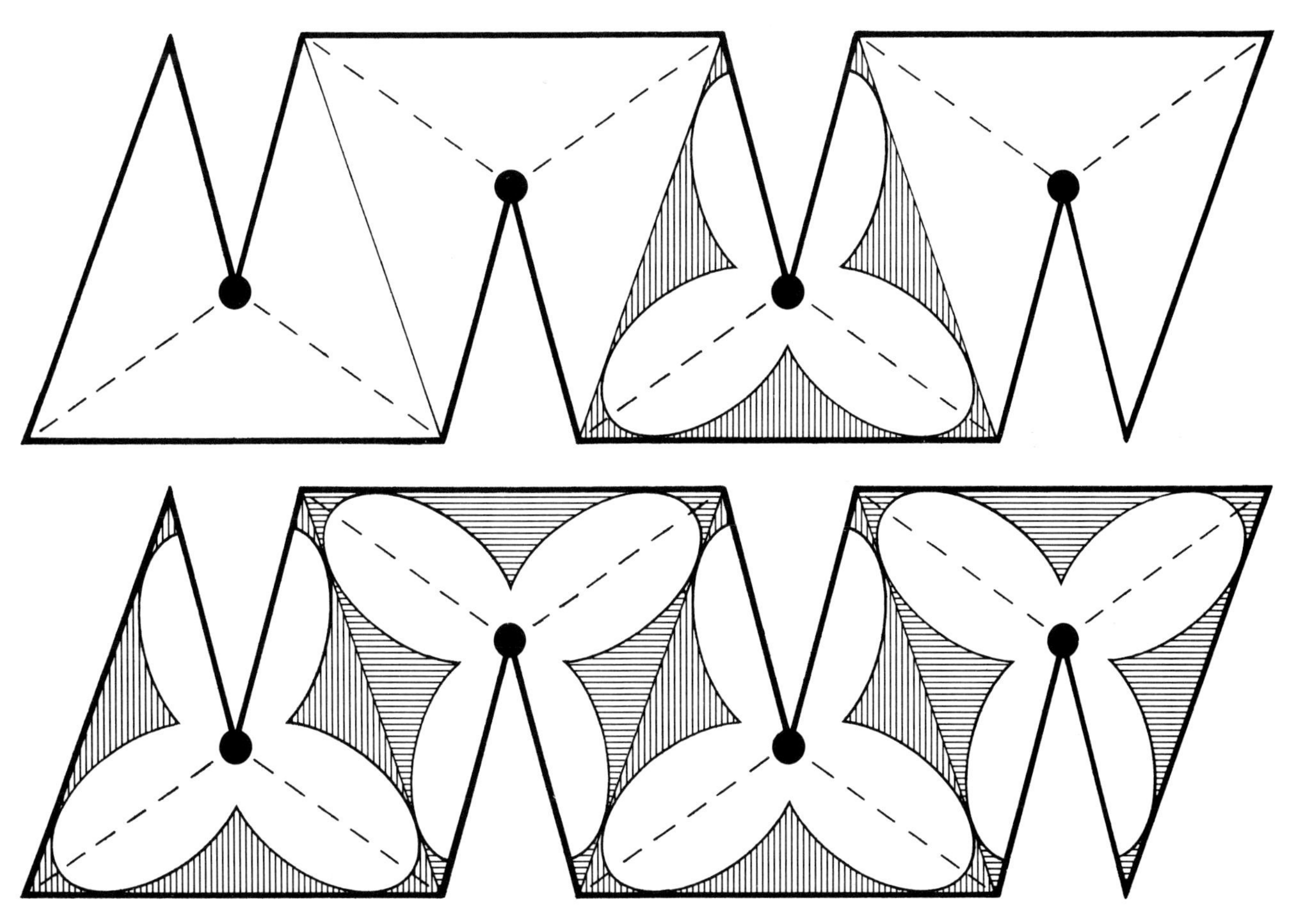

We call it the *Cooper Structure*: At its center is the nucleus of an $sp^3$ hybridized carbon. Projecting from the nucleus are 4 $sp^3$ orbitals, any 2 axes of which meet to form an angle of 109.5°. Carbon with this electron configuration occurs only when a C is bonded to 4 other atoms. Each $sp^3$ orbital will form a sigma bond by end-to-end overlap with an orbital from another atom. The axes of the resulting bonding orbitals meet also to form angles of 109.5° . . . To see the effect of tetrahedral bonding in $CH_3CH_3$ make 2 *Cooper Structures* and join them by running a piece of wire between the centers along the axis of an orbital from each carbon—best done if the ends of the 2 orbitals are cut or split to achieve overlap . . . Three or more *Cooper Structures* so joined will show other geometrical effects of $sp^3$ hybridization.

metric carbon before it can exist in the form of diastereoisomers. Where there are two asymmetric carbons in one molecule, stereoisomers result if the configuration at one of the asymmetric carbons is the same in both structures while the configuration at the second asymmetric carbon in one structure is the mirror image of the configuration of the substituents on the second asymmetric carbon in the other structure.

In order to give **R** and **S** designations for all four structures in #15–18, it is necessary only to determine the configurations for one isomer, and then all the others can be related back to this one. The configuration at C*2* in structure #15–18*a* is **S**, and that at C*2* is also **S**, as determined in #15–19.

**#15–19**

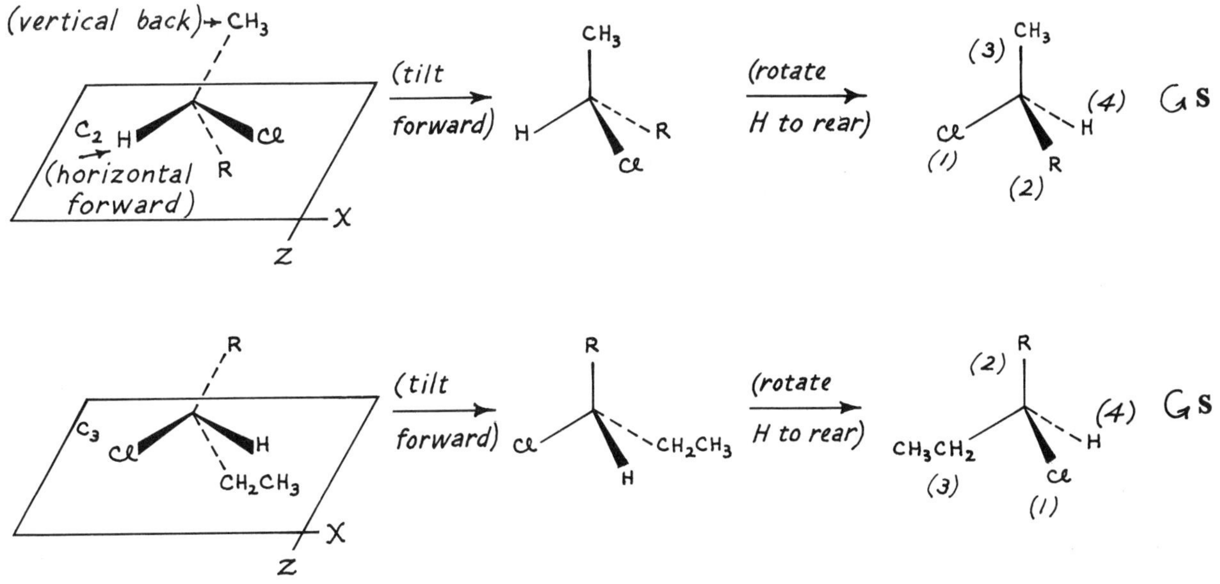

(vertical back) → $CH_3$; $C_2$ H, Cl, R; (horizontal forward); X; Z — (tilt forward) → $CH_3$, H, R, Cl — (rotate H to rear) → $CH_3$ (3), Cl (1), R (2), H (4) **S**

R; $C_3$ Cl, H, $CH_2CH_3$; X; Z — (tilt forward) → R, Cl, $CH_2CH_3$, H — (rotate H to rear) → R (2), $CH_3CH_2$ (3), Cl (1), H (4) **S**

(R = other part of the molecule)

Structure *a* in #15–18 is 2(**S**),3(**S**)-dichloropentane. Structure *b*, the enantiomer, is 2(**R**),3(**R**)-dichloropentane. In structure *c*, the configuration at C*2* is the same as it is in *b*. This gives 2(**S**),3(**R**)-dichloropentane for isomer *c*, while *d* is 2(**R**),3(**S**)-dichloropentane. Thus, the four isomers represent every possible combination of **R** and **S** configurations at the two asymmetric carbons.

Consider 2,3-dichlorobutane. It has two asymmetric carbons and would be expected to form four optical isomers as indicated in figure #15–20. However, if we examine structures *c* and *d* very carefully, we find that by rotating *d* by 180° in the plane of the paper it can be superimposed on *c*. This means *c* and *d* are merely two different ways of writing the same structure.

We can also determine that the structure represented by *c* and *d*

#15–20

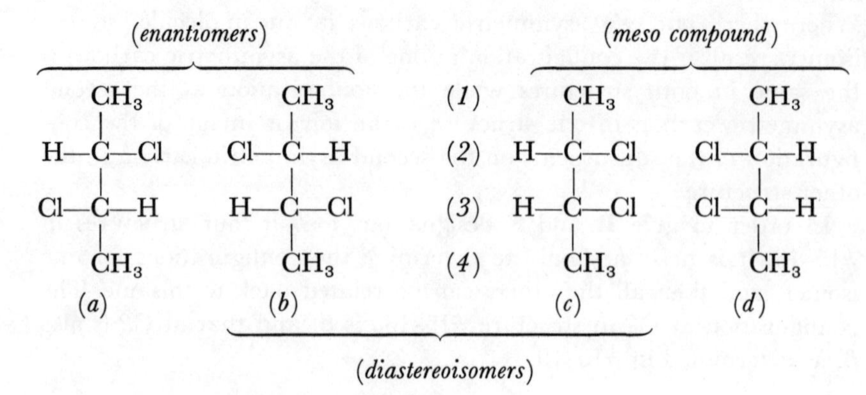


find that by rotating *d* by 180° in the plane of the paper it can be superimposed on *c*. This means *c* and *d* are merely two different ways of writing the same structure.

We can also determine that the structure represented by *c* and *d* is optically inactive. A plane bisecting the bond between C*2* and C*3* in #15–21*c* produces two "halves" of the molecule that are mirror

#15–21

$$\begin{array}{c} CH_3 \\ | \\ H\text{—}C\text{—}Cl \\ | \\ H\text{—}C\text{—}Cl \\ | \\ CH_3 \end{array} \qquad \begin{array}{c} CH_3 \\ | \\ H\text{—}C\text{—}Cl \\ | \end{array} \qquad \begin{array}{c} | \\ H\text{—}C\text{—}Cl \\ | \\ CH_3 \end{array}$$

images. The two fragments cannot be superimposed by rotation in the plane of the paper.

A plane which bisects a molecule and produces "halves" that are mirror images is called a *plane of symmetry*. The molecule is not actually broken into two pieces. The plane is merely a way to help us visualize the structural interrelationships within the molecule. When one half of a molecule is the mirror image of the other half, there can be no effect on plane polarized light. Whatever one asymmetric carbon might do to the polarization of light, the mirror-image "half" does the reverse. So the total effect is of no change. A molecule such as this one which contains asymmetric carbons but which is optically inactive because of internal compensation is called a *meso compound*. We can identify a meso compound by locating a plane of symmetry.

A plane of symmetry can bisect a bond as shown in #15–21 or it can bisect the molecule through a central carbon in the chain as shown in #15–22, where the molecule is again divided into two "halves" that are mirror images. This configuration of 2,4-pentanediol represents a meso compound.

#15–22

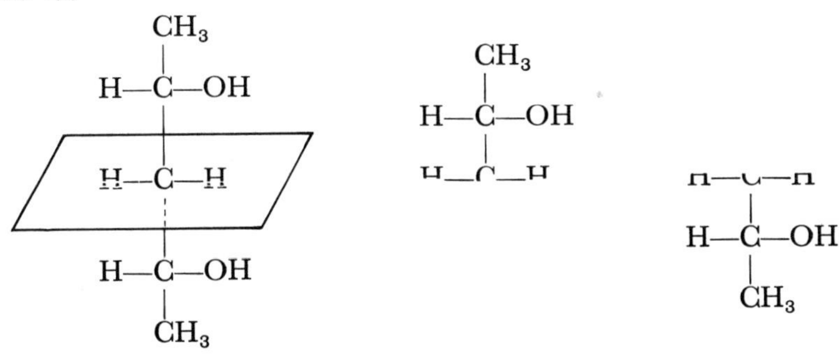


For contrast, compare the effect when a plane bisects the bond between C*2* and C*3* in structure *a* of #15–20. This bisection is diagrammed in #15–23. If one of the "halves" is mentally rotated in the

#15–23

$CH_3$
H—C—Cl
Cl—C—H
$CH_3$

$CH_3$
H—C—Cl

Cl—C—H
$CH_3$

plane of the page by 180°, it immediately becomes evident that the "halves" are identical, rather than mirror images as shown in #15–21 for the meso compounds. Both "halves" of this structure have the same effect on polarized light. The isomer with this arrangement of atoms is optically active. It has no plane of symmetry. The same is true for the configuration shown in #15–20*b*, which is the enantiomer (complete mirror image) of #15–20*a*. These two enantiomers are each diastereoisomers of the meso compound #15–20*c–d*. Convince yourself that this is true.

* * * * *

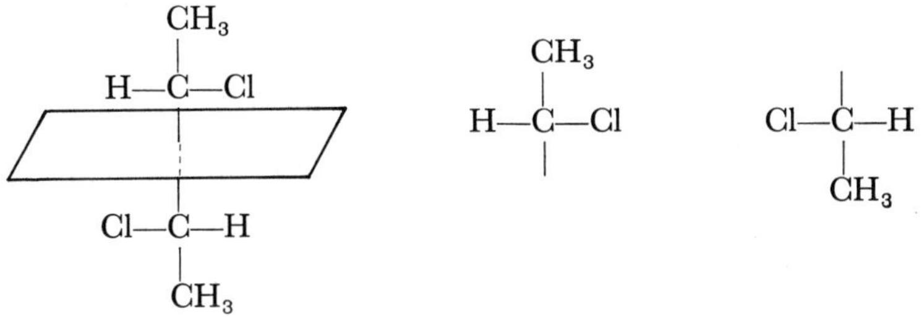

Cyclic compounds with asymmetric carbons may also possess a plane of symmetry. *Cis*-1,2-diiodocyclopentane is a meso compound. The "halves" of the ring on each side of the plane of symmetry are mirror images, as indicated in #15–24. The molecule is internally

#15–24

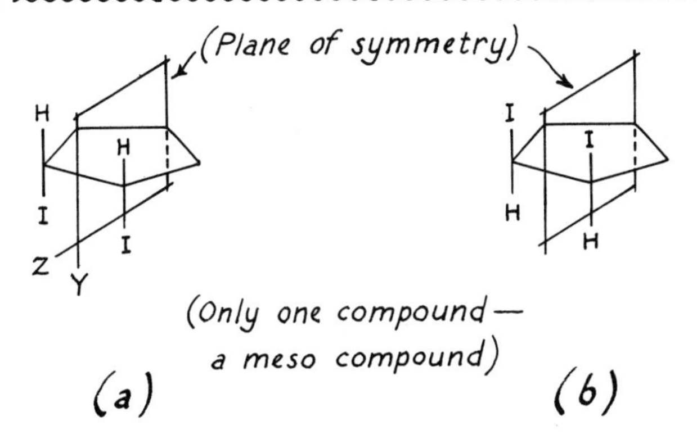

compensated and does not affect plane polarized light, even though it contains two asymmetric carbons. On the other hand, a similar plane bisecting *trans*-1,2-diiodocyclopentane is not a plane of symmetry. The two "halves" of the ring are not mirror images; they are identical, as indicated in #15–25. Thus the trans isomer can exist as enantiomers.

**#15–25**

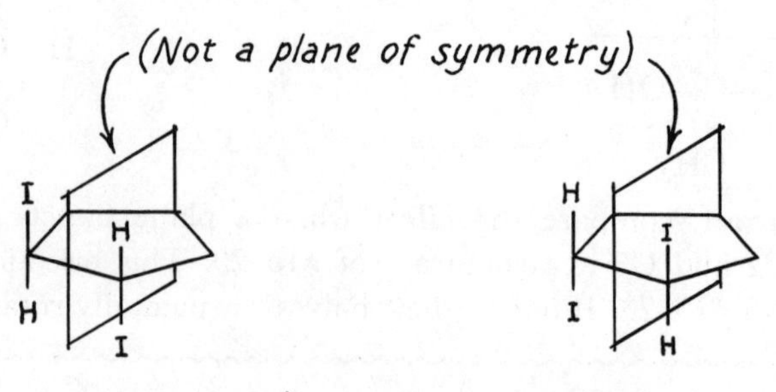


(Compounds are mirror images)

The figures in #15–25*a* and *b* represent the enantiomeric structures of two different substances, whereas #15–24*a* and *b* merely represent two different ways of drawing the same substance. If there is a plane of symmetry the compound is a meso compound—if no plane of symmetry can be located, the structure has at least one pair of enantiomeric forms.

We have not yet tried to draw a formula, given the name, so let us do so.

*Example 2*—Draw the formula of **R**-2-bromo-1-butanol.

*Step 1.* Draw the formula without regard for isomerism, just to identify the substituents on the asymmetric carbon.

**#15–26**

$$CH_3CH_2\overset{\displaystyle Br}{\overset{|}{C}}HCH_2OH$$

*Step 2.* Give numerical rating to the substituents on the asymmetric carbon.

$$H = 4 \qquad Br = 1 \qquad CH_3CH_2— = 3 \qquad —CH_2OH = 2$$

*Step 3.* Draw a tripod. Position the substituent of lowest priority at the rear and number the other three bonds from 1 to 3.

This compound is an **R**, so the numbering should proceed from the lowest to the highest in clockwise order, as shown in #15–27.

**#15–27**

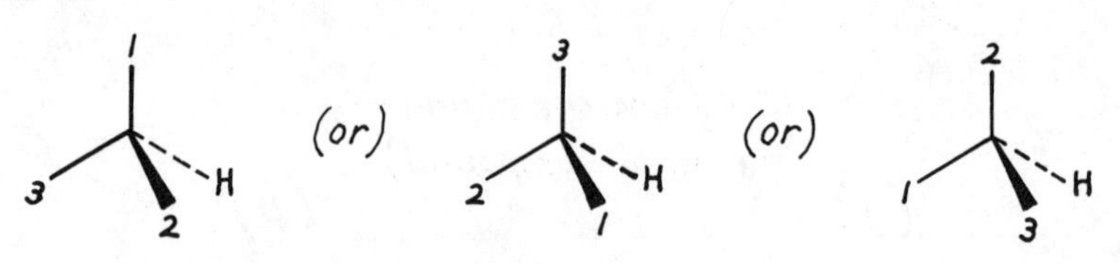

*Step 4.* Position the substituents according to the priority ratings assigned in Step 1, as in #15–28.

**#15–28**

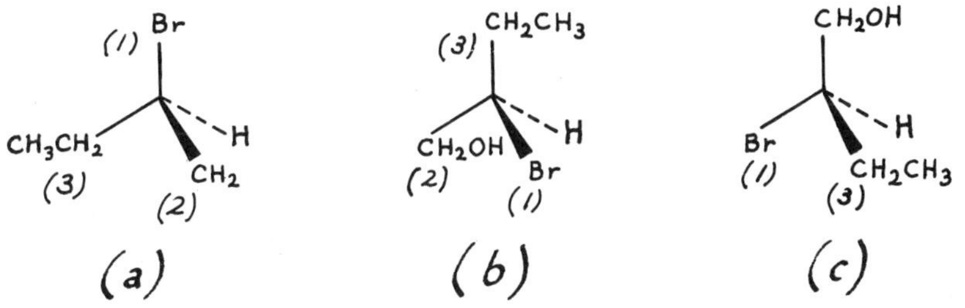

The order in which the three substituents are bonded at the asymmetric carbon is the same in each figure. The three are merely different drawings of the same structure.

*Step 5.* Draw the two-dimensional formula according to the rules established by Fischer.

Tilt the tripod from Step 4 backward so that two of the substituents are extending behind the plane of the paper and two are extending forward, as in #15–29. The substituents behind the plane of the paper,

**#15–29**

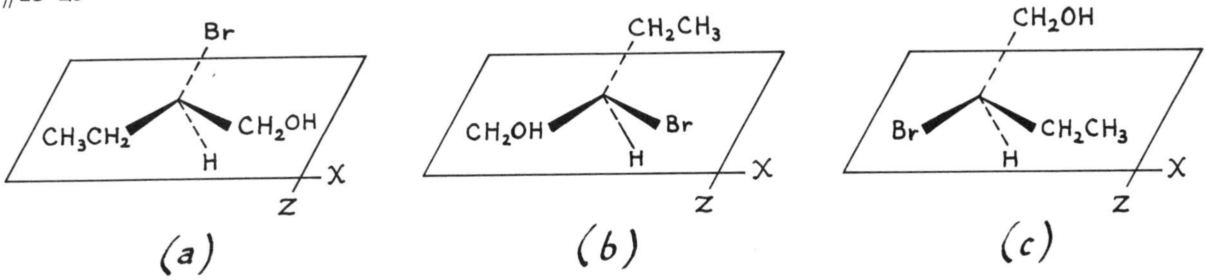

in reference plane YZ, are placed in the vertical position, and the substituents that extend forward, in reference plane XZ, are placed in the horizontal position in the two-dimensional formula. According to convention, formulas are usually written with the longest possible carbon chain in the horizontal position. Therefore, we will use structure *a* to determine the formula, although either *b* or *c* also provides the correct answer.

The formula for **R**-2-bromo-1-butanol is:

$$\begin{array}{c} Br \\ | \\ CH_3CH_2CCH_2OH. \\ | \\ H \end{array}$$

## § 16. Problems. Optical Isomerism

**1**) Given compound *a*, indicate whether each of the other structures is identical with *a* or is the mirror image of it, i.e., its enantiomer.

*a*) $$\begin{array}{rcl} & \mathrm{Cl} & \\ & | & \\ \mathrm{CH_3}— & \mathrm{C} & —\mathrm{CH_2CH_3} \\ & | & \\ & \mathrm{H} & \end{array}$$

*b*) $$\begin{array}{rcl} & \mathrm{H} & \\ & | & \\ \mathrm{CH_3CH_2}— & \mathrm{C} & —\mathrm{CH_3} \\ & | & \\ & \mathrm{Cl} & \end{array}$$

*c*) $$\begin{array}{rcl} & \mathrm{H} & \\ & | & \\ \mathrm{CH_3CH_2}— & \mathrm{C} & —\mathrm{Cl} \\ & | & \\ & \mathrm{CH_3} & \end{array}$$

*d*) $$\begin{array}{rcl} & \mathrm{Cl} & \\ & | & \\ \mathrm{CH_3}— & \mathrm{C} & —\mathrm{H} \\ & | & \\ & \mathrm{CH_2CH_3} & \end{array}$$

*e*) $$\begin{array}{rcl} & \mathrm{CH_3} & \\ & | & \\ \mathrm{CH_3CH_2}— & \mathrm{C} & —\mathrm{H} \\ & | & \\ & \mathrm{Cl} & \end{array}$$

*f*) $$\begin{array}{rcl} & \mathrm{H} & \\ & | & \\ \mathrm{Cl}— & \mathrm{C} & —\mathrm{CH_2CH_3} \\ & | & \\ & \mathrm{CH_3} & \end{array}$$

*g*) $$\begin{array}{rcl} & \mathrm{Cl} & \\ & | & \\ \mathrm{CH_3CH_2}— & \mathrm{C} & —\mathrm{CH_3} \\ & | & \\ & \mathrm{H} & \end{array}$$

*h*) $$\begin{array}{rcl} & \mathrm{H} & \\ & | & \\ \mathrm{CH_3}— & \mathrm{C} & —\mathrm{CH_2CH_3} \\ & | & \\ & \mathrm{Cl} & \end{array}$$

*i*) $$\begin{array}{rcl} & \mathrm{Cl} & \\ & | & \\ \mathrm{CH_3CH_2}— & \mathrm{C} & —\mathrm{H} \\ & | & \\ & \mathrm{CH_3} & \end{array}$$

*j*) $$\begin{array}{rcl} & \mathrm{CH_3} & \\ & | & \\ \mathrm{CH_3CH_2}— & \mathrm{C} & —\mathrm{Cl} \\ & | & \\ & \mathrm{H} & \end{array}$$

*k*) $$\begin{array}{rcl} & \mathrm{CH_3} & \\ & | & \\ \mathrm{Cl}— & \mathrm{C} & —\mathrm{CH_2CH_3} \\ & | & \\ & \mathrm{H} & \end{array}$$

*l*) $$\begin{array}{rcl} & \mathrm{CH_2CH_3} & \\ & | & \\ \mathrm{CH_3}— & \mathrm{C} & —\mathrm{H} \\ & | & \\ & \mathrm{Cl} & \end{array}$$

**2**) Indicate whether the formulas in each pair represent the same substance, mirror images, or diastereoisomers.

*a*) $$\begin{array}{rcl} & \mathrm{Cl} & \\ & | & \\ \mathrm{CH_3}— & \mathrm{C} & —\mathrm{CH_2Br} \\ & | & \\ & \mathrm{H} & \end{array} \qquad \begin{array}{rcl} & \mathrm{H} & \\ & | & \\ \mathrm{BrCH_2}— & \mathrm{C} & —\mathrm{CH_3} \\ & | & \\ & \mathrm{Cl} & \end{array}$$

*b*) $$\begin{array}{rcl} & \mathrm{Cl} & \\ & | & \\ \mathrm{CH_3}— & \mathrm{C} & —\mathrm{CH_2Br} \\ & | & \\ & \mathrm{H} & \end{array} \qquad \begin{array}{rcl} & \mathrm{H} & \\ & | & \\ \mathrm{CH_3}— & \mathrm{C} & —\mathrm{CH_2Br} \\ & | & \\ & \mathrm{Cl} & \end{array}$$

*c*) $$\begin{array}{rcl} & \mathrm{Br} & \\ & | & \\ \mathrm{HOCH_2}— & \mathrm{C} & —\mathrm{CH_3} \\ & | & \\ & \mathrm{H} & \end{array} \qquad \begin{array}{rcl} & \mathrm{Br} & \\ & | & \\ \mathrm{CH_3}— & \mathrm{C} & —\mathrm{CH_2OH} \\ & | & \\ & \mathrm{H} & \end{array}$$

*d*) $$\begin{array}{rcl} & \mathrm{CH_3} & \\ & | & \\ \mathrm{CH_3}— & \mathrm{C} & —\mathrm{CH_2OH} \\ & | & \\ & \mathrm{H} & \end{array} \qquad \begin{array}{rcl} & \mathrm{H} & \\ & | & \\ \mathrm{HOCH_2}— & \mathrm{C} & —\mathrm{CH_3} \\ & | & \\ & \mathrm{CH_3} & \end{array}$$

*e*) $$\begin{array}{rcl} & \mathrm{CH_3} & \\ & | & \\ \mathrm{H}— & \mathrm{C} & —\mathrm{Br} \\ & | & \\ \mathrm{Br}— & \mathrm{C} & —\mathrm{H} \\ & | & \\ & \mathrm{CH_2CH_3} & \end{array} \qquad \begin{array}{rcl} & \mathrm{CH_3} & \\ & | & \\ \mathrm{H}— & \mathrm{C} & —\mathrm{Br} \\ & | & \\ \mathrm{H}— & \mathrm{C} & —\mathrm{Br} \\ & | & \\ & \mathrm{CH_2CH_3} & \end{array}$$

*f*)

$CH_2CH_3$ / H—C—Cl / Cl—C—H / $CH_3$

$CH_3$ / Cl—C—H / H—C—Cl / $CH_2CH_3$

*g*)

OH H / $CH_3CH_2$—C——C—$CH_3$ / H OH

OH H / $CH_3$—C——C—$CH_2CH_3$ / H OH

*h*)

$CH_3$ / H—C—CN / $CH_2$ / H—C—CN / $CH_3$

$CH_3$ / CN—C—H / $CH_2$ / CN—C—H / $CH_3$

*i*)

$CH_3$ / $CH_3$

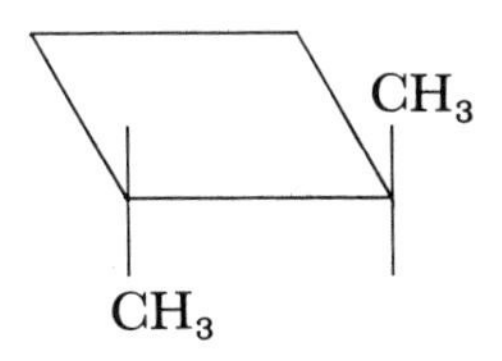

$CH_3$ / $CH_3$

*j*)

$CH_3$ $CH_3$

$CH_3$ $CH_3$

*k*)

H / Cl—C—$CH_2CH_3$ / $CH_3$

H / $CH_3CH_2$—C—Cl / $CH_3$

*l*)

Br / $CH_3CH_2$—C—COOH / H

H / $CH_3CH_2$—C—COOH / Br

*m*)

H Br / Br H

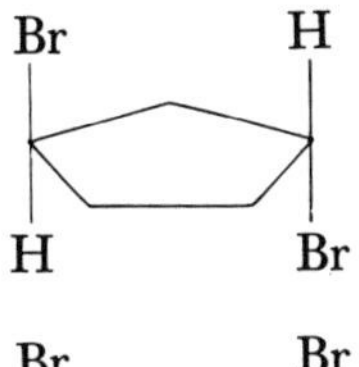

Br H / H Br

*n*)

H H / Br Br

Br Br / H H

*o*)
$$\begin{array}{ccc} & CH_2OH & \\ & | & \\ Br- & C & -Br \\ & | & \\ & CH_3 & \end{array} \qquad\qquad \begin{array}{ccc} & Br & \\ & | & \\ CH_3- & C & -CH_2OH \\ & | & \\ & Br & \end{array}$$

*p*)
$$\begin{array}{ccccc} & Br & & H & \\ & | & & | & \\ CH_3- & C & -CH_2- & C & -CH_3 \\ & | & & | & \\ & H & & Cl & \end{array} \qquad\qquad \begin{array}{ccccc} & H & & H & \\ & | & & | & \\ CH_3- & C & -CH_2- & C & -CH_3 \\ & | & & | & \\ & Br & & Cl & \end{array}$$

**3**) Determine which of the compounds in each group can exist in the form of optical isomers and then draw two-dimensional formulas for all the possible optical isomers of each of them. Label each set of enantiomers, each set of diastereoisomers, and each meso compound.

*a*) $CH_3CHOHCH_3$, $CH_3CHOHCH_2Cl$, $CH_3CH_2CH_2OH$
*b*) $CH_3CHCl_2$, $(CH_3)_2CBrCl$, $CH_3CHBrCl$
*c*) $CH_3CH_2CH_2OH$, $CH_3CH_2CDHOH$, $CH_3CDOHCH_3$
*d*) $CH_3CHBrCH_2CH_3$, $(CH_3)_2CBrCH_2CH_2CH_3$, $CH_3CH_2CHBrCH_2CH_3$
*e*) $CH_3CH_2NHCH_2CHClCH_3$, $CH_3OCH_2CHOHCH_3$, $CH_3CHCNCH_2CH_2CH_3$
*f*) $CH_3CHBrCH_2CHBrCH_3$, $CH_3CHBrCH_2CHBrCH_2CH_3$, $CH_3CBr_2CH_2CH_2CH_3$
*g*) $CH_3CHOHCHOHCH_2CH_3$, $CH_3CHOHCHOHCH_3$, $(CH_3)_2COHCH_2CHOHCH_3$

**4**) Draw three-dimensional "tripod" formulas to show the order of bonding of the substituents on the asymmetric carbon in each of these substances. Then write two-dimensional formulas for each compound following the Fischer convention.
*a*) **S**-2-butanol
*b*) 1-chloro-2(**R**)-bromobutane
*c*) 1-chloro-3(**S**)-bromobutane
*d*) **R**-2-methyl-1-butanol
*e*) **R**-2-bromopentanoic acid
*f*) **S**-2-bromobutanal
*g*) **R**-3-bromo-2-butanone
*h*) **S**-1-chloro-2-cyanopropane
*i*) **R**-3-methylhexane
*j*) **S**-3-phenyl-1-pentanol

**5**) Draw three-dimensional "tripod" formulas to show the order of bonding of the substituents at each asymmetric carbon in these compounds. Indicate whether the substituents are in the order **R** or **S**. Remember to keep Fischer's convention in mind when translating the two-dimensional into three-dimensional formulas.

*a*)
$$\begin{array}{ccc} & D & \\ & | & \\ H- & C & -NH_2 \\ & | & \\ & CH_3 & \end{array}$$

*b*)
$$\begin{array}{ccc} & CH_2Cl & \\ & | & \\ CH_3- & C & -CH_2CH_3 \\ & | & \\ & H & \end{array}$$

*c*) $$\mathrm{CH_2ClCH_2{-}\overset{\displaystyle I}{\underset{\displaystyle H}{C}}{-}CH_2CH_2Br}$$

*d*) $$\mathrm{HOCH_2{-}\overset{\displaystyle CH_3}{\underset{\displaystyle H}{C}}{-}\overset{\displaystyle O}{\overset{\|}{C}}H}$$

*e*) $$\mathrm{CH_3{-}\overset{\displaystyle OH}{\underset{\displaystyle D}{C}}{-}\overset{\displaystyle O}{\overset{\|}{C}}H}$$

*f*) $$\mathrm{CH_3O{-}\overset{\displaystyle H}{\underset{\displaystyle CH_2SCH_3}{C}}{-}CH_2OCH_3}$$

*g*) $$\mathrm{(OH)_2CH{-}\overset{\displaystyle H}{\underset{\displaystyle CH_2OH}{C}}{-}\overset{\displaystyle O}{\overset{\|}{C}}H}$$

*h*) $$\mathrm{CH_3{-}\overset{\displaystyle CH_2OH}{\underset{\displaystyle CHO}{C}}{-}COOH}$$

*i*) $$\mathrm{C_6H_5{-}\overset{\displaystyle CH_2Br}{\underset{\displaystyle CH_3}{C}}{-}CH_2CH_2Cl}$$

*j*) $$\mathrm{CH_2{=}CH{-}\overset{\displaystyle CH_3CH_2}{\underset{\displaystyle CH_2OH}{C}}{-}\overset{\displaystyle O}{\overset{\|}{C}}H}$$

**6**) Name by the IUPAC system, including the **R** and **S** notation, each of the structures that you drew in answer to question 3.

**7**) Draw two-dimensional formulas for the following:

*a*) 2(**R**),5(**R**)-dihydroxyhexane
*b*) 2(**R**),3(**S**)-dihydroxyhexane
*c*) 2(**S**),3(**R**)-dichloropentane
*d*) 2(**S**),3(**S**)-dichloropentane
*e*) 1(**R**),2(**S**)-diiodocyclopentane
*f*) 4(**S**)-chloro-2(**R**)-pentanol
*g*) 3(**R**)-chloro-5(**S**)-bromo-2(**S**)-hexanol
*h*) 5(**S**)-cyano-3(**S**)-hexanol
*i*) 4(**R**)-cyano-2(**S**)-pentanamine
*j*) (**RR**)-3,4-dichloro-3,4-dimethylhexane

**8**) For each of the following molecular formulas draw several two-dimensional structural formulas to illustrate the meaning of: enantiomer, diastereoisomer, and meso compound. If applicable include cyclic compounds as well as noncyclic.

*a*) $C_6H_{12}(CN)_2$ *b*) $C_5H_{10}O_2$ *c*) $C_6H_{10}Cl_2$
*d*) $C_6H_8Br_2$ *e*) $C_6H_{16}N_2$

## § 17. Summary Problems. Part I

Summary problems have been included for each Part to help you review the material presented in that area. Answers have not been provided. You should not need them for this type of drill. If you have trouble with any of them, go back to the original problems on the same topic.

**1**) Tables 17–1 and 17–2 list terms and concepts that you should know as the result of your work in Part I. Each term or concept is accompanied by the symbol that is used to represent it. These tables have been included as an aid to review. You should be able to cover the symbols and draw your own to go with each term or concept. Conversely, you should be able to express the concept or term represented by the symbol without looking at the other half of the table. When you understand the meaning of both the words and the symbols and can relate the two to each other, you have acquired some of the basic knowledge needed to solve organic chemistry.

**2**) Label every bond in each compound as to type, $\sigma$ or $\pi$. Indicate the kinds of electron, $sp^3$, $sp^2$, $sp$, etc., that make up the bonds in each compound. Assign bond angles. Diagram the orbitals for both the bonding and the nonbonding electrons in each compound. Make a model of each compound.

*a*) $(CH_3)_2CHCH_2CH_3$

*b*) $CH_3CH{=}CHCH_3$

*c*) $CH_3C{\equiv}CCH_3$

*d*) $CH_3\overset{\overset{\displaystyle O}{\|}}{C}CH_3$

*e*) $CH_3OCH_3$

*f*) $CH_3CH_2OH$

*g*) $CH_3COOCH_3$

*h*) $CH_3C_6H_5$

*i*) $CH_3NHCH_3$

*j*) $CH_2{=}CH{-}CH{=}CH_2$

**3**)

*a*) Draw all possible structures with the molecular formula $C_4H_8I_2$ that illustrate each of the following terms: position isomer, enantiomer, diastereoisomer, meso compound. Name each structure that you have drawn, including the **R** and **S** notation.

*b*) Repeat for $C_8H_{18}$.

**4**)

*a*) Draw both cyclic and noncyclic structures with the molecular formula $C_5H_{10}$ to illustrate the meaning of: cis-trans isomerism, optical isomerism, conformation. Name each structure you have drawn.

*b*) Repeat for $C_6H_{10}I_2$.

**5**) Draw one boat and two chair conformations for all isomers of dicyanocyclohexane. Label each as to whether it is cis or trans. Make a model of each.

**6**) Draw three staggered and three eclipsed conformations for each of the carbon-carbon bonds in 2,3-dichloropentane. Label the various conformers to indicate their structure. Make a model of each.

**7**) Draw structures with the molecular formula $C_6H_{12}O_2$ to illustrate the meaning of: alcohol, aldehyde, ketone, ether, carboxylic acid, cycloalkane, position isomerism, cis-trans isomerism (both cyclic and noncyclic), enantiomers, diastereoisomers, meso compound, racemic modification. Name each structure you have drawn, including the **R** and **S** notations.

**Table 17–1**

## Part I. Recapitulation of Structure

| *Conceptual* | *Structural* |
|---|---|
| electron orbital | 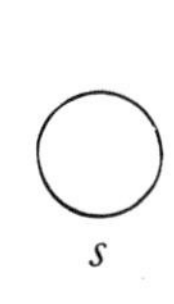  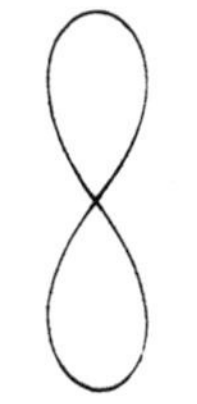 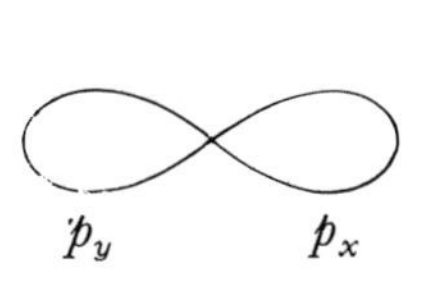  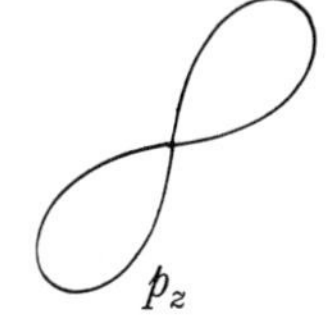  |
| hybridization | 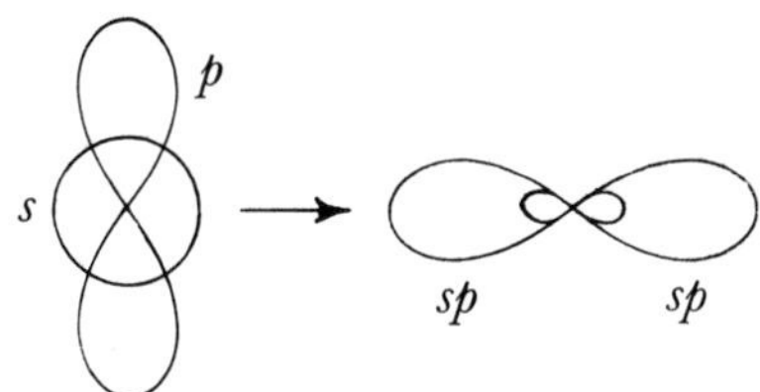  |
| sigma ($\sigma$) bond | C C → C C ≡ C—C |
| pi ($\pi$) bond | 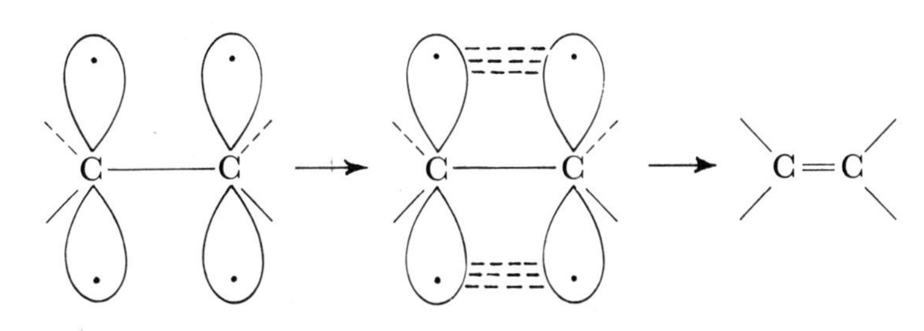  |
| $sp^3$ electron | Tetrahedral bonding<br>$\sigma$ bond 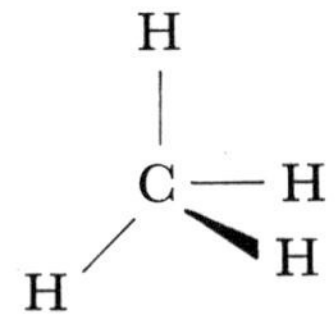  |
| $sp^2$ electron | bonding at 120°<br>$\sigma$ bond + $\pi$ bond 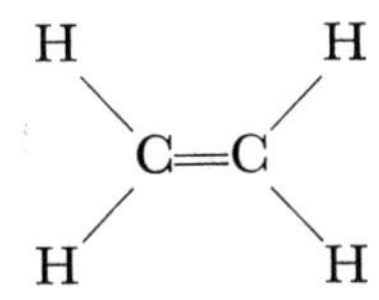  |
| $sp$ electron | bonding at 180°<br>$\sigma$ bond + 2 $\pi$ bonds H—C≡C—H |
| conformer | 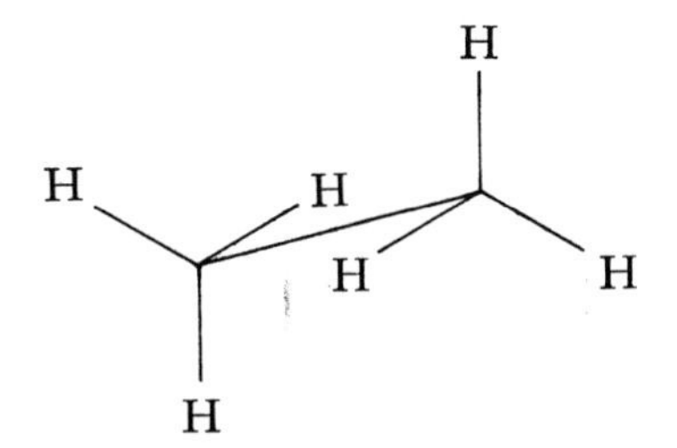 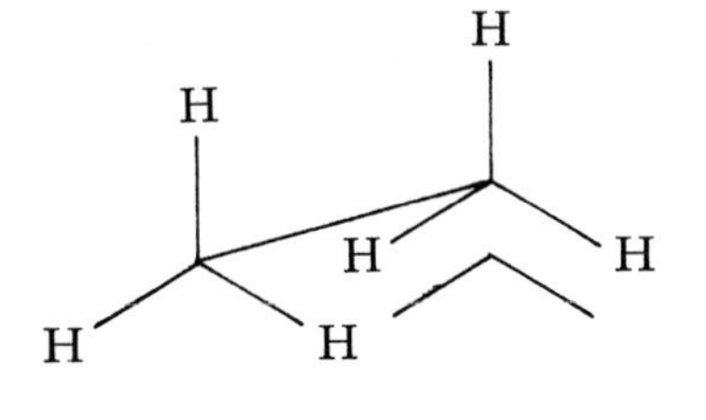 |

*Conceptual* *Structural*

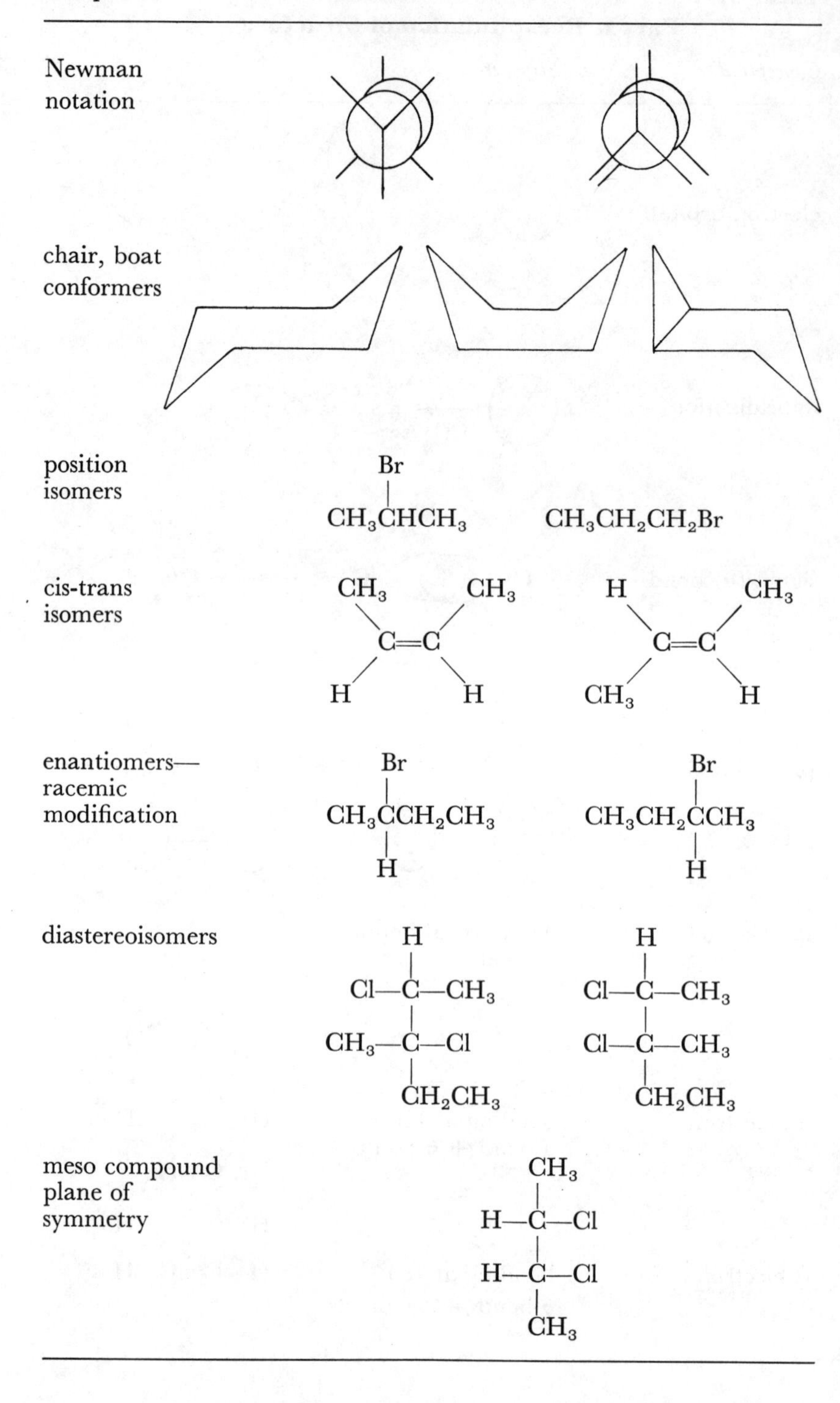

**Table 17–2**

## Part I. Recapitulation of Terms

| *Term* | *Symbol* |
|---|---|
| homologous series | $C_nH_{2n+2}$, $CH_4$, $CH_3CH_3$, $CH_3CH_2CH_3$, etc. |
| saturated compound | $CH_3CH_2CH_3$, $CH_3\overset{CH_3}{\overset{\vert}{C}}HCH_3$,  |
| unsaturated compound | 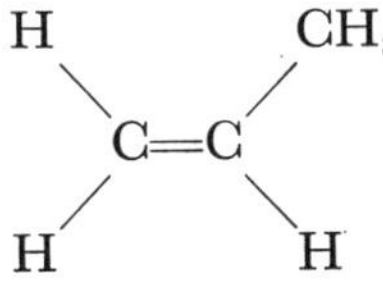 $\begin{matrix}H & & CH_3\\ & C{=}C & \\ H & & H\end{matrix}$, $H{-}C{\equiv}CCH_3$,  |
| alkane | $C_nH_{2m+2}$, $CH_3CH_2CH_3$, $CH_3CH_2CH_2CH_3$ |
| alkene | $C_nH_{2m}$, 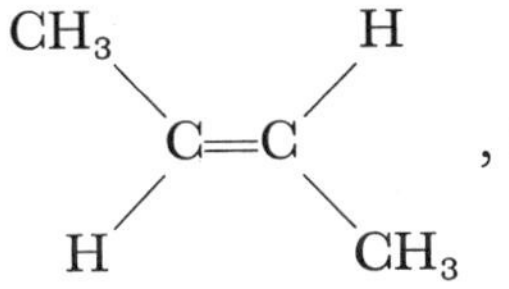 $\begin{matrix}CH_3 & & H\\ & C{=}C & \\ H & & CH_3\end{matrix}$, $CH_3CH{=}CH_2$ |
| alkyne | $C_nH_{2n-2}$, $H{-}C{\equiv}CCH_3$, $CH_3C{\equiv}CCH_3$ |
| alkyl group, **—R** | $C_nH_{2n+1}$, $-CH_3$, $-CH_2CH_3$ |
| cycloalkane | $C_nH_{2n}$,   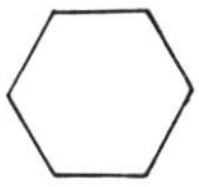 |
| alkyl halide | $RX$, $CH_3CH_2Cl$, $CH_3CH_2CH_2Br$ |
| alcohol | $ROH$, $CH_3OH$, $CH_3\overset{CH_3}{\overset{\vert}{C}}HOH$, $CH_3\underset{CH_3}{\underset{\vert}{\overset{CH_3}{\overset{\vert}{C}}}}OH$ |
| aldehyde | $R\overset{O}{\overset{\Vert}{C}}H$, $CH_3\overset{O}{\overset{\Vert}{C}}H$, $CH_3CH_2\overset{O}{\overset{\Vert}{C}}H$ |
| ketone | $R\overset{O}{\overset{\Vert}{C}}R'$, $CH_3\overset{O}{\overset{\Vert}{C}}CH_2CH_3$, $CH_3CH_2\overset{O}{\overset{\Vert}{C}}CH_2CH_3$ |
| carboxylic acid | $R\overset{O}{\overset{\Vert}{C}}OH$, $CH_3\overset{O}{\overset{\Vert}{C}}OH$, $CH_3CH_2\overset{O}{\overset{\Vert}{C}}OH$ |
| amine | $RNH_2$, $R_2NH$, $R_3N$, $CH_3NH_2$, $CH_3\overset{H}{\overset{\vert}{N}}CH_3$, $CH_3\overset{CH_3}{\overset{\vert}{N}}CH_3$ |
| ether | $ROR'$, $CH_3OCH_2CH_3$, $CH_3CH_2OCH_2CH_3$ |

# PART II

# The Geometry of the Electron Cloud. How to Predict and Depict Charge Distribution

The geometrical relationships between atoms and the factors determining these relationships were the main topics in Part I. The distribution of valence electrons in the molecular structure and the effect of this distribution on chemical reactivity are the main topics in Part II.

Each electron has a unit charge that is initially associated with a particular atom. When bonding occurs, the charge of the bonding electrons is distributed within a molecular orbital between the two bonded atoms. This charge distribution is due to the extremely rapid movement of the electron within a limited volume of space and relates to the statistical probability of finding that electron at any one point at any given time. The probability concept is sometimes difficult to visualize in concrete terms. As you try to understand variations in charge distribution it may help if you imagine that the "stuff", i.e., the charge which constitutes an electron, can be dispersed like the water vapor in a fog. Such an analogy provides you with a device for visualizing the effects produced by changes in the pattern of movement of an electron.

Each molecule can be described as surrounded by a *charge cloud*, also called an *electron cloud*. This charge cloud is generated by the movement of electrons within the molecule. The charge cloud is not necessarily either homogeneous in its composition (i.e., the charge is not necessarily evenly dispersed throughout) or symmetrical in its shape. *The pattern of charge distribution is unique to each substance.* There may be differences in charge density from one bond to another, as well as variations in the contours or shape of the cloud at different sites in the molecule. The geometry of any particular molecular orbital is determined by the geometry of adjacent orbitals, as well as by the nature of the bonded atoms, as you will learn through your study of Part II.

Density variations may be likened to differences in intensity of a fog. Although an entire terrain may be fogged in by a fine mist, there is usually a higher concentration of moisture in some sections than in other sections so that the visibility may vary from place to place. The charge density in an electron cloud may vary in a similar manner.

Differences in charge density are not haphazard but follow a predictable pattern. Just as one can anticipate that a fog will be denser over a body of water than over a land mass, so one can deduce that the charge density will be greater in the region of an atom with a negative formal charge (§18) than in the region of an atom with a positive formal charge.

The contours of an electron cloud can be visualized as varying like the contours of a cloud in the sky. In an electron cloud these variations are due to the underlying structure. You learned (#1–39) that the $\pi$ cloud of a multiple bond extends above and below the orbital of the $\sigma$ bond. Thus, a $\pi$ bond enlarges the electron cloud in the region of a double bond. You might even imagine a "bulge" at the site of a double bond.

Although we are concerned with the geometry of an electron cloud, it is not really practical to try to depict the specific cloud shape associated with each molecule. Of more importance to our study of chemical reactivity is an understanding of the differences that exist between the several sections of an electron cloud. These differences appear in the charge density and in the degree to which the electron cloud extends beyond the sigma bond framework that holds the atoms together.

Factors determining the characteristics of an electron cloud at specific sites include:

- The $\sigma$ bond framework.
- Inductive effects that redistribute charge density within the $\sigma$ bond framework.
- $\pi$ bonds which extend beyond the $\sigma$ bond framework.
- Resonance effects that redistribute charge density through the $\pi$ cloud.
- The presence of nonbonding electrons.
- The presence of atoms with a formal charge.

Each of these factors is studied individually in Part II. The role of each in determining the "structure" of the electron cloud is analyzed in terms of the physical basis for the action. The conventions developed specifically to symbolize in formulas, effects attributable to electron motility are given in detail. The relationships between chemical reactivity and charge density at specific sites are illustrated by acid-base reactions. When you finish your study of Part II, you should be able:

- to predict the sites of increased or decreased charge density in a given molecule;
- to represent by electron-dot formulas variations in charge density that are due to inductive and resonance effects;

- to relate chemical reactivity to "electronic geometry," and thereby to predict the kinds of reactions that a given substance will undergo.

## §18. Electron-Dot Formulas

In Part I we focused our attention on bonding electrons and on the roles they play in determining the relative positions of atoms in a molecule. There was only slight mention of the nonbonding electrons (electrons which in a given structure are not involved in bonding). However, chemical reactivity depends upon charge distribution within the electron cloud associated with a molecule; and all the valence electrons, both nonbonding and bonding, contribute to the electron cloud. Thus, we now need to consider both nonbonding and bonding electrons. As our interest expands to encompass the geometry of the electron cloud, the formulas we draw have to include a more specific representation of the valence electrons. Formulas such as those in #18–1 which employ dots, crosses, and/or small circles to represent electrons have proved useful. These are referred to as *electron-dot* or *Lewis formulas*. A clear understanding of this type of notation is essential to your study of organic chemistry.

**#18–1**

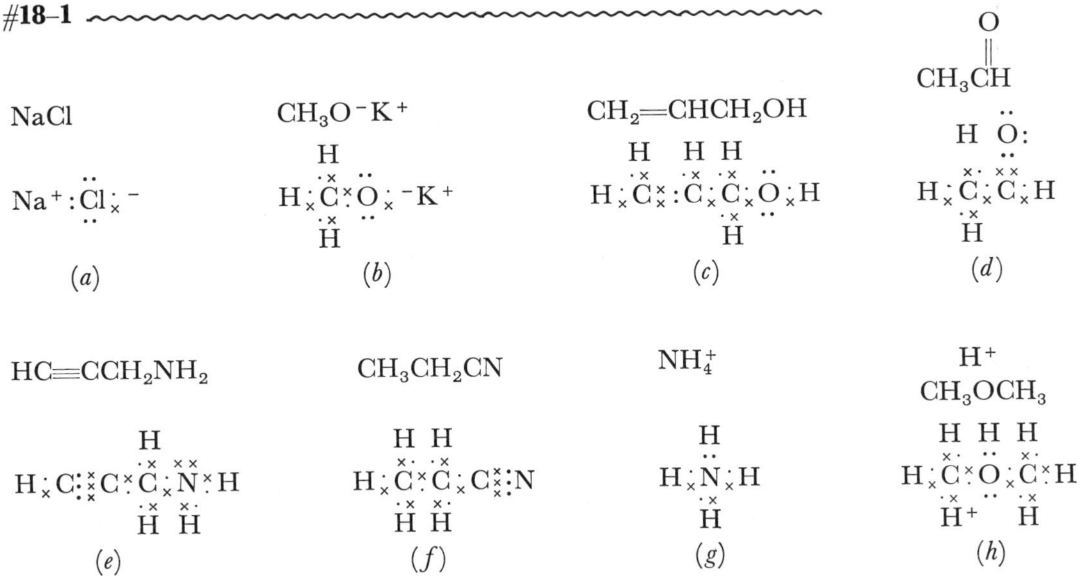

(a) NaCl

(b) $CH_3O^-K^+$

(c) $CH_2{=}CHCH_2OH$

(d) $CH_3CH{=}O$

(e) $HC{\equiv}CCH_2NH_2$

(f) $CH_3CH_2CN$

(g) $NH_4^+$

(h) $CH_3OCH_3$ with $H^+$

The formulas in #18–1 differentiate between the valence electrons contributed by the individual atoms in a molecule. Such differentiation has no significance in the actual molecule where the electrons in a given molecular orbital are indistinguishable. Assigning particular electrons to specific atoms in a formula is only a device which makes it easier to be certain that a proposed structure accounts in a reasonable manner for the total number of available valence electrons.

The number of valence electrons contributed by each element can be determined from its position in the Periodic Table, as illustrated in #18–2, which lists the elements most commonly encountered in organic chemistry. In this chart each dot represents a valence electron. A pair of dots at one side of a symbol indicates paired electrons in a single atomic orbital. A single dot at one side of a symbol indicates an unpaired electron in an atomic orbital.

**#18–2**

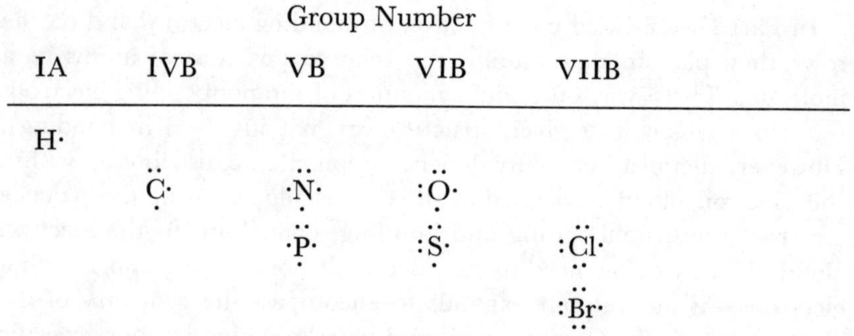

Group Number

| IA | IVB | VB | VIB | VIIB |
|---|---|---|---|---|
| $\mathrm{H}\cdot$ | | | | |
| | $\cdot\dot{\underset{\cdot}{\mathrm{C}}}\cdot$ | $\cdot\ddot{\underset{\cdot}{\mathrm{N}}}\cdot$ | $:\ddot{\underset{\cdot}{\mathrm{O}}}\cdot$ | |
| | | $\cdot\ddot{\underset{\cdot}{\mathrm{P}}}\cdot$ | $:\ddot{\underset{\cdot}{\mathrm{S}}}\cdot$ | $:\ddot{\underset{..}{\mathrm{Cl}}}\cdot$ |
| | | | | $:\ddot{\underset{..}{\mathrm{Br}}}\cdot$ |

The formulas in #18–1 include all of the types of bonding that you will encounter in your initial study of organic. These may be summarized as follows:

- *Ionic bonding* describes the electrostatic attraction between oppositely charged ions and serves to hold these ions in a constant position relative to each other. Sodium chloride (#18–1*a*) is a substance with ionic bonding. A sodium atom has donated an electron to a chlorine atom. The resulting ions, $Na^+$ and $:\ddot{\underset{..}{\mathrm{Cl}}}\underset{\times}{\cdot}^{-}$ are held in a crystal lattice because of electrostatic attraction. The relationship between the ions is shown by drawing them as separated units. The charge on each ion is included. One of the electrons on the chloride ion is represented by an x to indicate that it did not originally belong to the chlorine atom. The bonding between a methoxide ion and a potassium ion is ionic as indicated in #18–1*b*.
- A *covalent bond* results when two atoms share a pair of electrons in a common molecular orbital. There are single, double, and triple covalent bonds each of which can be described in terms of length, strength, and direction (see §1).
- A *single covalent bond* is formed when just one pair of electrons is shared between two atoms. Each of the atoms donates an unpaired electron to the bond. The single bonds in structures *c–f* in #18–1 are all this type of covalent bond.
- A *coordinate covalent bond* is a single bond that arises when both the bonding electrons are contributed by only one of the bonded atoms. As shown in #18–1*g* one of the four covalent N—H bonds in the ammonium ion has been formed by two electrons from the nitrogen atom. This is a coordinate covalent bond. There is no difference in the molecule between a covalent and a coordinate covalent bond. The distinction is one which helps us keep track

of the source of electrons as we study molecular structure. The bond between oxygen and hydrogen in the protonated ether of #18–1*h* is also a coordinate covalent bond.

- A *double covalent bond* is formed when two bonded atoms share four rather than two electrons. A double bond consists of one $\sigma$ and one $\pi$ bond (see §1). Examples are given in #18*c* and *d*.
- A *triple covalent bond* is formed when the two bonded atoms share three pairs or six electrons. A triple bond consists of one $\sigma$ bond and two $\pi$ bonds. Examples are given in #18–1*e* and *f*.

As discussed in §1, the electrons in carbon undergo hybridization so that carbon is tetravalent, i.e., forms four covalent bonds. The rest of the elements listed in #18–2 can form one covalent bond for each unpaired electron shown in the symbol. Electron-dot formulas for the neutral molecules formed between hydrogen and each of the Period 2 elements in #18–2 are shown in #18–3.

**#18–3**

```
   H           H           H
   ·×          ·×          ·×          ··
H ·× C ×· H  H ·× N ×· H  H ·× O :    H ×· Cl :
   ·×          ··          ··          ··
   H
  CH4         NH3         H2O         HCl
```

Except for hydrogen each of the atoms in these formulas "has a complete octet of electrons." This statement means that each has a share of eight different electrons. In methane four of the electrons come from the hydrogen atoms and four from the carbon, giving a total of eight shared electrons in four covalent bonds. Each hydrogen "has a duet of electrons," i.e., has a share of two different electrons.

Although the carbon shares in the charge from eight different electrons, the total charge available to the carbon is equivalent to that acquired from only four electrons (half of the eight shared electrons). Thus, we say carbon *has* four electrons and *shares* an octet of electrons. To understand this statement imagine two children each of whom has four candy bars all of different flavors. (We will ignore here the nutritional effects.) Each child breaks his candy bars in half and exchanges four halves, one of each kind, with his friend. Thus each child acquires candy in all eight flavors, yet he still has only a total of four candy bars. Similarly, there is an octet of electrons around the carbon, but carbon "owns" only a total of four electron charges.

Nitrogen in ammonia (#18–3*b*) shares three pairs of electrons with three hydrogens and has one pair of nonbonding electrons that is not shared. Thus, nitrogen has a share in a complete octet and "owns" five electrons (half of three pairs plus a complete pair). Nitrogen is in Group 5 and needs five electrons to equalize the positive charge on its nucleus. Nitrogen, therefore, does not carry a charge in this molecule.

An atom that "owns" more or fewer than enough electrons to

offset its nuclear charge carries either a negative or a positive charge. This charge is referred to as a *formal charge.*

The nitrogen in a molecule of hydrogen nitrate ($HNO_3$) has a positive formal charge of +1, while one of the oxygens has a negative formal charge of −1. Formal charges can be calculated as shown in #18–4. The number of electrons "owned" by an atom is equal to half the shared electrons plus the nonbonding electrons. The formal charge is the difference between the actual electron charge ("owned electrons") and that needed to neutralize the charge on the nucleus. The formal charges on atoms in a neutral molecule cancel out. However, there is an excess of negative formal charge in an anion and of positive formal charge in a cation. Note that nitrogen with a positive formal charge in #18–4 has contributed a pair of electrons to a coordinate covalent bond. Similarly, nitrogen with a positive formal charge in the ammonium ion and oxygen with a positive formal charge in the protonated ether (#18–1*h*) have both contributed a pair of electrons to a coordinate covalent bond.

**#18–4**

| Structure | | Atom | | Electrons | | Charge |
|---|---|---|---|---|---|---|
| | | H | "owns" | 1 *e* | *charge* = | 0 |
| H | | N | ,, | 4 *e*'s | ,, | +1 |
| ×·<br>×O×<br>× × | (*1*) | O*1* | ,, | 6 *e*'s | ,, | 0 |
| ·× ××<br>N:O×<br>•• ×× | (*2*) | O*2* | ,, | 7 *e*'s | ,, | −1 |
| ××<br>×O×<br>× × | (*3*) | O*3* | ,, | 6 *e*'s | ,, | 0 |
| | | | | | *Total charge* | 0 |

The most stable electron configuration for any molecular structure is the one which results in a complete octet for as many atoms as possible, with the exception of hydrogen which needs only a duet of electrons. When one is writing electron-dot formulas, the electrons associated with each atom are positioned so as to produce as many complete octets as possible. The electron arrangements given for the ground state atoms in #18–2 do not have to be maintained: single electrons may become paired; multiple bonds may be formed; a variety of positions are possible just so long as the dots representing the valence electrons of a particular atom remain adjacent to the symbol for that atom in the formula.

As an example of the process consider how to write a reasonable Lewis formula for carbon dioxide, $CO_2$.

Each oxygen atom has six valence electrons and the carbon has four giving a total of 16 electrons to be positioned in the formula. The three atoms could be fitted together as in #18–5, where one unpaired electron on each oxygen atom forms a bond by becoming paired with a single electron from carbon. However, none of the atoms in this formula shares in a complete octet and each atom has one or more unpaired, nonbonding electrons. Both situations represent unstable structures.

**#18–5**

·O:C:O· (electron-dot structure)

Alternatively, the electrons could be positioned as in #18–6. This configuration eliminates the problem of unpaired electrons but still leaves carbon without a complete octet. Carbon would have a formal charge of +2 in this structure while neither oxygen would carry a charge. This state of affairs could not produce a neutral molecule.

**#18–6**

:O:C:O: (electron-dot structure)

If a pair of electrons on each of the oxygens is shifted so as to form a double bond between each oxygen and carbon the structure in #18–7 results. Here each atom shares in an octet of electrons; carbon "owns" four electrons and oxygen "owns" six; there are no unpaired nonbonding electrons; and all 16 electrons have been accounted for.

**#18–7**

O::C::O (electron-dot structure) O=C=O

At first, drawing electron-dot formulas will involve just such a shifting around of the electrons as you seek the most stable electron configuration. However, with practice you will find that it takes fewer tries to come up with a reasonable arrangement. The following rules can serve as a guide to you in this process:

- Each atom contributes all of its valence electrons to a structure and all must be shown: hydrogen 1, carbon 4, nitrogen and phosphorous 5, oxygen and sulfur 6, the halogens 7.
- In a neutral molecule the total number of bonding and nonbonding electrons is equal to the sum of the valence electrons from all the atoms. For an anion this number is increased by the ionic charge, and for a cation it is decreased by the ionic charge.
- Each pair of dots in the formula represents a pair of electrons. A pair of dots between two atoms represents a single covalent bond; two pairs of dots indicate a double bond; and three pairs a triple bond.
- A pair of dots associated with an atom but not between two atoms represents a pair of nonbonding electrons.
- A single nonbonding electron leads to instability and should be included only when there is no other more satisfactory way to allocate electrons.
- As many atoms as possible except hydrogen should have a share in a complete octet of electrons. A hydrogen atom only needs a duet of electrons to complete an energy shell.
- An electron configuration that places a formal charge on two or more atoms in a neutral molecule represents a less stable structure than one in which none of the atoms carries a formal charge.

The following Examples show how to apply these rules.

*Example 1*—Write an electron-dot formula for acetaldehyde ($CH_3CHO$) and methylamine ($CH_3NH_2$).

*Step 1.* Write a skeletal formula for each without bonds.

The formula as written for acetaldehyde tells us that there are three hydrogens around one carbon, an oxygen and a hydrogen at the other carbon, and the two carbons are bonded to each other (see §3).

The formula for methylamine shows three hydrogens bonded to carbon, two hydrogens bonded to nitrogen, and a bond between the carbon and the nitrogen.

Thus, the two skeletal formulas without bonds are as shown in #18–8.

**#18–8**

```
  H O          H H
H C C H      H C N H
  H            H
```

*Step 2.* Indicate the $\sigma$ bond framework of the molecule by positioning a pair of electrons between each pair of adjacent atoms.

Initially assume that each $\sigma$ bond has been formed by the contribution of one electron by each of the bonded atoms. The result of such an assignment of electrons is given in #18–9.

**#18–9**

```
 H O          H H
 .. ..        .. ..
H:C:C:H      H:C:N:H
 ..           .. ..
 H            H
```

*Step 3.* Determine how many valence electrons remain to be allocated after Step 2 has been completed.

The number of electrons still to be assigned can be determined by subtracting the number of valence electrons shown in #18–9 from the total number of valence electrons in the molecule. This calculation is illustrated in #18–10.

**#18–10**

| *Acetaldehyde* | | |
|---|---|---|
| 2C | 2 × 4 = | 8 |
| 4H | 4 × 1 = | 4 |
| 1O | 1 × 6 = | 6 |
| | Total | 18 |
| Already used in step 2 | | 12 |
| Remaining | | 6 |

| *Methylamine* | | |
|---|---|---|
| 1C | 1 × 4 = | 4 |
| 5H | 5 × 1 = | 5 |
| 1N | 1 × 5 = | 5 |
| | Total | 14 |
| Already used in step 2 | | 12 |
| Remaining | | 2 |

*Step 4.* Position the remaining electrons so as to give each atom its full quota of valence electrons and so as to form as many complete octets of electrons as possible.

There are 6 electrons still be be assigned to the acetaldehyde structure in #18–9. Neither the oxygen nor the carbon bonded to the oxygen has a full complement of valence electrons. As shown in #18–11 there are several different ways in which six electrons could be positioned around these two atoms. However, of the four configurations only *d* provides a complete octet of electrons for all atoms other than hydrogen. In *d* each hydrogen has a duet of electrons and no atom carries a formal charge. Thus, *d* is the most reasonable formula for acetaldehyde.



**#18–11**

```
   ..           ..           
 H O:        H :O:        H :O:          H:O:
 .. ..        .. ..        ..  ..          .. ..
H:C:C:H     H:C :C:H     H:C::C:H       H:C:C:H
 .. ..        ..           ..             ..
 H            H            H              H
 (a)          (b)          (c)            (d)
```

The structure for methylamine in #18–9*b* accounts for all but two of the valence electrons. Nitrogen is the only atom in the formula with an incomplete octet. Therefore, as shown in #18–12, the remaining pair of electrons should be assigned to nitrogen as a nonbonding pair.

**#18–12**

```
 H H
 .. ..
H:C:N:H
 .. ..
 H
```

*Example 2*—Draw all the reasonable non-cyclic structures for a molecule composed of 2 carbons, 6 hydrogens, and 1 oxygen. Base your conclusions as to the reasonableness of the proposed structures on the assignment of electrons in the molecules.

*Step 1.* Draw the ground state symbol (hybridized for carbon) for each different atom other than hydrogen and determine whether there are unpaired electrons that could form $\sigma$ bonds between these atoms.

**#18–13**

```
          .        ..
H·       ·C·      :O·
          .        .
```

As shown in #18–13 each carbon has four unpaired electrons and can form four covalent bonds, while oxygen has two unpaired electrons.

The two carbons can bond together as in #18–14*a* and the oxygen can bond with a carbon as in #18–14*b*.

**#18–14**

```
 .  ×          .  ××
·C ×. C×      ·C ×. O×
 .  ×          .  ××
  (a)           (b)
```

*Step 2.* Draw all possible combinations of the three atoms joined by $\sigma$ bonds as described in #18–14.

One of the carbons can bond with a carbon and an oxygen as in #18–15*a* or the oxygen can bond with each of the carbons as in #18–15*b*.

**#18–15**

```
 .  ×  ..          ×  ..  ×
·C ×. C ×. O·     ×C ×. O ×. C×
 .  ×  ..          ×  ..  ×
    (a)               (b)
```

*Step 3.* Determine whether the number of hydrogens in the molecule corresponds to the number of unpaired electrons in the structures drawn in Step 2.

Each of the structures in #18–15 has 6 unpaired electrons and the molecular formula states that there are 6 hydrogens present. Therefore, the formulas in #18–15 do not need to be modified.

*Step 4.* Position a hydrogen with its unpaired electron wherever there is an unpaired electron available for bond formation in the structures drawn for Step 2.

#18–16

```
   H H                      H     H
   ·× ·× ··                 ·×  ··  ·×
H×C:C×O×H               H×C×O×C×H
   ·× ·× ··                 ·×  ··  ·×
   H H                      H     H
   (a)                         (b)
```

This operation leads to the two structures in #18–16. In each of them the carbons and the oxygen share in a complete octet of electrons; each carbon "owns" four electrons and the oxygen "owns" six, so that none of these atoms has a formal charge; each hydrogen shares a duet of electrons; there are no unpaired nonbonding electrons. Thus, each structure is a reasonable one. Both of them exist. The compound in #18–16*a* is ethanol ($CH_3CH_2OH$) and that in #18–16*b* is dimethyl ether ($CH_3OCH_3$).

*Example 3*—Draw reasonable structures for all non-cyclic compounds with the molecular formula $C_3H_6O$. Base your judgment of reasonableness on the electron configurations of the molecule.

*Step 1.* Determine the possible non-cyclic combinations in which three carbons and one oxygen can be joined by $\sigma$ bonds.

This operation involves pairing a single electron on one of the atoms with a single electron on another and fitting all of the atoms together in as many different sequences as possible. As shown in #18–17 there are three different positions for oxygen relative to the carbon atoms. Thus, there are three fundamental molecular frameworks.

#18–17

```
                                                  ··
                                                 ·O:
 ×   ·   ×  ··           ·   ×  ··   ×          ·  ·×  ·
×C×C×C×O:              ·C×C×O×C×             ·C×C×C·
 ×   ·   ×  ·            ·   ×  ··   ×          ·  ×  ·
    (a)                        (b)                  (c)
```

*Step. 2.* Determine whether the number of hydrogens to be positioned is the same as the number of unpaired electrons shown in each structure drawn for Step 1.

Draw the structures so as to emphasize the number of unpaired electrons available for bonding. Do this by using a line for the $\sigma$ bonds between the atoms and a dot for each of the remaining electrons as shown in #18–18. This rewrite clearly indicates eight unpaired electrons in each of the three structures.

#18–18

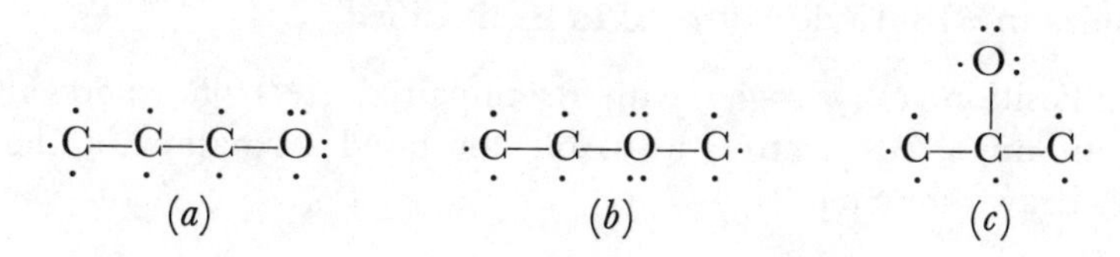

The molecular formula shows only six atoms in addition to carbon and oxygen. Thus, two electrons would remain unpaired no matter how we positioned the hydrogens. Unpaired electrons are so reactive that any structure with unpaired electrons immediately reacts with any available source of electrons to form another substance. Unpaired nonbonding electrons should not appear in a structure unless no other configuration is possible and the substance is, therefore, a free radical.

*Step 3.* Determine whether it is possible to reposition electrons in the three structures of #18–18 so as to eliminate two of the unpaired electrons.

Whenever unpaired electrons appear on adjacent atoms they may be shifted so as to form a second bond (representing a $\pi$ bond) between the two atoms. This operation is shown in #18–19 for each of the structures in #18–18. The arrows indicate electron shifts. For the moment the possibilities of cis-trans isomerism can be ignored because they can be accounted for in the final structures.

**#18–19**

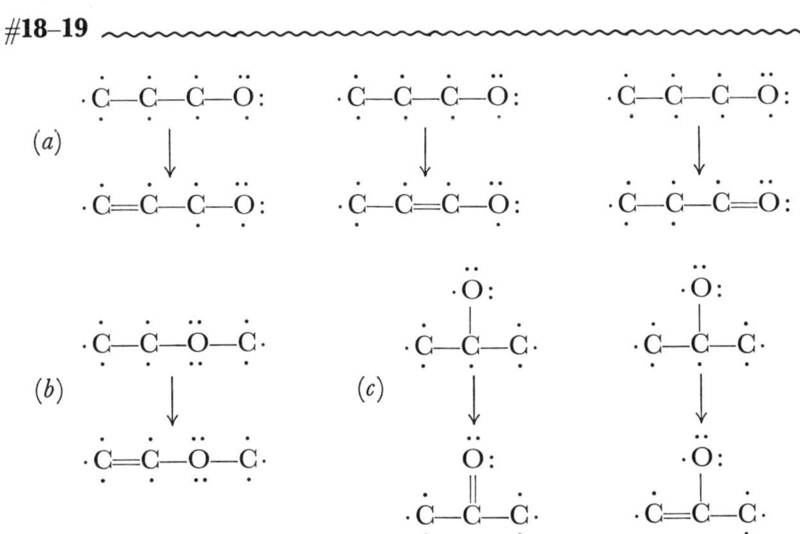

The electrons can be shifted only if there are unpaired electrons on adjacent carbons. Thus, the structure #18–18*a* gives three molecular frameworks with just the six unpaired electrons needed for the six hydrogens; #18–18*b* gives only one; and #18–18*c* gives two.

*Step 4.* Position a hydrogen with its unpaired electron wherever there is an unpaired electron available for bonding in each of the structures drawn for Step 3.

**#18–20**

H H H / H:C::C:C:O: / H H    H H H / H:C:C::C:O: / H H    H H H / H:C:C:C::O: / H H

$CH_2{=}CHCH_2OH$    $CH_3CH{=}CHOH$ (*unstable*)    $CH_3CH_2CH{=}O$

$$\begin{array}{ccc} \begin{matrix} H\;\;H\;\;\;\;\;\;H \\ H:\ddot{C}::\ddot{C}:\ddot{O}:\ddot{C}:H \\ \;\;\;\;\;\;\;\;\;\;\;\;\;\;H \end{matrix} & \begin{matrix} H\;\;\ddot{O}:H \\ H:\ddot{C}:\ddot{C}:\ddot{C}:H \\ H\;\;\;\;\;\;H \end{matrix} & \begin{matrix} H \\ H:\ddot{O}:H \\ H:\ddot{C}::\ddot{C}:\ddot{C}:H \\ H \end{matrix} \\ CH_2{=}\overset{H}{\overset{|}{C}}\ddot{O}CH_3 & CH_3\overset{:\ddot{O}}{\overset{\|}{C}}CH_3 & CH_2{=}\overset{:\ddot{O}H}{\overset{|}{C}}CH_3 \\ & & (unstable) \end{array}$$

The six compounds resulting from this operation are given in #18–20. In each of them all the atoms other than hydrogen share a complete octet; each carbon "owns" four electrons and each oxygen "owns" six so there are no formal charges; each hydrogen has a share in a duet of electrons and there are no unpaired nonbonding electrons. Thus, all are reasonable structures according to the criteria we have established. However, compounds in which an hydroxyl group is bonded to an unsaturated carbon tend to be unstable and to rearrange to the more stable isomer with the double bond between carbon and oxygen. This process is called tautomerism. Thus, $CH_3CH{=}CHOH$ becomes $CH_3CH_2\overset{:\ddot{O}}{\overset{\|}{C}}H$ and $CH_2{=}\overset{OH}{\overset{|}{C}}CH_3$ becomes $CH_3\overset{:\ddot{O}}{\overset{\|}{C}}CH_3$. Our answer, therefore, includes four compounds: allyl alcohol ($CH_2{=}CHCH_2\ddot{O}H$), propionaldehyde ($CH_3CH_2\overset{\ddot{O}}{\overset{\|}{C}}H$), acetone, ($CH_3\overset{:\ddot{O}}{\overset{\|}{C}}CH_3$) and vinylmethyl ether ($CH_2{=}CH\ddot{O}CH_3$).

As you learn to work with electron-dot formulas you will eventually use a line for the pair of electrons in all bonds except those of particular interest in a specific compound as in #18–21. However, you should always include the nonbonding electron pairs even if the presence of all the other valence electrons is merely indicated by the way in which the atoms are positioned in a condensed formula.

**#18–21**

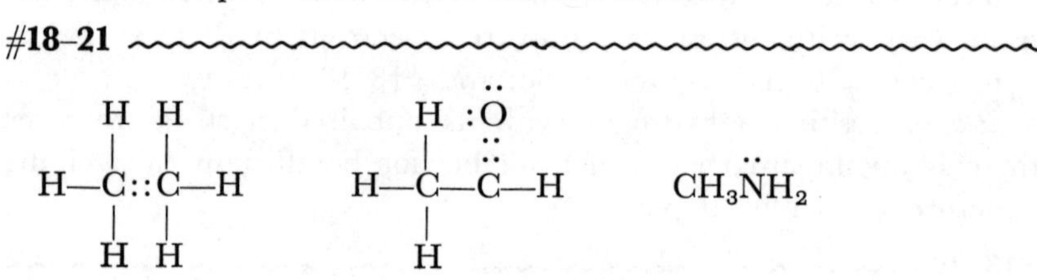


## §19. Problems. Electron-Dot Formulas

**1)** Each atom in a substance can be characterized by one or more of the statements below. Some of these statements relate to the intrinsic nature of the element itself, e.g., the number of valence electrons. Other statements relate to the nature of the bonding of that element in the particular structure, e.g., formal charge. You are to determine which of the statements apply to each atom in each structure. For example, consider $CH_3OH$. Statements 2, 6, 9 and 21 apply to the carbon; statements 4, 6, 11, 17, and 22 apply to the oxygen; statements 1 and 8 apply to all the hydrogens. Statements to be considered:

*The atom has:*

*1*) 1 valence electron
*2*) 4 valence electrons
*3*) 5 valence electrons
*4*) 6 valence electrons
*5*) 7 valence electrons
*6*) a complete octet
*7*) an incomplete octet
*8*) a complete duet
*9*) 4 electrons to balance against the nuclear charge
*10*) 5 electrons to balance against the nuclear charge
*11*) 6 electrons to balance against the nuclear charge
*12*) 7 electrons to balance against the nuclear charge
*13*) 8 electrons to balance against the nuclear charge
*14*) a formal charge of $-1$
*15*) a formal charge of $+1$
*16*) one pair of nonbonding electrons
*17*) two pairs of nonbonding electrons
*18*) three pairs of nonbonding electrons
*19*) a pi electron
*20*) contributed a pair of electrons to a coordinate covalent bond

*The orbital geometry at this atom is:*

*21*) tetrahedral
*22*) modified tetrahedral
*23*) that of $sp^2$ hybridization
*24*) that of $sp^3$ hybridization

*The substances in which the atoms are to be analyzed in terms of these statements are:*

*a*) $CH_2{=}CHCH_2OH$
*b*) $CH_3\overset{+}{C}HCH_3$
*c*) $CH_3F$
*d*) $CH_3PH_2$
*e*) $(CH_3NH_2CH_3)^+Br^-$
*f*) $(CH_3\overset{+}{O}H_2)NO_3^-$
*g*) $HC{\equiv}CCH_2SCH_3$
*h*) $(CH_3\overset{+}{\underset{|}{O}}CH_2Cl)Cl^-$ (with H attached to O)
*i*) $CH_3C{\equiv}C{:}^-Na^+$
*j*) $H_2C{=}CH\overset{O}{\overset{\|}{C}}O^-K^+$

**2**) Draw an electron-dot formula for each of the following substances using the rules presented in §18. Distinguish between the electrons donated by each atom so that you can easily keep track of the source of the electrons in each molecule.

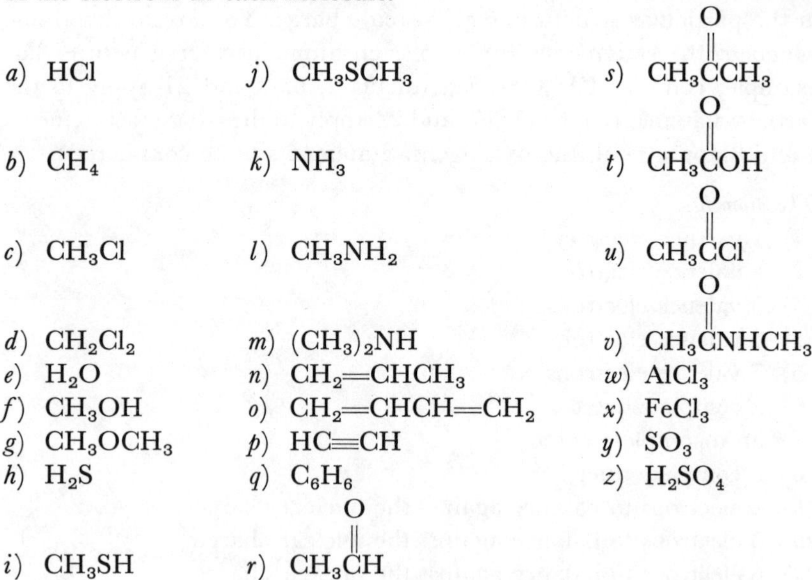


**3**) Draw an electron-dot formula for each of the following substances. Mark each coordinate covalent bond with a check and determine the formal charge on each atom in each structure.

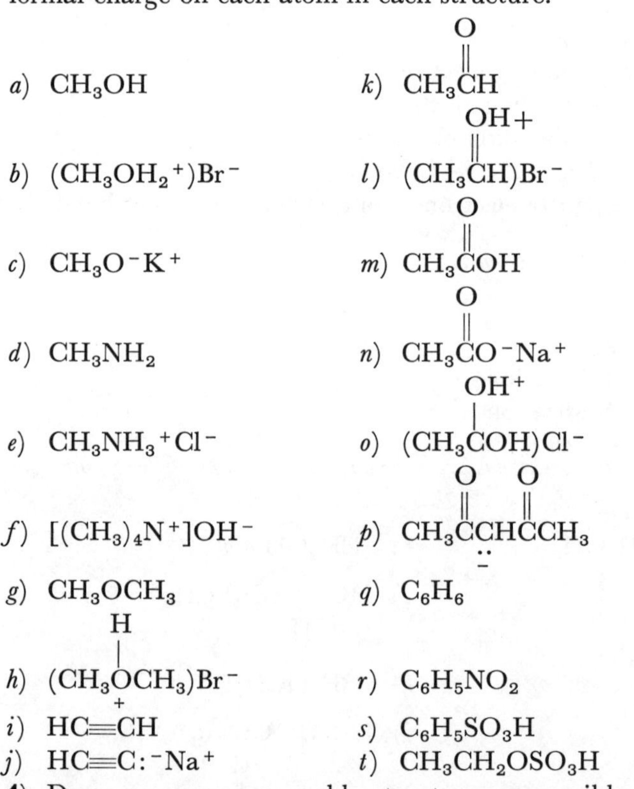


**4**) Draw as many reasonable structures as possible for each of the following molecular formulas. Follow the procedure described in Examples 2 and 3. Base your conclusions as to the reasonableness of the structures on the rules you have learned relating to the assignment of electrons within a structure. Do not include cyclic structures.

*a*) $C_3H_8O$ *b*) $C_3H_4Cl_2$ *c*) $C_3H_9N$ *d*) $C_4H_8O$
*e*) $C_4H_8O_2$ *f*) $C_4H_6O$ *g*) $C_4H_5Br$ *h*) $C_4H_7N$
*i*) $C_3H_7NO_2$ *j*) $C_4H_8S$ *k*) $C_4H_5OCl$ *l*) $C_4H_4O_2$

## §20. Formal Charge. Carbonium Ions, Carbanions, and Free Radicals

An atom carries a formal charge when in a given molecular structure it does not have a full share of the exact number of electrons needed to neutralize the charge on the nucleus (see §18). A formal charge may be either positive or negative. This electronic feature is not manifested in the shape of the molecular orbitals but does effect the density of the electron cloud in the region of the atom with the formal charge. In writing formulas we do not normally identify the formal charges on atoms in a neutral molecule. An ion, either an anion or a cation results when the formal charges on the atoms in a molecule do not cancel out. This fact is noted by a plus or a minus sign next to the entire unit as shown in #20–1. The particular atom with the formal charge may or may not be identified by the position of the plus or minus sign depending upon whether the charge may be considered as dispersed over the whole unit as in the ammonium ion or concentrated at a particular site as in the carbonium ion.

**#20–1**

```
  H             H       [ :Ö:    :Ö: ]-        H
  |             |       [    \   /   ]         |
H—N—H+        H—C+      [      N     ]       H—C:−
  |             |       [      ‖     ]         |
  H             H       [     Ö:     ]         H
```

The carbonium ion, the carbon free radical, and the carbanion are three important reactive organic species. They are rarely isolatable, having only a transitory existence as intermediates. Yet they play a fundamental role in determining the structure of reaction products.

**#20–2**

```
  R         H          H
  ..        ..         ..
R:C+      H:C+      CH₃C+
  ..        ..         ..
  R         H          H
```

These three species have the following characteristics:

- A *carbonium ion* is a chemical entity in which, as shown in #20–2, one of the carbon atoms has a share in only six electrons. The carbon shares three pairs of electrons with three other atoms, either carbon or hydrogen (designated by **R** in #20–2). Thus, the charge on the carbon nucleus is offset by the charge from only three electrons (half of the shared six), and the carbon atom has a formal charge of +1. This formal charge is localized on the carbon. There are only three filled orbitals around this carbon, which means that the geometry of the bonding is the geometry of three tangent spheres (#1–31). The unit is planar and the bonding angles are 120°.

## Table 20-1

### Reactive Particles

| | *Carbonium Ion* | *Free Radical* | *Carbanion* |
|---|---|---|---|
| Methyl | $\mathrm{H{-}\overset{H}{\underset{H}{C}}{+}}$ | $\mathrm{H{-}\overset{H}{\underset{H}{C}}\cdot}$ | $\mathrm{H{-}\overset{H}{\underset{H}{C}}:^{-}}$ |
| Primary 1° | $\mathrm{CH_3{-}\overset{H}{\underset{H}{C}}{+}}$ | $\mathrm{CH_3{-}\overset{H}{\underset{H}{C}}\cdot}$ | $\mathrm{CH_3{-}\overset{H}{\underset{H}{C}}:^{-}}$ |
| Secondary 2° | $\mathrm{CH_3{-}\overset{CH_3}{\underset{H}{C}}{+}}$ | $\mathrm{CH_3{-}\overset{CH_3}{\underset{H}{C}}\cdot}$ | $\mathrm{CH_3{-}\overset{CH_3}{\underset{H}{C}}:^{-}}$ |
| Tertiary 3° | $\mathrm{CH_3{-}\overset{CH_3}{\underset{CH_3}{C}}{+}}$ | $\mathrm{CH_3{-}\overset{CH_3}{\underset{CH_3}{C}}\cdot}$ | $\mathrm{CH_3{-}\overset{CH_3}{\underset{CH_3}{C}}:^{-}}$ |
| Allylic | $\mathrm{CH_2{=}CH\overset{H}{\underset{H}{C}}{+}}$ | $\mathrm{CH_2{=}CH\overset{H}{\underset{H}{C}}\cdot}$ | $\mathrm{CH_2{=}CH\overset{H}{\underset{H}{C}}:^{-}}$ |
| Benzylic | $\mathrm{C_6H_5{-}\overset{+}{\underset{H}{C}}{-}C_6H_5}$ | $\mathrm{C_6H_5{-}\overset{+}{\underset{H}{C}}{-}C_6H_5}$ | $\mathrm{C_6H_5{-}\overset{\ddot{-}}{\underset{H}{C}}{-}C_6H_5}$ |

$$\mathrm{R:\overset{R}{\underset{R}{\ddot{\underset{..}{C}}}}\cdot} \qquad \mathrm{H:\overset{H}{\underset{H}{\ddot{\underset{..}{C}}}}\cdot} \qquad \mathrm{CH_3\overset{H}{\underset{H}{\underset{..}{C}}}\cdot}$$

#### #20-3

- A *carbon free radical* is a chemical entity in which, as shown in #20-3, one of the carbon atoms shares three pairs of electrons with three other atoms (either carbon or hydrogen) and also has a single, unpaired nonbonding electron. Although this carbon has an incomplete octet, there is the charge from four valence electrons (half of six plus the nonbonding electron) to neutralize the nuclear charge. A free radical is very reactive because of the unpaired electron, but it does not have a charge. It is not an ion. The geometry is similar to that of $sp^2$ hybridization with the three filled orbitals in one plane and the orbital of the nonbonding electron perpendicular to this plane.

**#20–4**

```
   R          H           H
   ..         ..          ..
 R:C:⁻      H:C:⁻     CH₃C:⁻
   ..         ..          ..
   R          H           H
```

- A *carbanion* is a chemical entity in which, as shown in #20–4, one of the carbon atoms shares three pairs of electrons with three other atoms, either carbon or hydrogen and also has a pair of non-bonding electrons. The charge from five electrons (half of the shared six plus the pair of nonbonding electrons) can be assigned to this carbon. A charge of five is one more than necessary to neutralize the carbon nucleus. Thus, the significant carbon in a carbanion has a formal charge of −1. There are four filled orbitals around this carbon, which leads to the geometry of tetrahedral bonding.

Carbanions, carbon free radicals, and carbonium ions are classified according to the number of carbons bonded directly to the carbon that determines the nature of the species. This classification is:

- *Primary, 1°*—There is only one carbon bonded to the carbon atom with the distinctive electron configuration as in #20–5.

**#20–5**

```
     H             H             H
  |  |          |  |          |  |
 —C—C+         —C—C·         —C—C:⁻
  |  |          |  |          |  |
     H             H             H
```

- *Secondary, 2°*—There are two carbons bonded to the carbon atom with the distinctive electron configuration as in #20–6.

**#20–6**

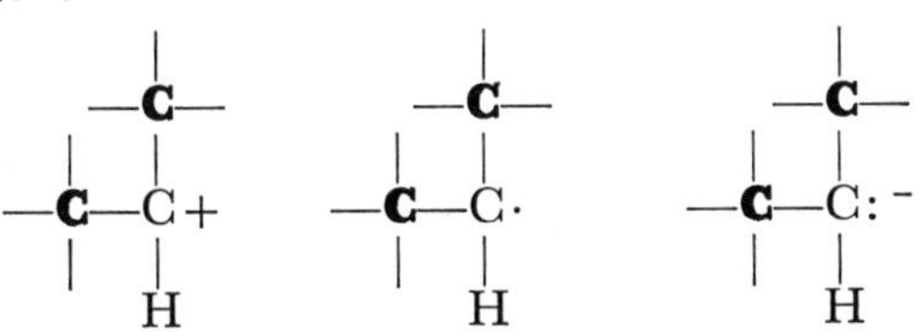

- *Tertiary, 2°*—There are three carbons bonded to the carbon atom with the distinctive electron configuration as in #20–7.

**#20–7**

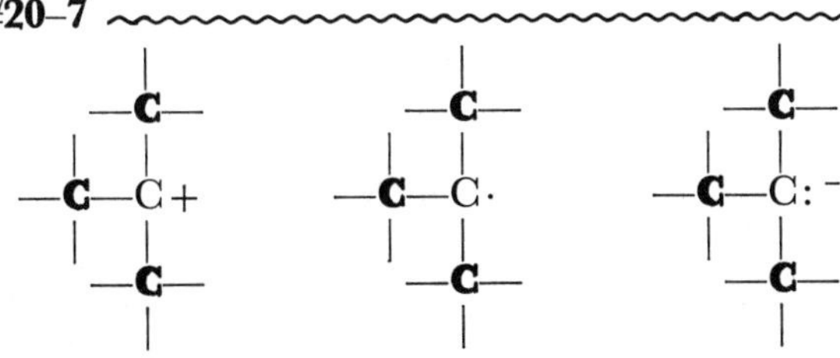

A carbonium ion can be formed by removing a hydride ion, H:$^-$, from an alkane carbon. The removal of a hydride ion, H:$^-$, from a 1° carbon requires more energy than the removal of a hydride ion, H:$^-$, from a 3° carbon. Thus, it is harder to form a 1° than a 3° carbonium ion.

A general principle is that the ion or radical that is made most easily (i.e., by the reaction that requires the least energy input) is the most stable. Conversely, the most stable ion or radical is the one that is made most easily. Thus, when we discuss ease of formation we are really talking about stability and vice versa. The terms are employed in a comparative rather than an absolute sense. "Ion B is less stable than ion A" means that more of ion A will be formed under the same reaction conditions. The statement also means that B will not remain in the ionic form for as long a time as A under the same conditions. The charge on B will be neutralized more easily. The term "less stable" does not imply that the entire molecular structure will be disrupted as B falls apart, but refers rather to the *comparative* stability of the two structures.

When we say that it requires more energy to form a 1° carbonium ion that a 3° we are also saying that a 3° carbonium ion is more stable than a 1° carbonium ion. The listing in Table 20–1 of 1°, 2°, and 3° carbonium ions places them in order of increasing stability.

It takes more energy to remove a hydrogen free radical, H·, from a 1° carbon than it does to remove a hydrogen free radical, H·, from a 3° carbon. Again the stability order of the particle is 3° > 2° > 1°. The reverse stability order is found for the carbanions. The 1° ion is more stable than the 2° ion which in turn is more stable than the 3°. The differences in stability are not as great for the carbanions as for the carbonium ions.

We need to understand the factors which cause these differences in stability:

Large inequities in the density of an electron cloud at different sites in a molecular structure represent an unstable distribution of electrons. It takes energy to keep the electron cloud piled up at one place, or to keep it very sparse at another, just as it takes energy to produce waves and troughs between waves on a body of water. Thus, the greater the charge localization, the less stable the charged particle is. Any structural feature which can neutralize or disperse the charge contributes to increased stability, whereas any feature which increases the localization of the charge leads to a decrease in stability. The term "charge" here refers to either + or −.

For a 3° carbonium ion to be more stable than a 1° carbonium ion, the portion of the positive charge actually concentrated at the electron-deficient carbon must be less for the 3° than for the 1° structure. The only structural difference between the two ions is found in the number of alkyl groups. Therefore, the alkyl groups must be helping to delocalize or reduce the formal charge on the carbon with an incomplete octet. Similarly, the alkyl groups must be helping to disperse the high density of the electron cloud associated with the single nonbonding electron in the free radical.

In order to understand how an alkyl group can contribute towards a more even distribution of the electron cloud in both a carbonium ion and a free radical, we need to examine the geometry of the electron orbitals in both species.

Consider first the free radical and compare an ethyl radical with an isopropyl radical to determine the reason for the increased stability of the 2° radical as compared to the 1°.

The oribtal geometry of the ethyl radical is shown in #20–8. There are two carbons, C*1* and C*2* joined by a σ bond; three sigma-bonded hydrogens form tetrahedral bond angles with C*2*; the two hydrogens at C*1* have the geometry of $sp^2$ hybridization with the two σ C*1*—H bonds in the same plane as the C*1*—C*2* bond and the orbital of the nonbonding electron perpendicular to this plane.

**#20–8**

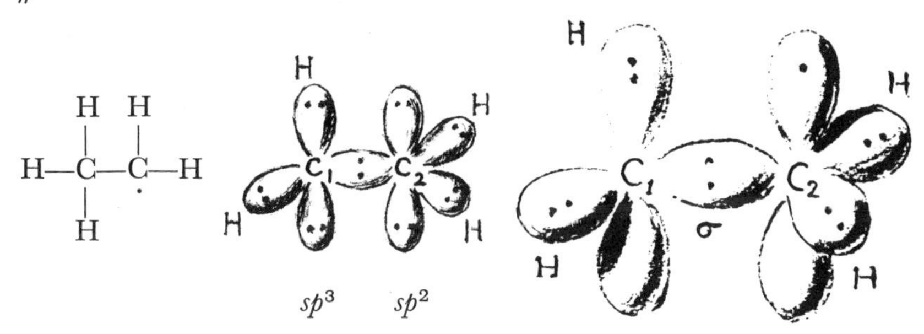


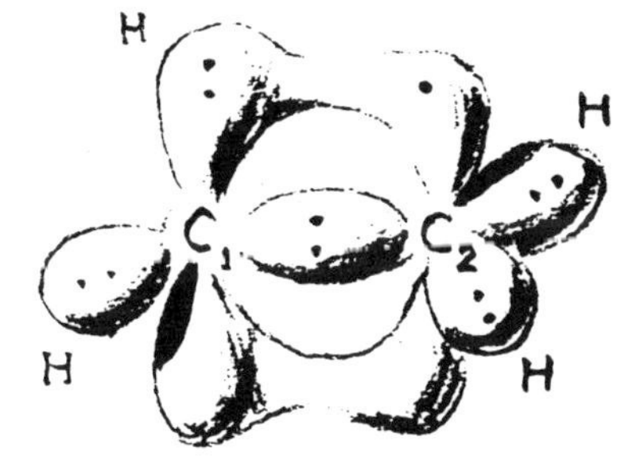


Overlapping of a C*2*—H σ orbital and the orbital of the nonbonding electron on C*1* occurs whenever the two become aligned. Orbital overlap leads to a charge redistribution. The charge density of the σ orbital decreases and the charge density of the half-filled orbital increases as the electrons in the overlapping orbitals redistribute their charge. An orbital with only one electron is designated as half-filled and represents an unstable situation. Any factor that increases the charge density in an unfilled orbital leads to greater stability.

The overlap between a σ orbital and the orbital of the nonbonding electron is not as extensive as the overlap of two *p* orbitals in the formation of a double bond. There is consequently a lesser electron exchange. However, the exchange does occur between each of the three C*2*—H σ orbitals and the orbital of the nonbonding electron on *C1*. The overall effect of the overlap tends to equalize the charge density in the electron cloud at the two carbons.

As stated above, it takes energy to maintain differences in charge density within an electron cloud. Any factor which tends to equalize charge distribution also contributes to the stability of the structural unit. There are two methyl groups adjacent to the carbon with the unpaired electron in the isopropyl radical. Two methyl groups are more effective than one in equalizing the charge density. Therefore, the 2° isopropyl radical is more stable than the 1° ethyl radical.

The redistribution of charge density as the result of σ orbital overlap in the manner described above is called *hyperconjugation.* Hyperconjugation is a dynamic effect that is very difficult to show in a simple two-dimensional formula. A specific symbolism has been developed for representing this property of alkyl groups. The exchange between the electron pair in the σ orbital of a C*2*—H bond and the unpaired electron on C*1* is indicated by a double bond between the two carbons as in #20–9. The resulting decrease in charge density in the C*2*—H orbital is represented by placing a single dot in place of the C*2*—H bond. A different formula is drawn for each carbon-hydrogen bond and a double-headed arrow is placed between the formulas. Such arrows mean that none of the structures actually exists, but all of the formulas are needed to show the structural characteristics of the actual substance. This set of formulas is a device for indicating that some of the electron "stuff" holding the hydrogens in place has been diverted to the orbital of the unpaired, nonbonding electron.

**#20–9**

```
 ┌   H H              H H                H H                H H    ┐
 │   | |              · |                | |                | |    │
 │ H—C—C—H  <-->  H—C=C—H  <-->  H·C=C—H  <-->  H—C=C—H │
 │   | ·              |                  |                  ·      │
 └   H                H                  H                  H      ┘
```

* * * * *

Carbonium ions can also be stabilized by hyperconjugation and the same stability order results: 3° > 2° > 1°. This effect can be understood if you visualize an "empty orbital" in place of the half-filled one we discussed for the ethyl radical. An ethyl carbonium ion with its "empty" orbital is diagrammed in #20–10.

**#20–10**

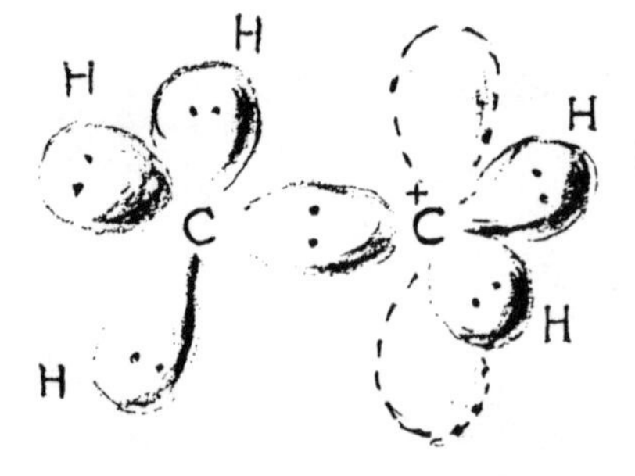


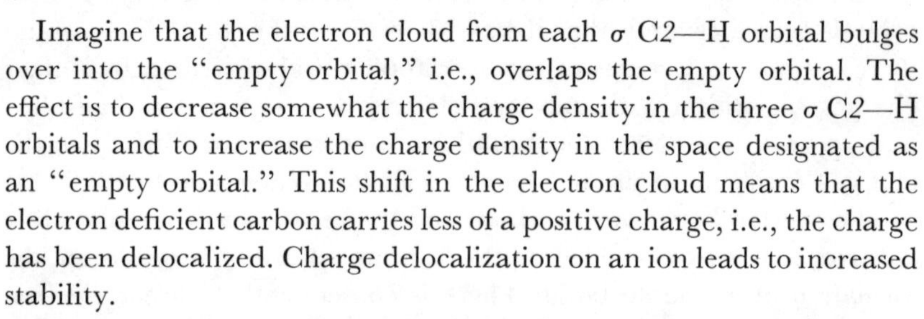

Imagine that the electron cloud from each σ C*2*—H orbital bulges over into the "empty orbital," i.e., overlaps the empty orbital. The effect is to decrease somewhat the charge density in the three σ C*2*—H orbitals and to increase the charge density in the space designated as an "empty orbital." This shift in the electron cloud means that the electron deficient carbon carries less of a positive charge, i.e., the charge has been delocalized. Charge delocalization on an ion leads to increased stability.

There are twice as many σ C—H bonds available for overlap in an isopropyl carbonium ion ($CH_3\overset{+}{C}HCH_3$) as in an ethyl carbonium ion ($CH_3CH_2+$). Thus, the former is more stable than the latter. Similarly the charge on a 3° carbonium ion is delocalized to a greater extent than the charge on a 2° carbonium ion, leading to still greater stability.

The formulas used to represent the effects of hyperconjugation in a carbonium ion are given in #20–11. The double bond between C*1*

**#20–11**

```
 ┌   H H              H H                H H                H H    ┐
 │   | |              + |                | |                | |    │
 │ H—C—C—H  <-->  H—C=C—H  <-->  H+C=C—H  <-->  H—C=C—H │
 │   | +              |                  |                  +      │
 └   H                H                  H                  H      ┘
```

and C*2* indicates the redistribution of charge density from C*2* to C*1* and the + at each alkyl hydrogen represents the slight electron deficiency generated in these sigma bonds as the result of this sharing.



* * * * *

Hyperconjugation does not play a significant role in determining the stability of a carbanion. The position of the orbital with the non-bonding electron pair is that of tetrahedral bonding as indicated in #20–12. Thus, this pair of electrons is further away from the C—H sigma bonds on the adjacent carbon than is the single nonbonding electron in the corresponding free radical. Additionally, with all the orbitals completely filled there is no opportunity to disperse the negative charge from the nonbonding electron pair.

**#20–12**

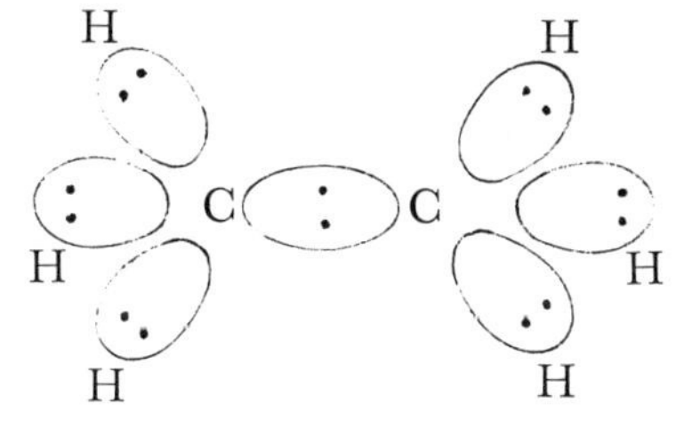


Alkyl groups release electron charge by *induction* (§24). This +**I** effect supplements hyperconjugation in a carbonium ion and offsets it in a free radical. For both species, induction is a minor effect compared to hyperconjugation and so was omitted from the discussion of relative stability. However, induction, although a minor effect for alkyl groups, is the main factor determining the relative stability of 1°, 2°, and 3° alkyl carbanions.

Electron release tends to increase the charge density at the carbon carrying the negative charge and thereby to decrease the stability of the anion. The + inductive effect increases with the number of alkyl groups present. Thus, the stability order for carbanions is 1° > 2° > 3°. The effect is a small one and the differences between these simple alkyl carbanions are small and hard to measure due to the great difficulty in abstracting a proton ($H^+$) from the parent compound.

* * * * *

The hydrogens on a carbon adjacent to a benzene ring are called *benzylic hydrogens*. The hydrogens on a carbon adjacent to an unsaturated, doubly-bonded carbon are called *allylic hydrogens*. The species formed when one of these hydrogens is removed (as $H:^-$, $H\cdot$, or $H^+$) is called (respectively) a benzylic or an allylic *carbonium ion*, or a *free radical*, or a *carbanion*. These units are each considerably more stable than the corresponding 3° unit.

In both the benzylic ($C_6H_5\overset{+}{C}R_2$,* $C_6H_5\dot{C}R_2$, $C_6H_5\ddot{C}R_2^-$) and allylic ($CH_2{=}CH\overset{+}{C}R_2$, $CH_2{=}CH\dot{C}R_2$, $CH_2{=}CH\ddot{C}R_2^-$) structures, there is extensive charge delocalization through the $\pi$ cloud which, because of resonance (§26), extends above and below the entire unit. A benzylic carbonium ion rather than a 3° carbonium ion is formed if there is the opportunity for both products. A vinyl carbonium ion, a free radical, or a carbanion is one where the +, ·, or $\ddot{-}$ is located on an unsaturated carbon of a double bond, as in $CH_3\overset{+}{C}{=}CH_2$ or

* **R** = hydrogen, alkyl, or aryl.

$CH_3CH{=}CH\cdot$. These particles are very unstable. The total stability order for carbonium ions and free radicals is: benzyl=allyl > 3° > 2° > 1° > methyl vinyl.

* * * * *

Note that we have *not* discussed the stability of carbonium ions relative to free radicals or to carbanions. Such a comparison does not provide generally useful information. However, an understanding of the stability of one carbonium ion relative to another carbonium ion or of one free radical relative to another free radical enables us to predict the main product of many reactions, as illustrated in Examples 1 and 2.

*Example 1*—A proton can bond at either C*1* or C*2* in isobutylene. Which reaction occurs most easily?

**#20–13**

$$H^+ + CH_2{=}\overset{\overset{\displaystyle CH_3}{|}}{C}{-}CH_3 \longrightarrow H{-}\underset{\underset{\displaystyle H}{|}}{\overset{\overset{\displaystyle H}{|}}{C}}{-}\underset{+}{\overset{\overset{\displaystyle CH_3}{|}}{C}}{-}CH_3$$

*Step 1.* Write equations showing the two possible products.

**#20–14**

$$H^+ + CH_2{=}\overset{\overset{\displaystyle CH_3}{|}}{C}{-}CH_3 \longrightarrow H{-}\underset{+}{\overset{\overset{\displaystyle H}{|}}{C}}{-}\underset{\underset{\displaystyle H}{|}}{\overset{\overset{\displaystyle CH_3}{|}}{C}}{-}CH_3$$

*Step 2.* Compare the structure of the two species to determine whether one is more stable than the other.

A 3° carbonium is formed when the proton bonds to C*1* as in equation #20–13.

A 1° carbonium ion is formed when a proton bonds to C*2* as in equation #20–14.

A 3° carbonium is more stable than a 1° carbonium ion. Therefore, the carbonium ion in equation #20–13 is formed more easily than the carbonium ion in equation #20–14. In other words it requires less energy input to make the *tert*-butyl carbonium ion than to make the isobutyl carbonium ion.

Whenever there is the possibility that one or the other of these two ions is the intermediate in a given reaction the *tert*-butyl carbonium ion will be the main species formed. The main product of the reaction will be the one derived from the more stable 3° intermediate.

*Example 2*—One step in the free radical chlorination of propane involves the formation of a carbon free radical intermediate. Propane contains both primary and secondary hydrogens and can therefore form two different free radicals. Which of the two possible free radicals is formed more easily?

*Step 1.* Write equations showing the formation of the free radicals.

Differentiate between the kinds of hydrogens in the formula of propane so as to identify the hydrogen to be removed by each reaction as shown in #20–15, #20–16.

**#20–15**

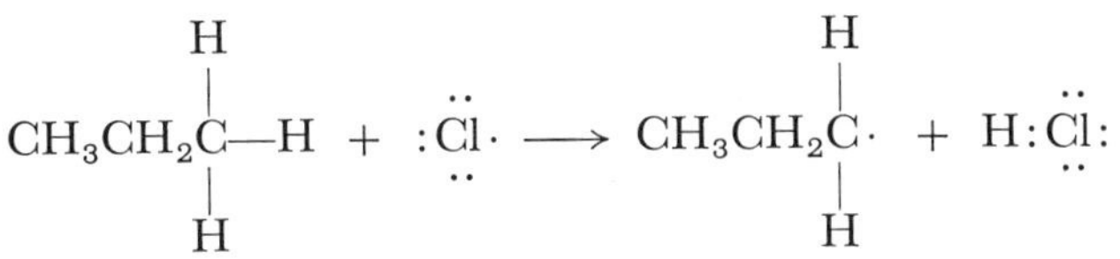

$$\mathrm{CH_3CH_2}\overset{\displaystyle \mathrm{H}\atop|}{\underset{|\atop\displaystyle \mathrm{H}}{\mathrm{C}}}\mathrm{{-}H} + :\underset{\cdot\cdot}{\overset{\cdot\cdot}{\mathrm{Cl}}}\cdot \longrightarrow \mathrm{CH_3CH_2}\overset{\displaystyle \mathrm{H}\atop|}{\underset{|\atop\displaystyle \mathrm{H}}{\mathrm{C}}}\cdot + \mathrm{H}:\underset{\cdot\cdot}{\overset{\cdot\cdot}{\mathrm{Cl}}}:$$

**#20–16**

$$\mathrm{CH_3}\overset{\displaystyle \mathrm{H}\atop|}{\underset{|\atop\displaystyle \mathrm{H}}{\mathrm{C}}}\mathrm{CH_3} + :\underset{\cdot\cdot}{\overset{\cdot\cdot}{\mathrm{Cl}}}\cdot \longrightarrow \mathrm{CH_3}\overset{\cdot}{\underset{|\atop\displaystyle \mathrm{H}}{\mathrm{C}}}\mathrm{CH_3} + \mathrm{H}:\underset{\cdot\cdot}{\overset{\cdot\cdot}{\mathrm{Cl}}}:$$

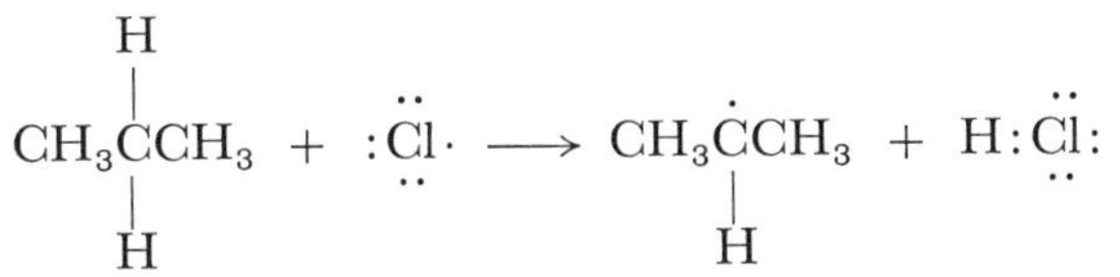

*Step 2.* Compare the structure of the two species to determine whether one is more stable than the other.

Removal of a H· from C*1* produces the *n*-propyl free radical as shown in equation #20–15. The *n*-propyl free radical is a 1° free radical.

Removal of a H· from C*2* produces the isopropyl free radical (#20–16). The isopropyl free radical is a 2° free radical.

An isopropyl free radical is formed more readily than a *n*-propyl free radical.

The problems in §21 give you practice in evaluating the relative stability of a variety of ions and free radicals. The objective is to prepare yourself to use the concept of relative stability in determining reaction products. Remember, in these examples, that relative stability relates to comparative degree of charge delocalization within the electron cloud.

## §21. Problems. Carbonium Ions, Carbanions, and Free Radicals

**1**) Name each species; indicate which member of each pair is more stable; state the reason for your choice.

*a*) $\mathrm{CH_3\overset{+}{C}HCH_3}$, $\mathrm{CH_3CH_2CH_2+}$

*b*) $\mathrm{(CH_3)_3C\cdot}$, $\mathrm{(CH_3)_2\dot{C}H}$

*c*) $\mathrm{C_6H_5\overset{+}{C}HCH_3}$, $\mathrm{CH_3\overset{+}{C}HCH_3}$

*d*) $\mathrm{C_6H_5\overset{\overset{-}{\cdot\cdot}}{C}HCH_3}$, $\mathrm{C_6H_5CH_2CH_2:^-}$

*e*) $\mathrm{CH_3+}$, $\mathrm{CH_3CH_2+}$

*f*) $\mathrm{CH_2{=}CHCH_2\!\cdot}$, $\mathrm{\cdot CH{=}CHCH_3}$

*g*) $\mathrm{(CH_3)_3C+}$, $\mathrm{CH_3CH_2CH_2+}$

*h*) $\mathrm{CH_3C_6H_5CH_2+}$, $\mathrm{CH_3C_6H_5CH_2CH_2+}$

*i*) $\mathrm{C_6H_5\dot{C}HCH_3}$, $\mathrm{CH_3\dot{C}HCH_3}$

*j*) $\mathrm{CH_3:^-}$, $\mathrm{CH_3\overset{\overset{-}{\cdot\cdot}}{C}HCH_3}$

*k*) $\mathrm{CH_3CH_2\dot{C}HCH_3}$, $\mathrm{CH_3C_6H_5\dot{C}HCH_3}$

*l*) $\mathrm{CH_3\overset{+}{C}{=}CHCH_3}$, $\mathrm{CH_3CH{=}CHCH_2+}$

**2**) Arrange the members of each series in order of increasing ease of formation, i.e., place first the one that is most difficult to form, and last the one that is easiest to form.

*a*) $CH_3CH_2+$ $(CH_3)_3C+$ $(CH_3)_2CH+$
*b*) $(CH_3)_3C\cdot$ $CH_3CH_2\cdot$ $CH_2{=}CHCH_2\cdot$
*c*) $(CH_3)_2CH:^-$ $C_6H_5CH_2:^-$ $CH_3CH_2:^-$
*d*) $CH_3\cdot$ $(CH_3)_2CH\cdot$ $CH_3CH_2\cdot$
*e*) $C_6H_5\overset{+}{C}HCH_3$ $(C_6H_5)_2CH+$ $C_6H_{13}\overset{+}{C}HCH_3$

**3**) Draw the formulas needed to represent hyperconjugation in the following species:

*a*) $CH_3CH_2\cdot$ $(CH_3)_2CH\cdot$ $(CH_3)_3C\cdot$
*b*) $CH_3CH_2+$ $(CH_3)_2CH+$ $(CH_3)_3C+$
*c*) $CH_3CH_2CH_2\cdot$ $(CH_3CH_2)_2CH\cdot$ $(CH_3CH_2)_3C\cdot$
*d*) $CH_3CH_2CH_2+$ $(CH_3CH_2)_2CH+$ $(CH_3CH_2)_3C+$

**4**) One step in the free radical halogenation of a hydrocarbon involves the formation of a carbon free radical intermediate according to the equation in #21–1. The ease of removal of a specific H· depends upon the stability of the free radical that is formed.

**#21–1**

$$-\overset{|}{\underset{|}{C}}-\overset{|}{\underset{|}{C}}:H + :\ddot{Cl}\cdot \longrightarrow -\overset{|}{\underset{|}{C}}-\overset{|}{\underset{|}{C}}\cdot + H:\ddot{Cl}:$$

Write the formula for each carbon free radical that can be formed from each of the following compounds by the reaction in #21–1. There will be a different free radical for each kind of hydrogen in the molecule (§5). List the free radicals for each molecule in the order of ease of formation with the one that is formed most easily being first on the list. Do not try to differentiate between two 1° or two 2° free radicals. Merely differentiate as to benzyl, allyl, 3°, 2°, 1°, methyl, and vinyl free radicals.

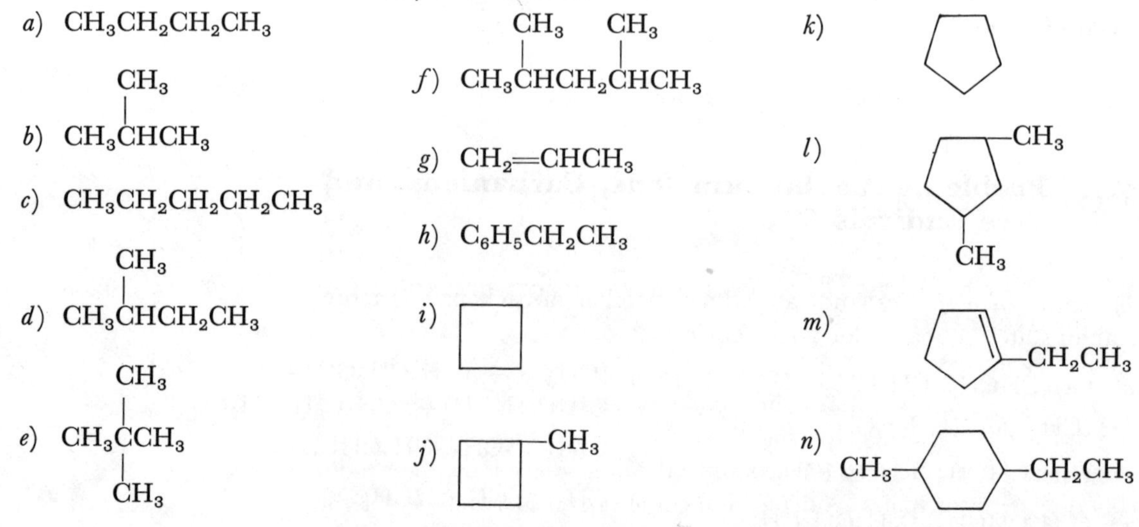


**5**) A proton can bond with an olefinic carbon forming a carbonium ion according to equation #21–2. The proton can presumably bond with either of the two olefinic carbons. However, the two carbonium ions are not formed in equal proportion. The major product is the more stable carbonium ion.

**#21-2**

$$-\overset{|}{C}\overset{..}{\underset{..}{}}\!\!=\!\overset{|}{C}- \ + \ H^+ \longrightarrow -\underset{+}{\overset{|}{C}}-\underset{|}{\overset{|}{C}}:H$$

Give the formula for each of the carbonium ions that can be formed when each of the following olefins reacts with a proton ($H^+$) as in equation #21-2. Indicate with a check mark the carbonium ion that is formed as the main product of this intermediate reaction.

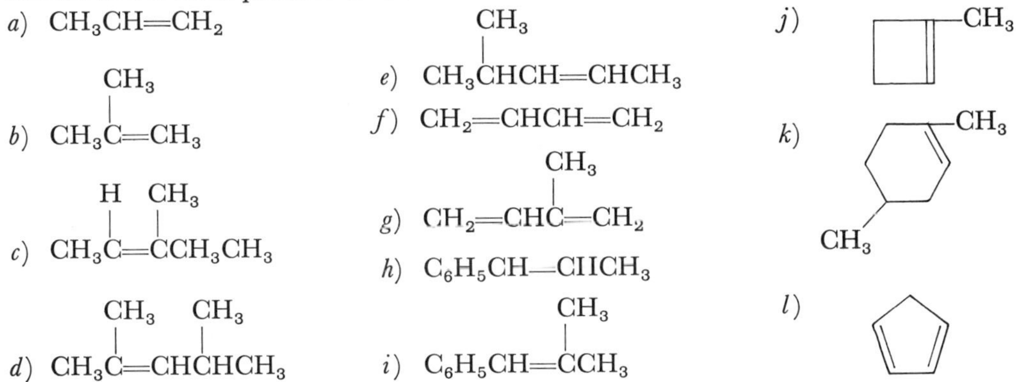


**6**) A bromine atom can react with an olefinic bond forming a carbon free radical as shown in equation #21-3. The bromine can presumably bond with either of the olefinic carbons. However, the reaction which produces the most stable free radical takes place more easily. Give the formula for the free radicals that can be formed from each of the olefins in question 5 when it reacts according to equation #21-3. Indicate with a √ the free radical that is formed most easily in each case.

**#21-3**

$$-\overset{|}{C}\overset{..}{\underset{..}{}}\!\!=\!\overset{|}{C}- \ + \ :\ddot{\underset{..}{Br}}\cdot \longrightarrow -\underset{\cdot}{\overset{|}{C}}-\underset{|}{\overset{|}{C}}:\ddot{\underset{..}{Br}}:$$

## §22. Nonbonding Electron Pairs. Proton Exchange Reactions

In §18 you learned that a molecular structure may include paired electrons that are not specifically involved in bonding two atoms together. In this section we examine the role played by these nonbonding electron pairs in determining the chemical reactivity of the structures in which they occur.

Nonbonding electron pairs occur in neutral atoms such as alcohols

($CH_3\ddot{O}H$), ethers ($CH_3\ddot{O}CH_3$), amines ($CH_3\ddot{N}H_2$), and alkyl halides ($CH_3\ddot{Cl}$:). Nonbonding electron pairs may also be responsible for the formal charge on an anion such as is found in a carbanion ($CH_3\ddot{C}H_2^-$), an alkoxide ion ($CH_3\ddot{O}:^-$), or an amide ion ($CH_3\ddot{N}H^-$).

A substance with a pair of nonbonding electrons that can be donated to the formation of a coordinate covalent bond with an electron-deficient atom is called a *Brönsted* base. Typically a Brönsted base can donate a pair of electrons to a coordinate covalent bond with a proton, $H^+$. Protons do not exist as such in the reaction medium but are obtained from another substance as shown in #22–1. Thus, such reactions are referred to as proton exchange reactions.

**#22–1**

$$\ddot{N}H_3 + H:\ddot{Cl}: \rightleftharpoons NH_4^+ + :\ddot{Cl}:^-$$

$$CH_3\ddot{O}H + H_3\ddot{O}^+ \rightleftharpoons CH_3\ddot{O}H_2^+ + H_2\ddot{O}:$$

Substances vary in their tendency to share nonbonding electrons. A substance with a strong tendency to donate a pair of nonbonding electrons to a coordinate covalent bond is classified as a *strong Brönsted base*; a substance with a slight tendency to donate an electron pair is *weak Brönsted base.*

In terms of electronic geometry, the orbitals of nonbonding electrons create regions within the electron cloud where the electrons are potentially available for bonding. The orbitals of nonbonding electrons are less constrained than the orbitals of bonding electrons. This difference between orbitals can be represented diagrammatically by drawing the orbitals of nonbonding electrons as somewhat larger than the orbitals of bonding electrons. In this book the orbitals of nonbonding electrons are represented by a different shape. The ellipse is reserved for bonding electrons and the less symmetrical figure with one end more pointed than the other, as shown in #22–2, designates the orbitals of nonbonding electrons.

The orbital of the nonbonding electron pair on nitrogen has been drawn larger in #22–2 than the orbitals of the nonbonding electron

**#22–2**

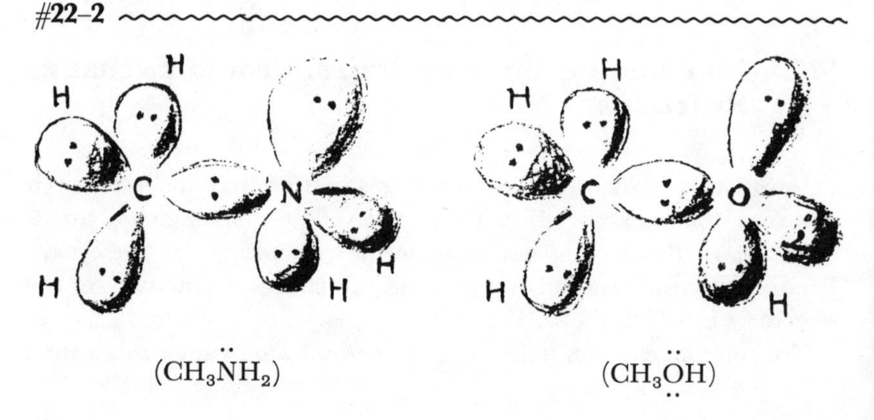

pairs on oxygen to represent a significant difference between the two atoms. Oxygen is more electronegative than nitrogen. This statement means that the oxygen nucleus has a stronger attraction for an electron than does the nitrogen nucleus. Thus, a nonbonding electron pair on oxygen is held more tightly than a nonbonding electron pair on nitrogen. This difference can be crudely represented by drawing the orbitals of the nonbonding electrons on nitrogen as a larger figure.

The greater attraction of oxygen for a nonbonding electron pair means that the pair is not as available for forming a coordinate covalent bond as is the pair on nitrogen. Consequently methyl amine ($CH_3\dot{\dot{N}}H_2$) is a stronger Brönsted base than methyl alcohol ($CH_3\ddot{\underset{..}{O}}H$).

The relative strength of a Brönsted base is also determined by the charge density in the region of the nonbonding electron pair. A greater charge density makes the electron pair more available for bonding and results in a stronger Brönsted base. You will learn about the two major factors that affect charge distribution in §24 and §26. However, the effect of differences in charge density can be illustrated with the concept of formal charge which you have already studied §(18).

**#22-3**

$$CH_3\ddot{\underset{..}{O}}:^- + H_3\ddot{O}^+ \rightleftharpoons CH_3\ddot{\underset{..}{O}}H + H\ddot{\underset{..}{O}}H$$

$$CH_3\ddot{\underset{..}{O}}H + H_3\ddot{O}^+ \rightleftharpoons CH_3\overset{H}{\overset{..+}{\underset{..}{O}}}H + H\ddot{\underset{..}{O}}H$$

The equations in #22-3 include reactions in which a methoxide ion ($CH_3\ddot{\underset{..}{O}}:^-$) and methanol ($CH_3\ddot{\underset{..}{O}}H$) each reacts as a Brönsted base. Experiments show that the methoxide ion is a much stronger base than methanol. The oxygen atom has a formal charge of $-1$ in the methoxide ion and is neutral in methanol. Therefore, the charge density is greater in the region of the oxygen atom in the methoxide ion than in methanol. The electron pair is more available for bonding and so the methoxide ion is a stronger base.

The difference in electronic geometry responsible for the difference in chemical reactivity of these two substances can be roughly dia-

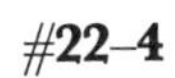
**#22-4**

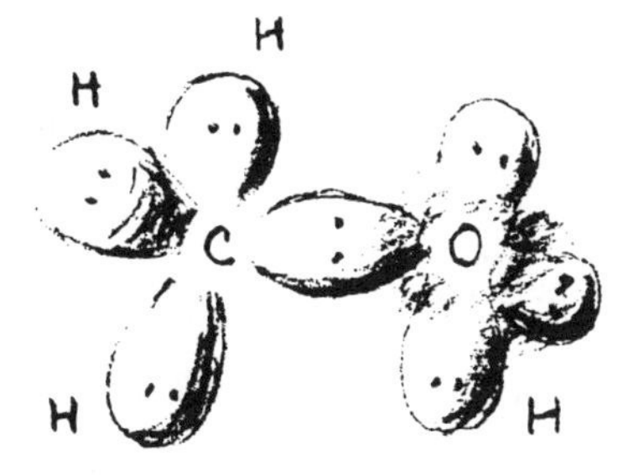


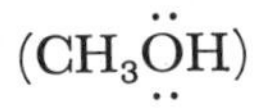
($CH_3\ddot{\underset{..}{O}}H$)

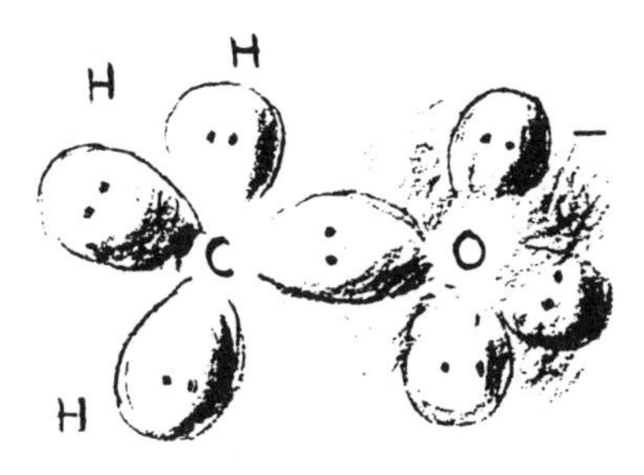


($CH_3\ddot{\underset{..}{O}}:^-$)

grammed as in #22–4, where density variations are indicated by shading in some areas.

* * * * *

The structure that provides the proton in a proton exchange reaction is called a *Brönsted acid.* Examples of some proton exchange reactions are given in #22–5. Note that a proton leaves behind the electrons by which it was bonded in the acid. Presumably any structure that contains a hydrogen could function as a Brönsted acid. However, there are great differences in the ease with which various structures give up a proton. It may take a very strong base such as an amide ion ($\ddot{N}H_2^-$) to remove the proton from an extremely weak acid, such as acetylene ($HC{\equiv}CH$), whereas a strong acid such as HCl will readily give up a proton to a weak base such as $H_2O$. (Write equations for these two reactions.)

**#22–5**

| *Brönsted base* | *Brönsted acid* |
|---|---|

$$CH_3\ddot{N}H_2 + CH_3\overset{\overset{:\ddot{O}}{\|}}{C}\ddot{O}H \rightleftharpoons CH_3\overset{H^+}{\ddot{N}}H_2 + CH_3\overset{\overset{:\ddot{O}}{\|}}{C}\ddot{O}:^-$$

$$CH_3\ddot{O}H + H:\ddot{Cl}: \rightleftharpoons CH_3\overset{H^+}{\ddot{O}}H + :\ddot{Cl}:^-$$

$$CH_3\ddot{O}CH_3 + H_3\ddot{O}^+ \rightleftharpoons CH_3\overset{H^+}{\ddot{O}}CH_3 + H\ddot{O}H$$

$$HC{\equiv}C:^- + C_6H_5\ddot{O}H \rightleftharpoons HC{\equiv}CH + C_6H_5\ddot{O}:^-$$

$$C_6H_5\ddot{O}:^- + CH_3CH_2\ddot{O}H \rightleftharpoons C_6H_5\ddot{O}H + CH_3CH_2\ddot{O}:^-$$

A substance that gives up a proton more easily than a hydronium ion ($H_3O^+$) does is classified as a *strong Brönsted acid.* A substance that gives up a proton less readily than a hydronium ion does is classified as a *weak Brönsted acid.* Differences in acidity and basicity can be related to the geometry of the electron cloud. These relationships are considered in detail in §24 and §26.

* * * * *

Any chemical structure in which there is an atom with an incomplete octet can accept a share of a pair of electrons in a coordinate covalent bond. Substances that need a pair of electrons to provide a complete octet for all atoms (except hydrogen) are called *Lewis acids.* Examples of some Lewis acids are given in #22–6.

The substance that donates a pair of electrons to form a coordinate covalent bond with a Lewis acid is called a *Lewis base.* Examples of some Lewis acid-base reactions are given in #22–7. A Lewis base has

#22–6

```
H.  .H        .Cl.  .Cl.          .O×  ×O.
×    ×         .×    ×.            .×    ×.
  C+              Al                  S            H+
  .×              .×                  ××
  H              :Cl:                :O:
```

#22–7

```
     CH3                           CH3
     |          ..                 |
CH3—C+   +  :Cl:-   ⟶   CH3—C—Cl
     |          ..                 |
     CH3                           CH3
```

essentially the same structure as a Brönsted base. Both have a pair of nonbonding electrons to donate. The classification of a base depends upon the nature of the acid with which it reacts, rather than on the structure of the base itself. The same substance can function as either a Lewis or a Brönsted base as illustrated in #22–8.

#22–8

```
                                                   H+
                ..              ..                 ..               ..
(Brönsted   CH3NCH3  +  H:Cl:  ⇌  CH3NCH3  +  :Cl:-
 base)          |               ..                 |                ..
                H                                  H
                                                                ..
                           .F. .F.             CH3:F:
                           .×   ×.              |    ×.
                ..                                      ..
(Lewis      CH3NCH3  +      B      ⟶  CH3N  :  B×F:
 Base)          |                               |       ..
                H           ×.                  H   ×.
                           :F:                     :F:
                            ..                      ..
```

The term *electrophile* is applied to any substance that can accept electrons. Electrophile means electron-loving. According to this definition a Lewis acid is an electrophile. The term *nucleophile* is applied to substances with electrons to donate. These substances need a nucleus to which they can donate electrons. Thus, a Brönsted base and a Lewis base are nucleophiles.

The problems in §23 are to help you recognize the presence and the importance of nonbonding electron pairs. They also provide practice in writing proton exchange reactions and in identifying Lewis and Brönsted acid and bases.

## §23. Problems. Nonbonding Electron Pairs. Proton Exchange Reactions

**1**) Fill in the nonbonding electron pairs (if any) that have not been included in the following formulas.

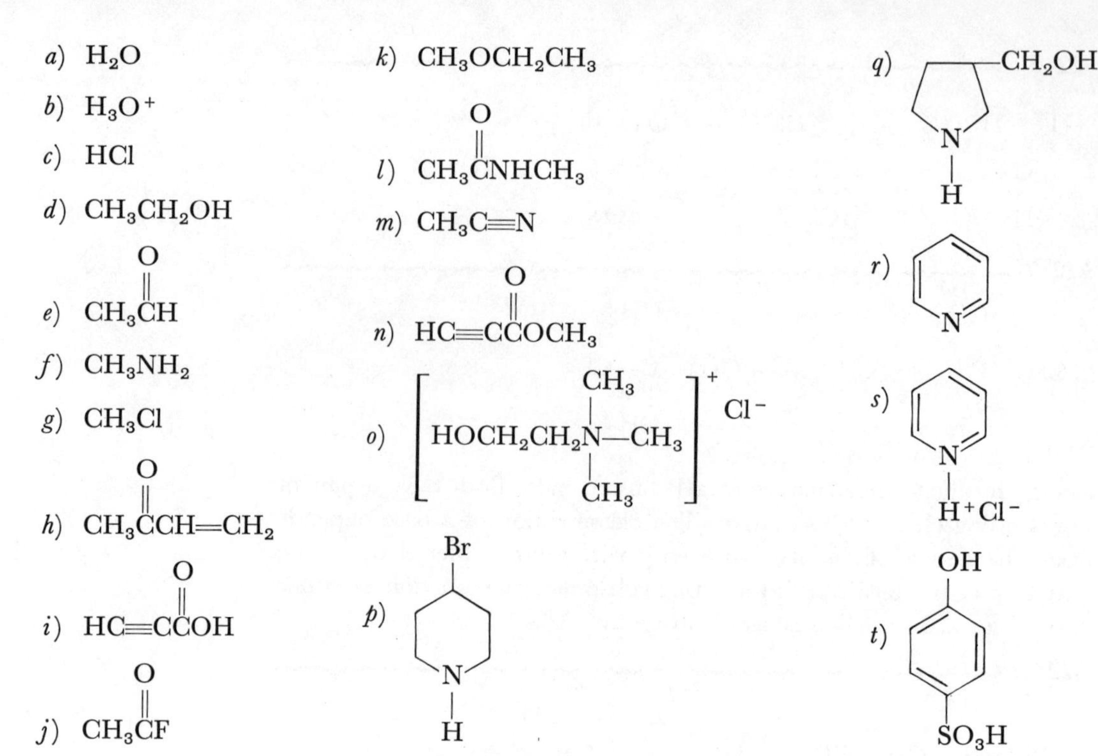


**2**) A substance is said to be *protonated* when one of its atoms donates a pair of nonbonding electrons to the formation of a coordinate covalent bond with a proton. There are two groups of compounds among those listed in problem 1 that are readily protonated. These are the compounds in which nitrogen has a pair of nonbonding electrons and those in which oxygen has two pairs of nonbonding electrons. The nonbonding pair of electrons remaining on oxygen after protonation does not readily react with another proton. The nonbonding electron pairs on a halogen atom (except for fluorine) are also not readily shared with a proton.

With the above information in mind, draw the formula of the protonated product for each substance listed in problem 1 that can undergo protonation. Calculate the formal charge on the protonated atom in each instance.

**3**) Classify each of the following substances as having the potential to function as a *1* Brönsted base, *2* Brönsted acid, *3* Lewis base, *4* Lewis acid. Some of the substances may fall into more than one category. Do not consider as potentially acidic those hydrogens bonded directly to carbon or to nitrogen, but do include those bonded to oxygen.

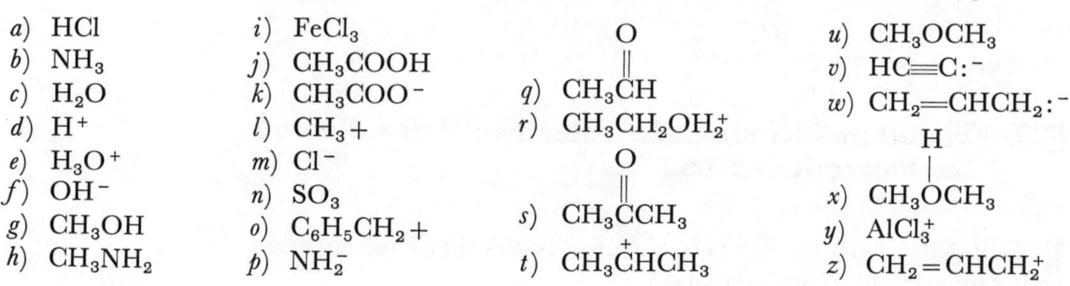

**4**) Write an equation for the reaction between HCl and each of the substances you listed in your answer to problem 2 as a potential Brönsted base.
**5**) Write an equation for the reaction between a hydroxide ion ($OH^-$) and each of the substances you listed in your answer to problem 2 as a potential Brönsted acid.
**6**) Write an equation for a possible reaction between water and each of the substances in the following parts of problem 3: *a, c, d, j, l, n, o, r, t, x,* and *z*. Indicate whether water is functioning as a Lewis or a Brönsted base in each reaction. Some of these will be equilibrium reactions (§28) but for the moment do not concern yourself with this aspect of the reaction.

## §24. Inductive Effect. Partial Charge

The diagram in #24–1 was designed to indicate roughly charge distribution in a molecule of propane. When a σ bond is formed, the charge from two electrons is concentrated mainly between the two bonded atoms. This effect is indicated in #24–1.

**#24–1**

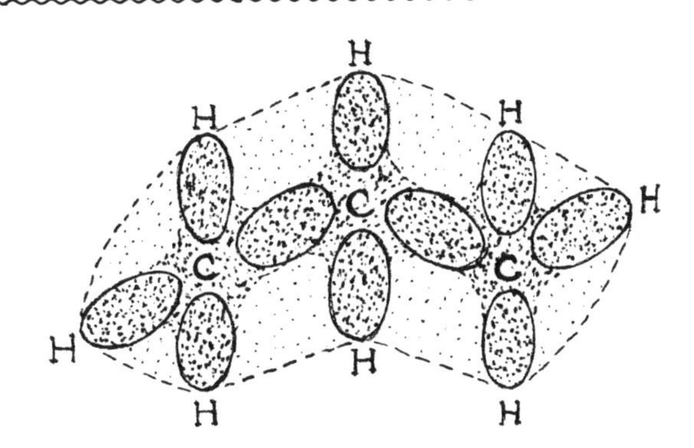


In propane all the bonds are between identical atoms (carbon–carbon) or between atoms of equivalent *electronegativity** (carbon–hydrogen). Under these circumstances the available charge is distributed evenly between the two bonded atoms. However, when one of the bonded atoms is more electronegative than the other, the charge from the bonding electrons is not distributed evenly. The more electronegative atom attracts more than half the available charge. For example, in HCl the high electronegativity of chlorine as compared to hydrogen causes more of the charge to concentrate at the chlorine atom.

* Electronegativity is a measure of the attraction of an atom of an element for an electron. A highly electronegative element has a strong affinity for an electron. Pauling electronegativity values are listed in Table 24–1 for the elements most commonly encountered in organic chemistry.

**Table 24–1**

**Electronegativity Values According to Pauling**

| | | | | | |
|---|---|---|---|---|---|
| *Electronegativity increases across a period.* | | | | | |
| *It decreases* | H 2.1 | | | | |
| *down* | | C 2.5 | N 3.0 | O 3.5 | F 4.0 |
| *a group* | | | P 2.1 | S 2.5 | Cl 3.0 |
| | | | | | Br 2.8 |

The extra share of negative charge acquired by the chlorine atom means that this atom has more than enough negative charge to offset the positive charge on its nucleus. The chlorine atom has not acquired a complete unit of extra charge and so does not have a formal negative charge as defined in §18. The chlorine atom has only acquired part of a charge and is described as having a *partial negative charge* designated by δ−.

As the result of electron withdrawal from the hydrogen atom by the chlorine atom, the hydrogen atom has been left without its full quota of negative charge and so has a *partial positive charge* designated by δ+. Such unequal charge distribution may be indicated as part of a formula as shown in #24–2.

**#24–2**

(δ+)(δ−)
H:C̈l:

When a highly electronegative atom such as chlorine is bonded to a carbon chain, the effect of electron withdrawal by the chlorine atom may extend along the chain through as many as three or four carbons. The carbon adjacent to the electronegative atom has the largest partial charge. The degree of electron deficiency decreases with each succeeding carbon. An attempt has been made in #24–3 to diagram this effect for *n*-propyl chloride ($CH_3CH_2CH_2Cl$). The orbitals of the nonbonding electrons on chlorine have been drawn larger than the bonding orbitals on *C3* to emphasize the increased charge concentration on the chlorine atom and also to indicate the greater freedom of motion associated with nonbonding as compared to bonding electron pairs.

**#24–3**

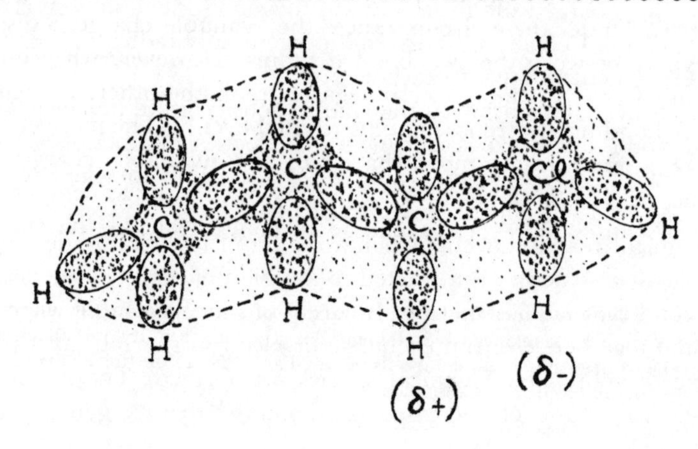

**Table 24–2**

**Inductive Effect**

*Classification of substituents, arranged in approximate order of decreasing strength*

| +I | | −I | |
|---|---|---|---|
| $-COO^-$ | $-NR_3^+$ | $-F$ | $-OH$ |
| $-CR_3$ | $-NH_3^+$ | $-Cl$ | $-C{\equiv}CR$ |
| $-CHR_2$ | $-NO_2$ | $-Br$ | Ar— |
| $-CHR$ | $-SO_2R$ | $-I$ | $-CH{=}CR_2$ |
| $-CH_3$ | $-CN$ | $-OR$ | |
| $-D$ | $-COOH$ | $-SH$ | |

Whenever there is a covalent bond between two unlike atoms, the more electronegative atom has a partial negative charge and the less electronegative atom has a partial positive charge. This statement applies to multiple as well as single covalent bonds. Thus, although there is an overall increase in the charge density in the region of a carbon–oxygen double bond, the carbon atom has a partial positive charge as indicated in #24–4 and the oxygen atom has a partial negative charge. The carbons α (i.e., adjacent) to the carbonyl group also have a partial positive charge but a lesser one than does the carbonyl carbon itself.

**#24–4**

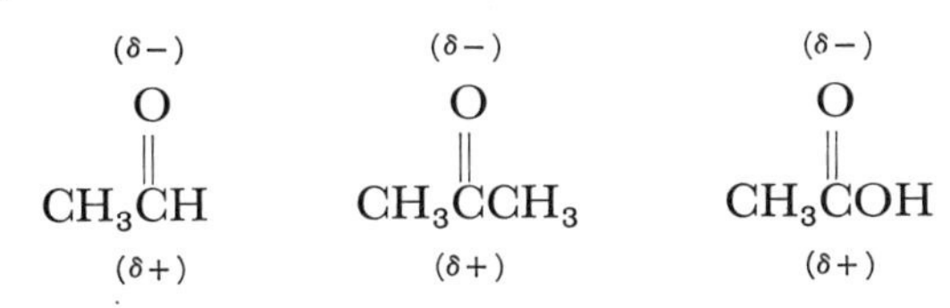


*Induction* is the term applied to the phenomenon described above. Electron withdrawal (or negative-charge withdrawal) that is due to differences in the electronegativity of bonded atoms occurs through the σ bond framework and is said to take place by induction.

A substituent that creates an unevenness in the charge density by withdrawing charge from nearby atoms through induction is described as exerting a **−I** effect. This designation generally indicates the effect of the substituent relative to carbon.

A substituent that increases the charge density at an adjacent atom by induction is described as exerting a **+I** effect. A number of common substituents are classified in Table 24–2 according to whether they exert a **+I** or a **−I** effect relative to carbon.

A chemical reaction occurs when a region of low charge density in one molecular structure is attracted to a region of high charge density in a second molecular structure. If conditions are favorable when the two come together, electron exchange or electron sharing takes place and one or more new chemical species are formed. The regions of increased (δ−) and of decreased (δ+) charge density generated by inductive effects provide centers of potential chemical reactivity. Problem 1 in §25 has been included to accustom you to thinking of molecular structures in terms of the presence or absence of such potentially reactive sites.

The concept that *chemical reactivity is related to the structure of the electron cloud* is fundamental to all of organic chemistry. Examples 1 and 2 illustrate how this concept provides criteria that you can apply whenever you examine two molecular structures with a view to determining whether or not they can be expected to react with each other under favorable conditions.

*Example 1*—Would you expect a reaction to occur between methyl amine ($CH_3\ddot{N}H_2$) and hydrogen chloride (HCl) under favorable conditions of temperature and concentration?

*Step 1.* Analyze the charge distribution pattern of one of the possible reactants to determine whether there is a potentially reactive site in the structure.

Methylamine:

Nitrogen is somewhat more electronegative than carbon (Table 24–1) and therefore exerts a −**I** effect creating a slight partial negative charge in the region of the nitrogen and a slight partial positive charge in the region of the carbon as represented in formula #24–5.

$$\underset{(\delta+)\ (\delta-)}{CH_3\ddot{N}H_2}$$

**#24–5**

The amino nitrogen also has a pair of nonbonding electrons. The presence of this nonbonding electron pair is the most important feature of the structure because these electrons create a region of available charge at the nitrogen—it is therefore a potentially reactive site.

*Step 2.* Similarly analyze the charge distribution pattern on the other possible reactant.

Hydrogen chloride:

The chlorine atom is considerably more electronegative than the hydrogen atom (Table 24–1). Thus, there is a significant difference in the charge density at these two atoms. The hydrogen carries a $\delta+$ charge and can be considered as electron deficient. The chlorine carries a $\delta-$ charge. Either atom is a potentially reactive site, as is indicated in #24–6.

$$\overset{(\delta+)(\delta-)}{H:\ddot{\underset{..}{Cl}}:}$$

**#24–6**

*Step 3.* Compare the results obtained in Steps 1 and 2 to determine whether there is a region of decreased charge density on one reactant that can interact with a region of increased charge density on the other.

The nonbonding electron pair plus the $\delta-$ on the amino nitrogen indicates that this substance can function as a Brönsted or a Lewis base and as a nucleophile.

If the methylamine is to react there has to be a region of electron deficiency in the other reactant. Hydrogen chloride does have an electron-deficient hydrogen and can function as a Brönsted acid.

On the basis of these observations, one can conclude that methyl amine will react with hydrogen chloride under favorable reaction conditions. The reaction would involve a proton exchange between a Brönsted acid and a Brönsted base.

The equation for the reaction is given in #24–7. This equation tells

**#24–7**

$$\mathrm{H{-}\overset{\displaystyle H}{\underset{\displaystyle H}{C}}{:}\underset{\displaystyle H}{\ddot{N}}{:}H + H{:}\ddot{Cl}{:} \rightleftharpoons CH_3{:}\overset{\displaystyle H}{\underset{\displaystyle H}{\ddot{N}}}{:}H + {:}\ddot{Cl}{:}^-}$$

$(\delta+\ \delta-)$

you that when the electron-deficient hydrogen of HCl approaches the electron-rich region near the amino nitrogen, the nonbonding electron pair from the nitrogen forms a coordinate covalent bond with the hydrogen. As this new bond is being formed, the original bond between the hydrogen and chlorine is broken. The hydrogen leaves behind its bonding electrons and a chloride ion with a formal negative charge is formed. Read the equation carefully and identify the symbols used to convey the above information.

*Example 2*—Would you expect a reaction to occur between methyl amine ($CH_3\ddot{N}H_2$) and methanol $CH_3\ddot{O}H$) under favorable conditions?

*Step 1.* Analyze the charge distribution pattern of the two possible reactants.

Methylamine:

As indicated in Example 1, there is a region of available charge on the amino nitrogen.

Methanol:

Oxygen has a **−I** effect so the oxygen will have a partial negative charge and the carbon will have a partial positive charge as indicated in #24–8.

**#24–8**

$$CH_3\ddot{O}H$$

$(\delta+)\ (\delta-)$

The oxygen also has nonbonding electron pairs. Thus, the pattern of electron distribution is essentially the same in the two molecules.

*Step 2.* Compare the two structures to determine whether there is a region of electron deficiency in one that could react with an electron-rich region in the other.

The main feature of each molecule is the electron-rich region with available electrons at the hetero atom.* Two electron-rich regions do not react together. Each of these substances can function as a Brönsted base, a Lewis base, a nucleophile. Neither structure has a region of sufficient electron deficiency to react with the electron-rich site on the

* A *hetero atom* is any atom that is neither carbon nor hydrogen.

other structure. Methanol and methyl amine would not be expected to react with each other.

You are not yet prepared to write complete equations for a variety of reactions. However, you should be able to detect molecular structures that have the potential of reacting together. Problem 2 in §25 provides you with practice in identifying potentially reactive pairs of compounds.

* * * * *

The significance of induction can be better understood if we examine the role played by induction in determining the relative acidity of the series of carboxylic acids in #24–9. These acids are arranged with the weakest acid first and the strongest acid last. A weak Brönsted acid is one that does not readily give up a proton to a base.

**#24–9**

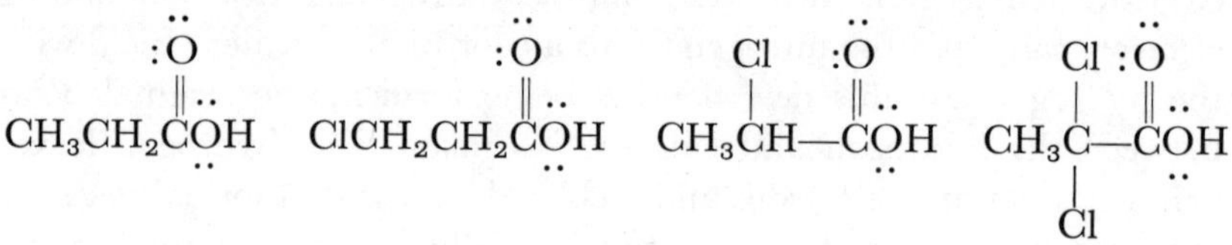

$CH_3CH_2C(=\ddot{O}:)\ddot{O}H \quad ClCH_2CH_2C(=\ddot{O}:)\ddot{O}H \quad CH_3CH(Cl){-}C(=\ddot{O}:)\ddot{O}H \quad CH_3C(Cl)_2{-}C(=\ddot{O}:)\ddot{O}H$

The relative acidity of a carboxylic acid is related to the amount of energy needed to form the carboxylate ion from the acid. It takes less energy to form a more stable ion than to form a less stable ion.

Consider the relative stability of the two carboxylate ions in #24–10. Chlorine, being highly electronegative, exerts a −**I** effect and reduces the charge density on the $-COO^-$ group. This charge delocalization increases the stability of the 2-chloropropanoate ion (#24–10b) over that of the unsubstituted propanoate ion (#24–10a). Thus, it is easier to form the 2-chloropropanoate ion and 2-chloropropanoic acid is a stronger acid than propanoic acid.

**#24–10**

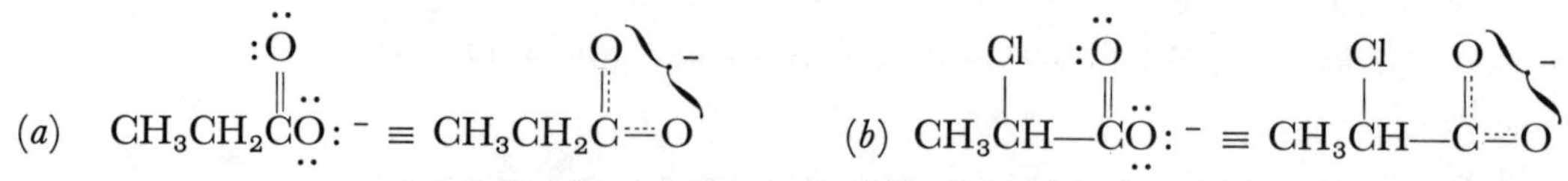

(*a*) $CH_3CH_2C(=\ddot{O}:)\ddot{O}:^- \equiv CH_3CH_2C(\cdots O)(\cdots O)^-$ (*b*) $CH_3CH(Cl){-}C(=\ddot{O}:)\ddot{O}:^- \equiv CH_3CH(Cl){-}C(\cdots O)(\cdots O)^-$

The effectiveness of induction diminishes as the distance between the —$COO^-$ group and the chlorine atom is increased. Charge delocalization occurs to a much lesser degree at the —$COO^-$ group in the 3-chloropropanoate anion than at the —$COO^-$ group in the 2-chloropropanoate anion. The latter anion is, therefore, more stable than the former and 2-chloropropanoic acid is a stronger Brönsted acid than 3-chloropropanoic acid.

As might be expected, two halogens result in a greater charge delocalization than one halogen. The strongest acid of the series in #24–9 is 2,2-dichloropropanoic acid.

* * * * *

Relative basicity is also determined by inductive effects. Consider the two reactions in #24–11 and –12. The difference in the lengths of the arrows in #24–11 indicate that the reaction proceeds quite far towards completion before equilibrium is established (§28). By contrast, the arrows in #24–12 indicate that equilibrium is established while there is still a fairly high concentration of reactants in the reaction mixture. This difference in reactivity indicates that ethyl amine is a stronger base than the fluoro derivative, an observation that can be explained in terms of induction.

**#24–11**

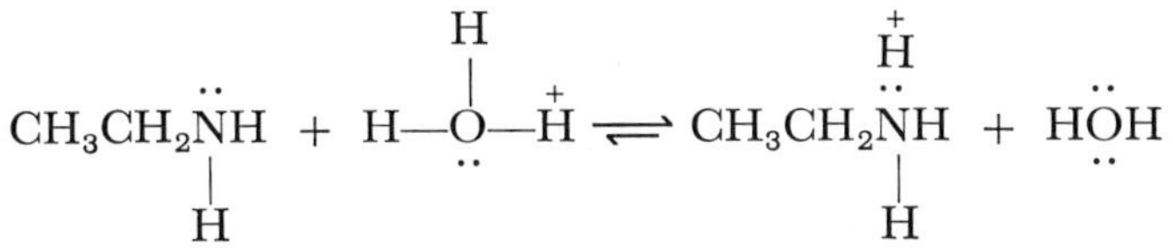

$$CH_3CH_2\ddot{N}H_2 + H_3\ddot{O}^+ \rightleftharpoons CH_3CH_2\ddot{N}H_3^+ + H\ddot{O}H$$

**#24–12**

$$FCH_2CH_2\ddot{N}H_2 + H_3\ddot{O}^+ \rightleftharpoons FCH_2CH_2\ddot{N}H_3^+ + H\ddot{O}H$$

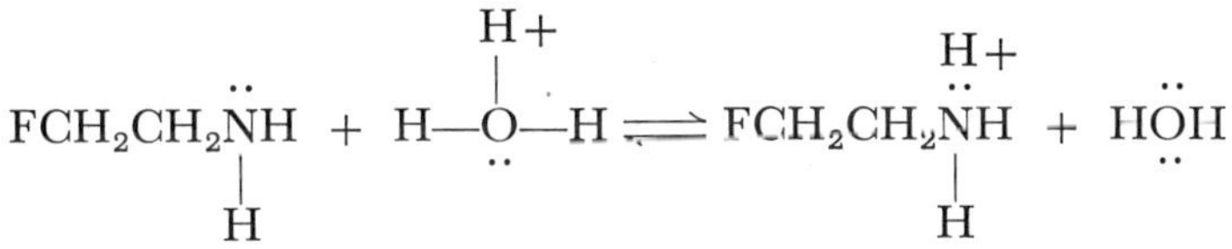

The electron withdrawing action of the halogen reduces basicity of the fluoro derivative in two ways: the $-\mathbf{I}$ effect reduces the charge density on the nitrogen in the amine and decreases the availability of the nonbonding electron pair; the $-\mathbf{I}$ effect intensifies the + charge on the nitrogen in the cation, reducing the stability of the ion which, therefore, is more difficult to make than the ethyl ammonium ion. Thus, a substituent with a $-\mathbf{I}$ effect strengthens a Brönsted acid and weakens a Brönsted base.

The following Examples illustrate how to apply the concepts discussed in this section.

*Example 3*—Given the two pairs of substances in #24–13, which member of each pair would you expect to be the more stable and why?

**#24–13**

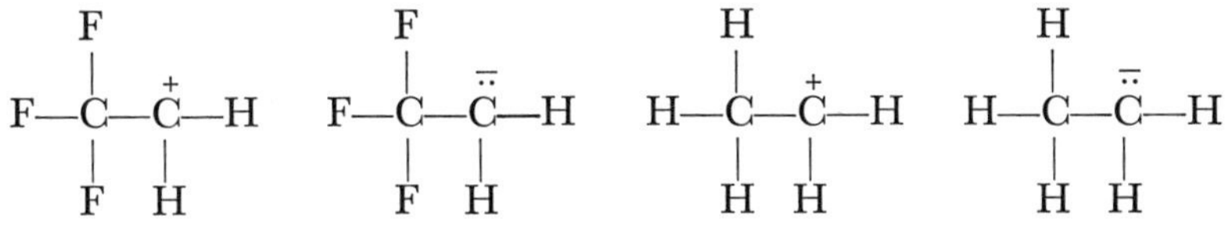

$$F_3C\text{—}\overset{+}{C}H_2 \qquad F_3C\text{—}\overset{\cdot\cdot}{C}H_2^- \qquad \textit{(a)}$$

$$H_3C\text{—}\overset{+}{C}H_2 \qquad H_3C\text{—}\overset{\cdot\cdot}{C}H_2^- \qquad \textit{(b)}$$

*Step 1.* Summarize the effects that would stabilize and destabilize each ion.

- The two carbonium ions will be stabilized by electron release and destabilized by electron withdrawal.
- The two carbanions will be stabilized by electron withdrawal and destabilized by electron release.

*Step 2.* Summarize the characteristics of the various substituents in each ion.

- Alkyl groups tend to be electron-releasing both by induction and hyperconjugation.
- Fluorine is a highly electronegative atom which withdraws electrons by induction.

*Step 3.* Combine both sets of data into a conclusion.

Of the two ions in #24–13a, the carbanion will be stabilized by the inductive electron withdrawal of the fluorine, whereas the carbonium ion will be rendered less stable.

Of the two ions in #24–13b, the electron release of the methyl group will help stabilize the carbonium ion, whereas it will decrease the stability of the carbanion.

*Example 4*—Arrange the three acids in #24–14 in order of increasing acidity.

**#24–14**

$FCH_2CH_2COOH$ $BrCH_2CH_2COOH$ $CH_3CHFCOOH$

*Step 1.* Consider the effect of each substituent in terms of its relative inductive effect.

Both F and Br exert a $-\mathbf{I}$ effect (Table 24–2). F is more electronegative than Br so that its effect is greater.

A $-\mathbf{I}$ effect stabilizes the acid anion and leads to increased acidity in a Brönsted acid. 3-fluoropropanoic acid is a stronger acid than 3-bromopropanoic acid.

*Step 2.* Consider the relative positions of the substituents.

The electron-withdrawing substituent is closer to the $—COO^-$ group in 2-fluoropropanoic acid than it is in 3-fluoropropanoic acid. The charge delocalization on the carboxylate group is greater in the 2-fluoro than in the 3-fluoro derivative because of the closer proximity of the $-\mathbf{I}$ substituent to the $-COO^-$ group.

Greater charge delocalization leads to increased anion stability, which means that the anion is easier to make and, therefore, less energy is needed to remove the proton from the —COOH group. The stronger acid is the one which gives up a proton more readily. Thus, 2-fluoropropanoic acid is a stronger acid than 3-fluoropropanoic acid.

*Step 3.* Arrange the three acids in series with the weakest acid first and the strongest acid last.

The acidity of the three substances increases:

$BrCH_2CH_2COOH$ $FCH_2CH_2COOH$ $CH_3CHFCOOH$

## §25. Problems. Inductive Effect. Partial Charge

**1**) Indicate the partial charge by $\delta-$ or $\delta+$ at each of the hetero atoms in the following compounds and at each carbon adjacent to a hetero atom.

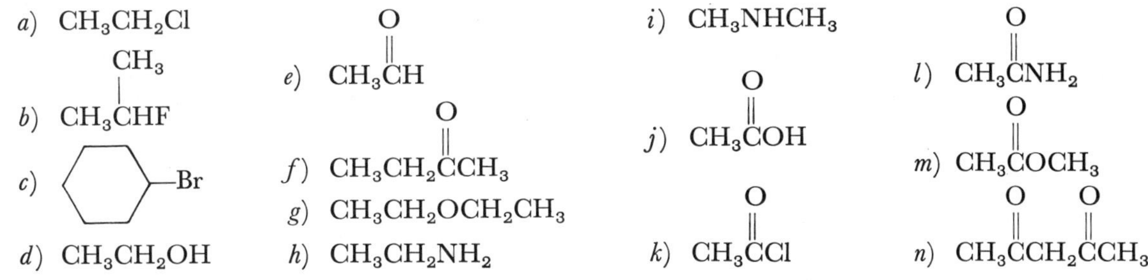


**2**) Some of the pairs of substances listed below react together under favorable conditions, whereas other pairs do not do so. You are to write an equation for each reaction that can occur.

At this stage of your study you have the following criterion to use in deciding whether or not a reaction can occur: A reaction takes place when an electron-rich site on one substance is attracted to an electron-deficient site on another substance and the structures are such that the electron-rich species can donate electrons to or share electrons with the electron-deficient species.

There are two parts to this criterion: the attraction of oppositely charged sites and the availability of electrons for bonding. To determine whether both requirements are met by a pair of proposed reactants, (*a*) indicate the regions of partial charge on each molecule by $\delta-$ and $\delta+$ and (*b*) add all nonbonding electron pairs to the formulas.

A pair of reactants that do satisfy both requirements can be expected to react together under appropriate conditions of temperature, pressure, and concentration. Write the equations for these reactions without concern as to how far the reaction proceeds before equilibrium is established (§28).

*a*) $CH_3OH + HBr$
*b*) $CH_3OH + H_2$
*c*) $CH_3NH_2 + NaOH$
*d*) $CH_3NH_2 + HCl$
*e*) $CH_3CH_2CH_3 + HF$
*f*) $CH_3SH + HF$
*g*) $CH_3OCH_3 + HCl$
*h*) $CH_3OCH_3 + H_2$
*i*) $(CH_3)_2CHCH_3 + HF$
*j*) $CH_3CHO + HI$
*k*) $CH_3CH_2Cl + HBr$
*l*) $CH_3COOH + NaOH$

**3**) Write the formula of the anion formed when each of the following carboxylic acids functions as a Brönsted acid. Arrange the anions in each series so that the least stable anion is placed first and the most stable is last. Then arrange the original acids in order of increasing acidity for each series, i.e., place the substance first that has the least tendency to give up a proton and function as a Brönsted acid. Explain the order you choose.

*a*) *1.* $BrCH_2CH_2CH_2COOH$ *2.* $CH_3CHBrCH_2COOH$
*3.* $CH_3CH_2CHBrCOOH$

*b*) *1.* $CH_3COOH$ *2.* $F_3CCOOH$
*3.* $FCH_2COOH$

*c*) *1.* $BrCH_2COOH$ *2.* $ICH_2COOH$
*3.* $ClCH_2COOH$

*d*) *1.* $HSCH_2CH_2COOH$ *2.* $CH_3CH_2COOH$
*3.* $HOCH_2CH_2COOH$

*e*) *1.* $CH_3COOH$ *2.* $CH_3CH_2COOH$
*3.* $(CH_3)_2CHCOOH$

**4**) Write the formula for the cation formed when each of the following amines functions as a Brönsted base and accepts a proton. Arrange the cations in each series so that the least stable cation is first and the most stable cation is last. Then arrange the amines in each series in order of increasing basicity, i.e., place the substance first that is least likely to donate a pair of electrons to a proton and to function as a Brönsted base. Explain why the basicity increases in the order you indicated for each series.

*a*) *1.* $BrCH_2CH_2CH_2NH_2$  *2.* $CH_3CHBrCH_2NH_2$  *3.* $CH_3CH_2CHBrNH_2$

*b*) *1.* $CH_3NH_2$  *2.* $CH_3CH_2CH_2NH_2$  *3.* $CH_3CH_2NH_2$

*c*) *1.* $ClCH_2CH_2NH_2$  *2.* $CH_3CH_2NH_2$  *3.* $Cl_2CHCH_2NH_2$

*d*) *1.* (pyrrolidine ring, N–H, with Br)  *2.* (pyrrolidine ring, N–H, with I)  *3.* (pyrrolidine ring, N–H, with Cl)

*e*) *1.* $CH_3NH_2$  *2.* $(CH_3)_2CHNH_2$  *3.* $ClCH_2CH_2NH_2$

**5**) An alcohol can function as an acid by the reaction shown in #25–1. Write the formula for the anion formed when each of the following alcohols functions as a Brönsted acid. Arrange the anions in

**#25–1**

$$CH_3CH_2\ddot{\underset{..}{O}}H + :\ddot{N}H_2^- \rightleftharpoons CH_3CH_2\ddot{\underset{..}{O}}:^- + \ddot{N}H_3$$

each series with the least stable in position one and the most stable in position three. Then arrange the alcohols in each series in order of increasing acidity. Explain the order of increasing acidity in each of the series you have set up.

*a*) *1.* $CH_3CH_2CH_2OH$  *2.* $CH_3CHClCH_2OH$  *3.* $ClCH_2CH_2CH_2OH$

*b*) *1.* $(CH_3)_3COH$  *2.* $(CH_3)_2CHCH_2OH$  *3.* $CH_3CH_2CH(OH)CH_3$

*c*) *1.* $F_2CHCH_2CH_2OH$  *2.* $FCH_2CH_2CH_2OH$  *3.* $F_3CCH_2CH_2OH$

*d*) *1.* 2-fluorocyclohexanol (F, ring, –OH)  *2.* 3-fluorocyclohexanol (F, ring, –OH)  *3.* F–(cyclohexane ring)–OH

*e*) *1.* $NO_2CH_2CH_2CH_2OH$  *2.* $CH_3CH(NO_2)CH_2OH$  *3.* $CH_3CH_2CH_2OH$

**6**) An alcohol can function as a Brönsted base when it reacts according to equation #25–2. Write the formula for the cation formed when each of the alcohols in problem 5 reacts as a base by the reaction in #25–2. Arrange the cations in each series with the least stable in position one and the most stable in position three. Then arrange the alcohols in each series in order of increasing basicity. Explain the basicity order in each of the series.

**#25–2**

$$CH_3CH_2\ddot{\underset{..}{O}}H + H:\ddot{\underset{..}{Cl}}: \rightleftharpoons CH_3CH_2\overset{H+}{\underset{..}{O}}H + :\ddot{\underset{..}{Cl}}:^-$$

**7**) Arrange the members of each of the following series in order of increasing stability and explain the order.

*a*) *1.* $ClCH_2CH_2CH_2+$ *2.* $CH_3CH_2CH_2+$ *3.* $CH_3\overset{+}{C}HCH_3$

*b*) *1.* $ClCH_2CH_2CH_2:^-$ *2.* $ClCH_2CH_2CH_2+$ *3.* $CH_3CHClCH_2:^-$

*c*) *1.* $CNCH_2\overset{-}{\ddot{C}}HCH_3$ *2.* $CNCH_2\overset{+}{C}HCH_3$ *3.* $CNCH_2\overset{-}{\ddot{C}}HCH_2CN$

*d*) *1.* $C_6H_5$—$\overset{+}{C}HCH_2CH_3$ *2.* $C_6H_5$—$\overset{+}{C}HCH_2CH_2F$ *3.* $C_6H_5$—$\overset{+}{C}HCHFCH_3$

*e*) *1.* $BrCH_2\overset{+}{C}(CH_3)_2$ *2.* $(CH_3)_3C+$ *3.* $(BrCH_2)_2\overset{+}{C}CH_3$

**8**) Arrange the members of the following series in order of increasing acidity and explain the order.

*a*) *1.* $CNCH_2COOH$ *2.* $CH_3CH_2COOH$ *3.* $CH_3COOH$

*b*) *1.* $CNCH_2CH_2COOH$ *2.* $CH_3CH(CN)COOH$ *3.* $(CH_3)_2CHCOOH$

*c*) *1.* $CH_3COOH$ *2.* $CH_3OCH_2COOH$ *3.* $HCOOH$

*d*) *1.* (ortho-F)$C_6H_4$—COOH *2.* (meta-F)$C_6H_4$—COOH *3.* F—$C_6H_4$—COOH

*e*) *1.* $NO_2$—$C_6H_4$—COOH *2.* $C_6H_5$—COOH *3.* (ortho-$NO_2$)$C_6H_4$—COOH

*f*) *1.* $(CH_3)_2CH_2OH$ *2.* $CH_3OH$ *3.* $CH_3CH_2OH$

*g*) *1.* $BrCH_2CH_2OH$ *2.* $FCH_2CH_2OH$ *3.* $ICH_2CH_2OH$

*h*) *1.* cyclopentyl—OH *2.* cyclopentyl—$CH_2OH$ *3.* 1-methylcyclopentyl—OH ($CH_3$)

**9**) Arrange the members of the following series in order of increasing basicity and explain the order.

*a*) *1.* $BrCH_2CH_2NH_2$ *2.* $FCH_2CH_2NH_2$ *3.* $ICH_2CH_2NH_2$

*b*) *1.* $BrCH_2CH_2OH$ *2.* $FCH_2CH_2OH$ *3.* $ICH_2CH_2OH$

*c*) *1.* $CNCH_2CH_2NH_2$ *2.* $CH_3CH_2CH_2NH_2$ *3.* $CH_3CH(CN)CH_2CH_2NH_2$

*d*) *1.* 3-bromopiperidine (Br; N—H) *2.* 4-bromopiperidine (Br; N—H) *3.* 2-bromopiperidine (Br; N—H)

*e*) *1.* $FCH_2CH_2NHCH_2CH_3$ *2.* $(FCH_2CH_2)_2NH$ *3.* $(CH_3CH_2)_2NH$

*f*) *1.* $CH_3OCH_2NH_2$ *2.* $CH_3CHINH_2$ *3.* $CH_3SCH_2NH_2$

*g*) *1.* $(CH_3)_2CH_2OH$ *2.* $CH_3OH$ *3.* $CH_3CH_2OH$

*h*) *1.* $CH_3CH_2OH$ *2.* $CH_3CH_2Cl$ *3.* $CH_3CH_2NH_2$

## §26. Resonance. Molecular Orbital Diagrams. Canonical Forms

Equation #26–1 represents a proton exchange reaction whereby acetic acid is converted into an acetate ion. The electron-dot formula for the anion in #26–1*a* shows a formal negative charge on *O2* and a double bond between *C1* and *O1*. However, in terms of completed electron octets it would be equally logical to place the formal charge on O*1* and the double bond between C*1* and O*2*, as shown in #26–1*b*.

**#26–1**

$$CH_3C(=\ddot{O}:)\ddot{O}H + :\ddot{O}H^- \longrightarrow CH_3C1(=\dot{\ddot{O}}\,1)(-\dot{\ddot{O}}\,2^-) \longleftrightarrow CH_3C1(-\dot{\ddot{O}}\,1^-)(=\dot{\ddot{O}}\,2) + H\ddot{O}H$$

(*a*) (*b*)

(*Italicized numbers are included to identify the particular carbon and oxygen atoms for the discussion*)

Experimental data reveal that both of the carbon–oxygen bonds have double bond character. They are equal in length, being somewhat shorter than a single bond and somewhat longer than a traditional double bond. Further experimentation reveals that, instead of there being one neutral oxygen and one oxygen with a formal charge of $-1$, each oxygen carries a partial negative charge, $\delta -$. How can we reconcile the formula and these observations? Does either formula really represent the structure properly? If not, how do we show the structure with symbols?

Let us try to visualize the electron orbitals in this anion by a detailed analysis of each individual carbon–oxygen bond and a consideration of how the two fit together in a complete unit. While this analysis has been based on structure #26–1*a*, the same statements can be made about structure #26–1*b* by merely interchanging O*2* and O*1*.

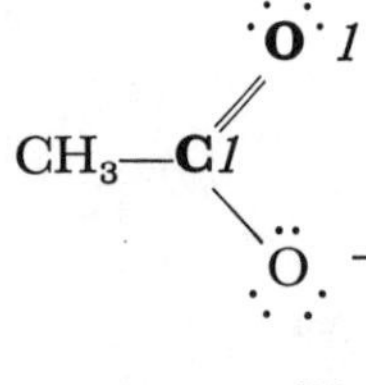


**#26–2**

The double bond between C*1* and O*1* involves four electrons, two from the carbon atom and two from the oxygen atom. The $\sigma$ bond is formed by the end-to-end overlap of two $sp^2$ orbitals and the $\pi$ bond is formed by the sideways overlap of two parallel $p$ orbitals, one from each atom. These relationships are diagrammed in #26–2.

The remaining four electrons on O*1* consist of two nonbonding electron pairs. The orbitals of these electrons are in the position of $sp^2$ hybridization relative to the carbon–oxygen double bond as indicated in #26–3. The axes of the two orbitals of the nonbonding electrons form angles of about 120° with the carbon–oxygen $\sigma$ bond and are in a plane perpendicular to the plane of the two original $p$ oribitals.

#26–3

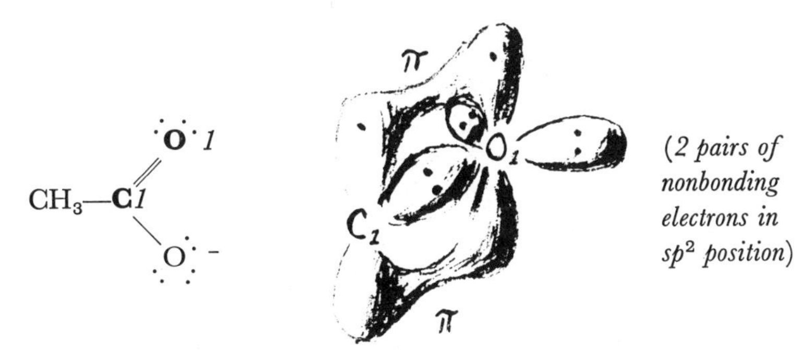


The two remaining electrons on C*1* are involved in $\sigma$ bonds to O*2* and to C*2*. These bonds are also in the position of $sp^2$ hybridization as indicated in #26–4.

#26–4

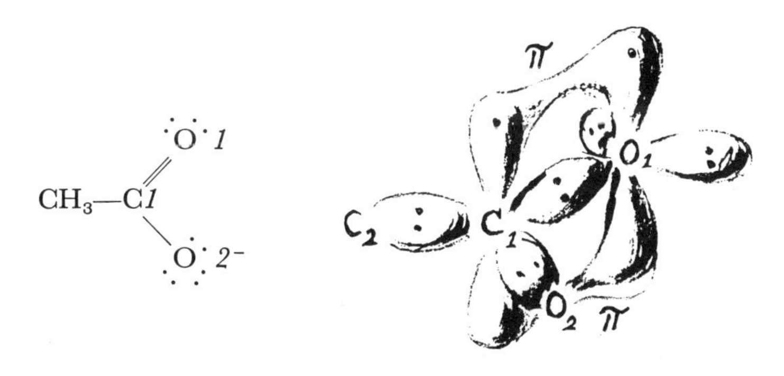


Consider next the bond between C*1* and O*2* as it is represented in #26–1*a*. The electron left behind by the proton was arbitrarily assigned to O*2*, placing a formal charge of −1 on this atom. One of these seven electrons is involved in the $\sigma$ bond with C*1*. The other six make up three pairs of nonbonding electrons which, in the absence of any outside influence, would be expected to occupy orbitals with tetrahedral geometry as indicated in #26–5. However, in this molecule there is a structural feature which modifies the geometry of these orbitals.

#26–5

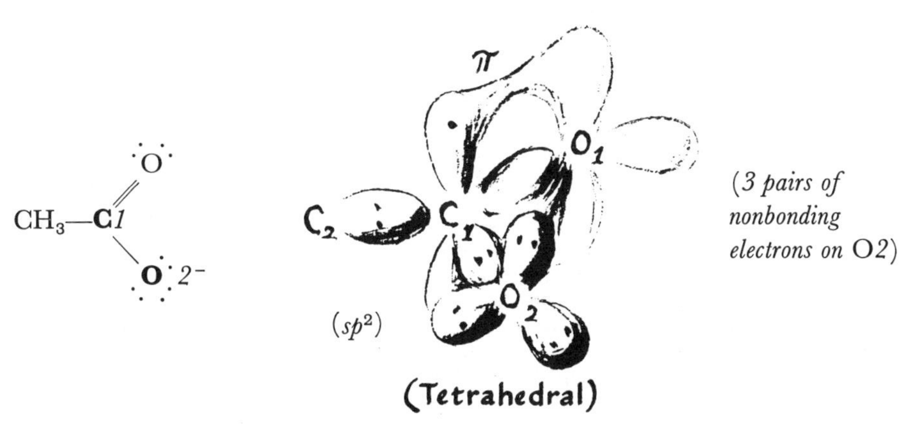

This structural feature is the *p* orbital on C*1*.* It is so positioned that overlap occurs between this *p* orbital and the orbital of a nonbonding electron pair on O*2*. Whenever there is orbital overlap, charge redistribution occurs which may lead to the formation of a new molecular orbital. The effect here can be likened to the formation of the $\pi$ cloud of a double bond. As a consequence of overlap, imagine an orbital rearrangement which aligns one of the nonbonding orbitals on O*2* with the *p* orbital on C*1* and places the other two nonbonding orbitals in the position of $sp^2$ hybridization. Then, because of sideways overlap, a $\pi$ cloud forms above and below the $\sigma$ bond between C*1* and O*2*. This $\pi$ cloud involves three rather than two electrons and contributes double bond character to the C*1*—O*2* bond. These relationships are diagrammed in #26–6.

**#26–6**

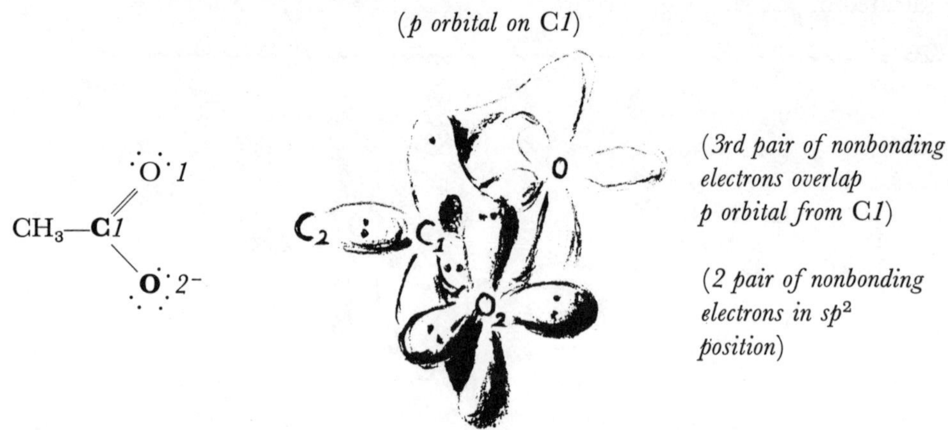


Thus far we have considered the electron configuration at bond C*1*—O*1* and at bond C*1*—O*2* as though they were independent of each other. This is a fallacy because the atom C*1* is common to both bonds and the *p* oribital on C*1* is involved in the $\pi$ cloud at each bond.

**#26–7**

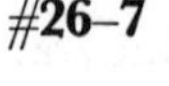

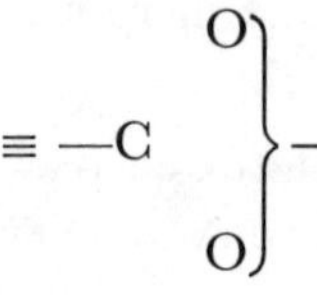


The structure of the complete —COO⁻ group is diagrammed in #26–7. There are three parallel orbitals, one from each of the atoms C*1*, O*1*, and O*2*. These orbitals are perpendicular to the plane through the three atoms. Each adjacent pair of orbitals overlaps side-

* The term *p* orbital actually refers to the electron configuration prior to bond formation. The electron becomes a pi electron in the $\pi$ orbital of the double bond which forms as soon as the atoms are bonded together. Reference to a *p* electron indicates the geometry of the electron orbital in terms of the configuration of a single carbon atom and is used to explain the origin of the pi electrons and their relative position rather than to indicate an actual orbital configuration in the molecule.

ways, forming a continuous row of three overlapping orbitals. Sideways overlap leads to the formation of a $\pi$ orbital over the three atoms and charge redistribution through a $\pi$ cloud above and below the plane of the three atoms.

There are four electrons in this $\pi$ orbital—two from O*2* and one each from O*1* and C*1*. Thus, the charge due to the "extra" electron has been distributed over all three atoms. This charge dispersal is not exactly even because the electronegativity of the two oxygen atoms causes the charge density to be somewhat greater at the two oxygens than at the carbon.

Charge redistribution through a $\pi$ cloud as described above for a carboxylate anion occurs whenever the molecular structure is such that the orbitals of pi and/or nonbonding electrons can overlap. This overlap is sideways rather than the end-to-end overlap of $\sigma$ bonding and may vary in degree in different structures. In order for such overlap to occur the pi and/or nonbonding electrons must be on adjacent atoms.

*Resonance* is the term used to describe the process of charge redistribution through a $\pi$ orbital formed by sideways orbital overlap, just as *induction* is the term used to describe charge redistribution through $\sigma$ orbitals because of differences in the electronegativity of the bonded atoms.

Resonance is a troublesome concept for some, but it really need not be. Just keep in mind the molecular orbital diagram and remember that "something" happens when there is an overlapping of pi and/or nonbonding electron orbitals from adjacent atoms. This "something" involves charge equalization and is described by the name of resonance. This "something" means that the electrons occupy different molecular orbitals than they would have occupied without the overlap.

Perhaps an analogy can serve to clarify the concept of resonance. Imagine a shallow box composed of two equal compartments with a removable partition between them. If you empty one glass of water into compartment A and two glasses of water into compartment B, the water in B will be twice as deep as the water in A. If you now remove the partition, the water will spread out to the same depth in both compartments. The water in compartment A will be deeper and the water in compartment B will be shallower than before the partition was removed because each compartment now contains the equivalent of one and a half glasses of water.

In your imagination equate one electron with one glass of water and picture an electron cloud with the fluidity of water. Then the tendency of water to seek its own level can be considered analogous to the redistribution of charge which leads to equalization of charge density in the electron cloud. The removal of the partition in the box, thus allowing for free flow of the water, corresponds to the orbital overlap that occurs when pi and/or nonbonding electrons are on adjacent atoms. It is this overlap that allows the electrons to intermingle and charge redistribution to occur.

Although resonance tends to equalize charge density, the presence of an atom more electronegative than carbon leads to a region of increased charge density in the $\pi$ cloud, just as a depression in the box results in deeper water in that area.

Reconsider what was said about charge delocalization in the —COO⁻ group in terms of the above analogy to see that you now understand better.

* * * * *

Whenever resonance plays a role in determining the charge distribution within a molecule, that particular molecular structure cannot be represented by just one electron-dot formula. It will be possible to draw two, three, or even more formulas in which all the atoms are in the same relative positions, but in which one or more pairs of electrons are in different positions. For example, the two formulas for the acetate ion shown in #26–8 differ only in the position of one pair of electrons.

**#26–8**

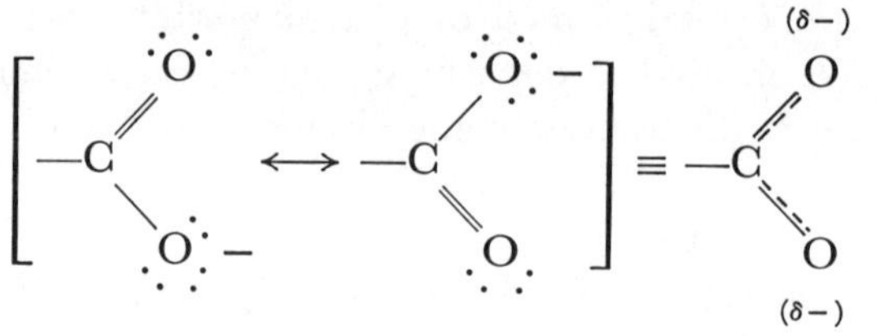

(δ−)

(δ−)

(*a*) (*b*) (*c*)

Not only will it be possible to draw these several different formulas; all of them will be needed to describe accurately the structural characteristics of the molecule or ion. The molecular orbital diagram (#26–7) of the —COO⁻ group indicates that a $\pi$ cloud extends over all three atoms so that both of the C—O bonds have some double bond character and each of the oxygens carries a partial negative charge. Neither #26–8*a* or #26–8*b* conveys this information. However, if you consider both formulas or if you combine them as in #26–8*c* you can obtain a clearer picture of the actual structure.

The structures represented by the various electron-dot formulas are referred to as the *canonical forms* of the molecule. Their relationship to each other is indicated by a double-headed arrow and by enclosing the set of formulas in square brackets. A canonical form does not exist as an isolatable entity. It serves merely to indicate one aspect, one set of structural characteristics that is found in the molecule. For example, structure #26–8*a* tells us that there is a negative charge on the oxygen atom numbered *1.* Structure #26–8*b* tells us that there is a negative charge on the oxygen numbered *2.* The two formulas together tell us that in the actual molecule there is a partial negative charge on each of the oxygens.

If two canonical forms are of equal stability they will contribute equally to the properties of the molecule. The canonical forms in #26–8*a* and *b* are *equivalent*, i.e., indistinguishable unless the atoms are numbered. Therefore, they are of equal stability. The degree of partial negative charge on each of the oxygens is the same. If two canonical forms are not equally stable, the less stable one will make a lesser contribution to the characteristics of the molecule. For example, the electron-dot formula for acetic acid (#26–9), which places a positive formal charge on the highly electronegative oxygen, represents a relatively unstable canonical form that makes only a slight contribution to the overall properties of the acetic acid molecule.

**#26–9**

$$\left[ CH_3\overset{\overset{:\ddot{O}}{\|}}{C}\!-\!\underset{\cdot\cdot}{\ddot{O}}H \longleftrightarrow CH_3\underset{+}{\overset{\overset{:\ddot{O}:^-}{|}}{C}}\!-\!\underset{\cdot\cdot}{\ddot{O}}H \longleftrightarrow CH_3\overset{\overset{:\ddot{O}:^-}{|}}{C}\!=\!\underset{+}{\ddot{O}}H \right]$$

$$\left[ CH_3\overset{\overset{:\ddot{O}}{\|}}{C}\!-\!\underset{\cdot\cdot}{\ddot{O}}:^- \longleftrightarrow CH_3\overset{\overset{:\ddot{O}:^-}{|}}{C}\!=\!\ddot{O}: \longleftrightarrow CH_3\overset{\overset{:\ddot{O}:^-}{|}}{C}\!-\!\underset{\cdot\cdot}{\ddot{O}}:^- \right]$$

When it is possible to draw two or more equivalent canonical forms for a substance, resonance plays a significant role in determining its characteristics. A structure in which charge delocalization occurs through resonance is more stable than one in which this effect is missing. The relative stability of two substances can be determined by comparing the number of equivalent canonical forms that can be drawn for each. For example, as shown in #26–9, two of the canonical forms for the acetate ion are equivalent, whereas none of the forms for acetic acid are equivalent. Thus, we say that the acetate ion has a higher level of resonance energy than acetic acid; or we say that the acetate ion has a higher degree of resonance stabilization than acetic acid.

The extent to which the charge on an anion is delocalized by resonance helps determine the acidity of the substance from which the anion is made. For example, while the alcoholate anion, $RCH_2\underset{\cdot\cdot}{\ddot{O}}:^-$ contains nonbonding electrons on the oxygen there are no pi or nonbonding electrons on the adjacent carbon to provide the opportunity for orbital overlap and charge delocalization through resonance. Thus, there is no real difference in stability between the alcohol and the alcoholate ion. It is much easier to convert a $R\overset{\overset{:\ddot{O}}{\|}}{C}\underset{\cdot\cdot}{\ddot{O}}H$ into $R\overset{\overset{:\ddot{O}}{\|}}{C}\underset{\cdot\cdot}{\ddot{O}}:^-$ than it is to convert $R\underset{\cdot\cdot}{\ddot{O}}H$ into $R\underset{\cdot\cdot}{\ddot{O}}:^-$, because $R\overset{\overset{:\ddot{O}}{\|}}{C}\underset{\cdot\cdot}{\ddot{O}}:^-$ has a higher degree of resonance stabilization than $R\overset{\overset{:\ddot{O}}{\|}}{C}\underset{\cdot\cdot}{\ddot{O}}H$, whereas $R\underset{\cdot\cdot}{\ddot{O}}H$ and $R\underset{\cdot\cdot}{\ddot{O}}:^-$ are of about equal stability. Thus, $\mathrm{R\overset{\overset{:\ddot{O}}{\|}}{C}\underset{..}{\ddot{O}}H}$ is a stronger acid than $\mathrm{R\underset{..}{\ddot{O}}H}$.

Although resonance can best be understood through visualization of the $\pi$ cloud as in a molecular orbital diagram, electron-dot formulas do provide a mechanical device for evaluating the relative degree of resonance stabilization in related structures. Therefore, you need to know how to draw canonical forms and how to evaluate their relative stability.

The starting point for determining the canonical forms of any substance is an electron-dot formula drawn according to the rules in §18. This formula must allocate a complete octet of electrons to as many atoms as possible.

The object is to draw as many different structures as possible merely by relocating electrons within the original structure. There is a set of rules which must be obeyed during the relocation process. These include:

- Only pi and nonbonding electrons may be relocated.
- A $\sigma$ bond may not be broken.
- An electron may not be moved away from the atom to which it is assigned in the original formula.
- An electron may merely be shifted into a different position on the atom to which it is originally assigned.
- The electrons in a pi pair may be moved individually.
- Several successive shifts may be made to achieve the dual objective of leaving as many atoms as possible with a complete octet and of creating a minimum of new formal charges.

The relocation of electrons on a particular atom does not change the number of electrons shared by this atom but does either create or destroy an octet on an adjacent atom. This kind of shift also brings about a change in the formal charge of the original atom and of the adjacent atom. This process can best be understood if we work through some problems. Keep in mind as you work that you are not trying to show the actual movement of electrons within the molecular structure. You are merely trying to determine, without making a molecular orbital diagram, whether or not orbital overlap can occur. The rules that have been set up enable you, through the symbolism of the formulas, to determine whether a particular structure does involve orbital overlap and is, therefore, resonance stabilized.

Orbital overlap (sideways) and varying degrees of charge redistribution through resonance occur whenever Y in the sequence of atoms $A = B - Y$ has an electronic configuration that includes one of the following:

- a pair of nonbonding electrons, $A = B - Y\!:$
- a pi electron, $A = B - Y = X$
- a single nonbonding electron, $A = B - Y\cdot$
- an incomplete octet. $A = B - Y+$

This statement can be confirmed by showing that equivalent canonical forms can be drawn for substances with each of these general formulas.

Examine first the structure A∸B—Y:. The general formula includes only the $\sigma$ bonds between the three atoms and the significant arrangement of pi and nonbonding electrons that indicates resonance. The dots representing electrons can be positioned in four different locations around the symbol of an element, to the right, to the left, above, and below. When drawing canonical forms from a given electron-dot formula you may:

- shift a pair of pi electrons from its original position into any of the three other positions around the symbol for either of the two bonded atoms;
- shift a pair of nonbonding electrons from its original position into any of the other three positions around the symbol of the atom to which it is assigned.

Such positioning of the pi electrons in A∸B—Y: is illustrated in #26–10*a–f*. Figures *a*, *b*, and *c* relocate the electron pair in the various positions on atom A, whereas *d*, *e*, and *f* relocate the pair on atom B. Application of the operations in #26–10*a–f* to an actual substance, the acetate ion, is given in #26–10*g–l*. Note that when an electron pair is moved from between two atoms onto one of the atoms, a formal charge is generated on each of the atoms. This kind of change in charge is referred to as *charge separation*. Electrons are usually shifted in pairs, although if the structure involves a free radical, a pair may be split and each member shifted to a different position (#26–15).

Each theoretically possible electron shift may not actually be possible for a specific structure and, even if possible, may not lead to a stable configuration. Examination of the structures formed by the operations in #26–10*g–l* immediately eliminates some of the proposed shifts for the acetate ion. Atom A has only nonbonding electron pairs so there is no significance to steps *h* and *i*. The symbol ::O has no physical meaning. There is no orbital arrangement that leads to the doubling up of nonbonding electron pairs. Step l also has to be eliminated because it produces a carbon with ten electrons ($H_3C$═). Thus, *g*, *j*, and *k* are left as possible shifts for the pi electrons.

There are also presumably three possible shifts for the nonbonding electron pair on Y. However, in this structure Y has only nonbonding electron pairs so there is only one possible shift, the shift which moves a pair of electrons from a nonbonding position on Y to a double bond between B and Y. This shift is shown in #26–11 for structures *g*, *j*, and *k* from #26–10. The corresponding general formulas are included with each acetate ion.

Of the six forms, #26–10*g*, *j*, and *k*, #26–11*g*, *j*, and *k*, only #26–11*g* gives every possible atom a complete octet and has only one atom with a formal charge. Thus, this form represents a more stable configuration than the others. This form is also equivalent to the original form given in #26–10. As discussed above, the fact that equivalent canonical forms can be drawn means that there is orbital overlap in the actual

#26–10

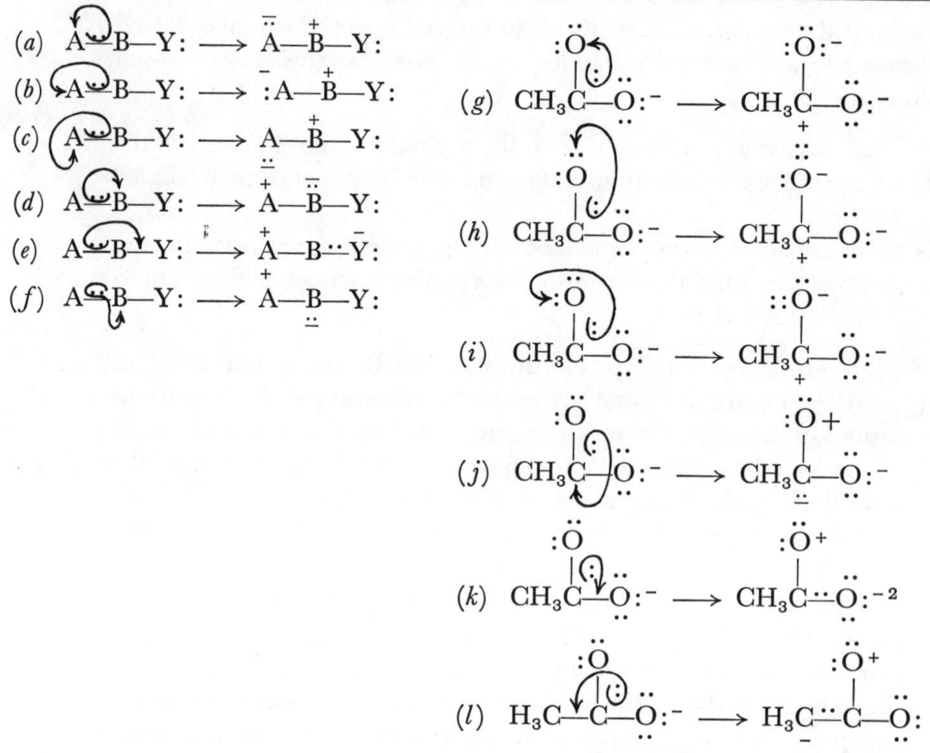

A∶B—Y:　　　　　$CH_3C(=O)$—O:⁻ (acetate ion, :Ö above C)

(a) A∶B—Y: ⟶ A—B—Y: (A with −, B with +)

(b) A∶B—Y: ⟶ :A—B—Y: (A with −, B with +)

(c) A∶B—Y: ⟶ A—B—Y: (B with +, electron pair below)

(d) A∶B—Y: ⟶ A—B—Y: (A with +, B with −)

(e) A∶B—Y: ⟶ A—B∶Y: (A with +, Y with −)

(f) A∶B—Y: ⟶ A—B—Y: (B with +, electron pair below)

(g) $CH_3C$—O:⁻ (with :O above) ⟶ $CH_3C$—O:⁻ (with :O:⁻ above, C with +)

(h) $CH_3C$—O:⁻ ⟶ $CH_3C$—O:⁻ (with :O⁻ above, C with +)

(i) $CH_3C$—O:⁻ ⟶ $CH_3C$—O:⁻ (with ::O⁻ above, C with +)

(j) $CH_3C$—O:⁻ ⟶ $CH_3C$—O:⁻ (with :O+ above, electron pair below C)

(k) $CH_3C$—O:⁻ ⟶ $CH_3C$∶O:⁻² (with :O+ above)

(l) $H_3C$—C—O:⁻ ⟶ $H_3C$∶C—O: (with :O+ above, C with −)

#26–11

(a) A—B—Y(:) ⟶ A—B∶Y⁺ (A with −, B with +)

(g) $CH_3C$—O:⁻ (with :O:⁻ above, C with +) ⟶ $CH_3C$∶O: (with :O:⁻ above)

(d) A—B—Y(:) ⟶ A—B∶Y⁺ (A with +, B with −2)

(j) $CH_3C$—O: (with :O+ above) ⟶ $CH_3C$∶O: (with :O+ above, C with −2)

(e) A—B∶Y(:) ⟶ A—B∷Y (A with +, B with −)

(k) $CH_3C$∶O: (with :O+ above, −2) ⟶ $CH_3C$∷O:⁻ (with :O+ above, C with −)

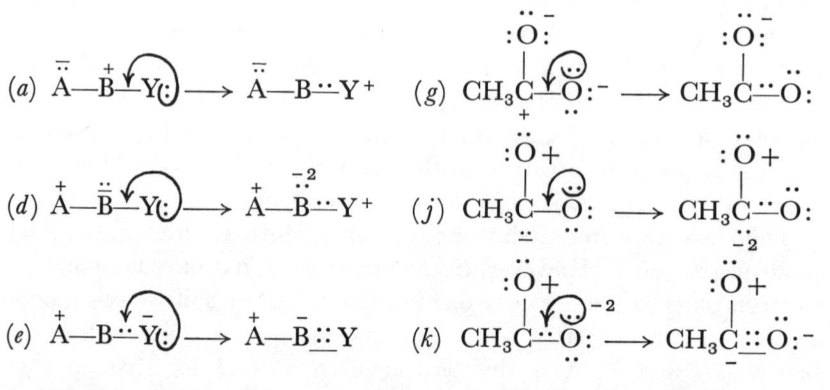

structure and resonance determines the charge distribution.

The sequence of electron shifts shown in #26–10*g* and #26–11*g* answers the question of whether or not equivalent canonical forms can be drawn for the acetate ion. The other sequences are not needed. Similarly, there is one essential sequence of electron shifts that provides the answer to this equation for each of the other three general formulas. The essential sequence is given below for each of the four general formulas. The presence of the structural relationship summarized by any of the general formulas provides the first indication of resonance in a substance. The production of equivalent canonical forms as the result of applying the essential sequence of electron shifts to the formula of the substance indicates a high level of resonance stabilization. The general formulas and how each is treated follow:

A∸B—Y:

Shift the pair of pi electrons from between A and B into a non-bonding position on A. Eliminate the incomplete octet that this move generates on B by shifting the nonbonding electron pair on Y into a double bond between Y and B. These operations are shown in #26–12 for both the general formula and the acetate ion. This diagram is a rewrite of the essential sequence from #26–10 and –11.

A∸B—Y∸X

**#26–12**

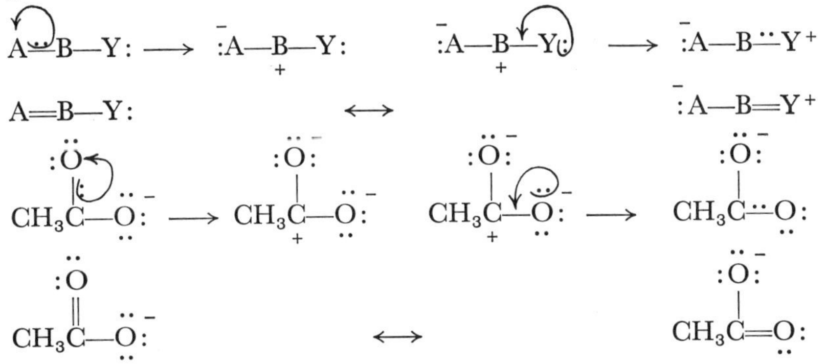

A∸B—Y: ⟶ $\bar{:}$A—$\underset{+}{\text{B}}$—Y:    $\bar{:}$A—$\underset{+}{\text{B}}$—Y:  ⟶ $\bar{:}$A—B∸$Y^+$

A=B—Y: ⟷ $\bar{:}$A—B=$Y^+$

$CH_3C(=\ddot{O}:)—\ddot{O}:^-$ ⟶ $CH_3\underset{+}{C}(—\ddot{O}:^-)—\ddot{O}:^-$    $CH_3\underset{+}{C}(—\ddot{O}:^-)—\ddot{O}:^-$ ⟶ $CH_3C(—\ddot{O}:^-)∸\ddot{O}:$

$CH_3C(=\ddot{O}:)—\ddot{O}:^-$ ⟷ $CH_3C(—\ddot{O}:^-)=\ddot{O}:$

The sequence of shifts shown for the general formula in #26–13 is essentially the same as that for A∸B—Y:. Move the pair of pi electrons from between A and B into a nonbonding position on A. Eliminate the sextet generated on B by shifting the pair of pi electrons from its position between Y and X into a double bond between Y and B.

**#26–13**

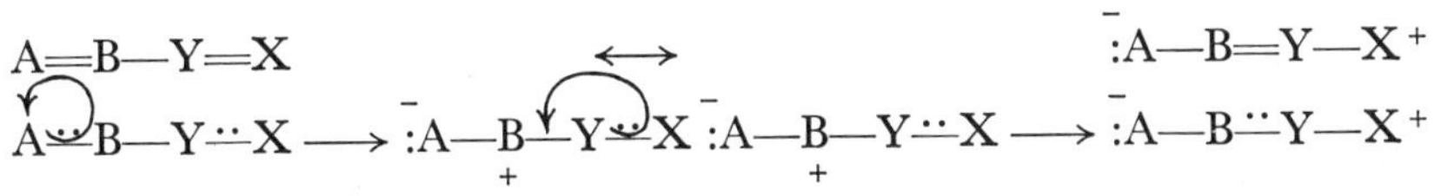

A=B—Y=X ⟷ $\bar{:}$A—B=Y—$X^+$

A∸B—Y∸X ⟶ $\bar{:}$A—$\underset{+}{\text{B}}$—Y∸X $\bar{:}$A—$\underset{+}{\text{B}}$—Y∸X ⟶ $\bar{:}$A—B∸Y—$X^+$

With two double bonds in the molecule, the question arises as to which pair of the doubly bonded atoms should be designated as A and B. Two sequences have to be carried out. Each set of double bonds has to be designated as A—B. This point is illustrated by the two sequences in #26–14, which shows how to draw the main canonical forms for dimethyl glyoxal. The two equivalent forms #26–14*b* and *c* indicate orbital overlap with charge distribution over four atoms.

Shift one of the pi electrons on A into a nonbonding position as shown in #26–15. Shift the second pi electron into position between B and Y so that it can form a double bond with the single nonbonding electron from Y.

#26-14

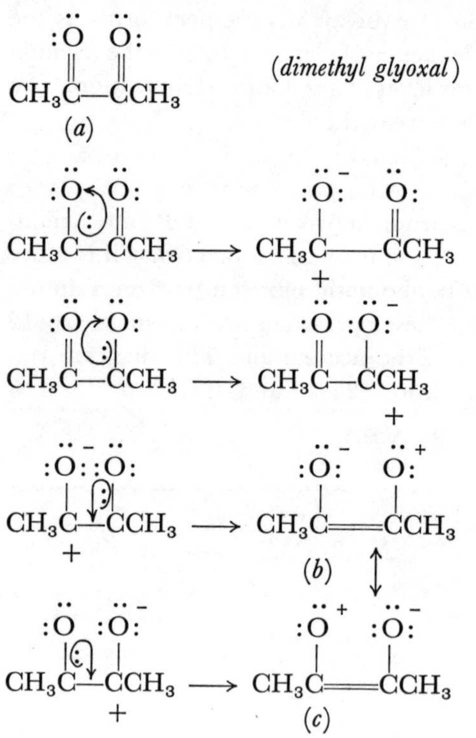

$$CH_3\overset{\overset{\ddot{O}}{\|}}{C}-\overset{\overset{\ddot{O}}{\|}}{C}CH_3 \quad (a)$$

(*dimethyl glyoxal*)

$$CH_3\overset{\overset{:\ddot{O}:}{\|}}{C}-\overset{\overset{\ddot{O}:}{\|}}{C}CH_3 \longrightarrow CH_3\overset{\overset{:\ddot{O}:^-}{|}}{\underset{+}{C}}-\overset{\overset{\ddot{O}:}{\|}}{C}CH_3$$

$$CH_3\overset{\overset{:\ddot{O}}{\|}}{C}-\overset{\overset{\ddot{O}:}{\|}}{C}CH_3 \longrightarrow CH_3\overset{\overset{:\ddot{O}}{\|}}{C}-\overset{\overset{:\ddot{O}:^-}{|}}{\underset{+}{C}}CH_3$$

$$CH_3\overset{\overset{:\ddot{O}:^-}{|}}{\underset{+}{C}}-\overset{\overset{:\ddot{O}:}{\|}}{C}CH_3 \longrightarrow \underset{(b)}{CH_3\overset{\overset{:\ddot{O}:^-}{|}}{C}=\overset{\overset{\ddot{O}:^+}{|}}{C}CH_3}$$

$$\updownarrow$$

$$CH_3\overset{\overset{:\ddot{O}}{\|}}{C}-\overset{\overset{:\ddot{O}:^-}{|}}{\underset{+}{C}}CH_3 \longrightarrow \underset{(c)}{CH_3\overset{\overset{:\ddot{O}^+}{|}}{C}=\overset{\overset{:\ddot{O}:^-}{|}}{C}CH_3}$$

$$A = B - Y\cdot$$

$$A{=}B{-}Y\cdot \longleftrightarrow \cdot A{-}B{=}Y$$

$$A{=}B{-}Y\cdot \longrightarrow \cdot A{-}B{\because}Y$$

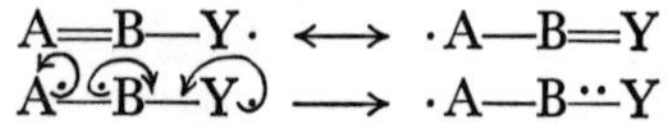

#26-15

An allyl free radical provides us with an example of this structural relationships as shown in #26-16.

#26-16

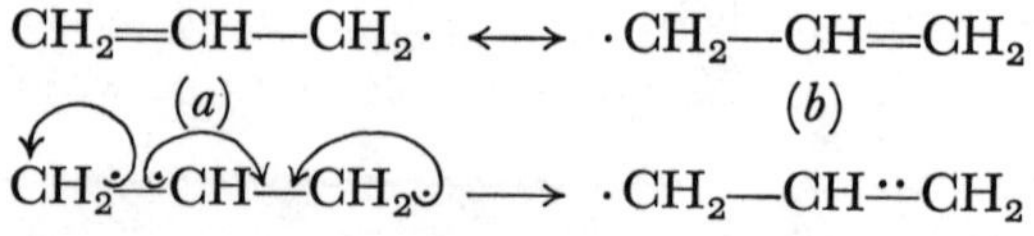

$$\underset{(a)}{CH_2{=}CH{-}CH_2\cdot} \longleftrightarrow \underset{(b)}{\cdot CH_2{-}CH{=}CH_2}$$

$$CH_2{=}CH{-}CH_2\cdot \longrightarrow \cdot CH_2{-}CH{\because}CH_2$$

The equivalent canonical forms are indicative of the relative ease with which the allyl free radical is formed. The more stable a free radical is, the easier it is to make.

$$A = B - Y+$$

Shift the pair of pi electrons on B so that they help form a double bond between B and Y, as shown in #26-17. This operation effectively shifts the electron deficiency from Y to A just as the operations in #26-15 shifted the nonbonding electron from Y to A.

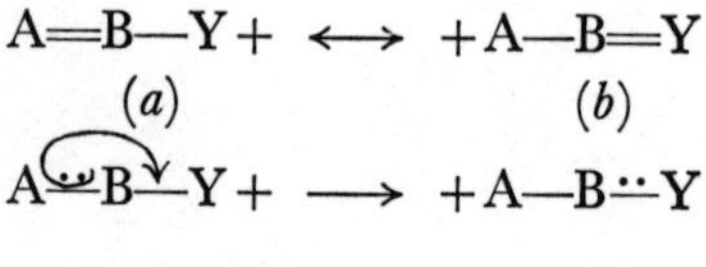

$$\underset{(a)}{A{=}B{-}Y+} \longleftrightarrow \underset{(b)}{+A{-}B{=}Y}$$

$$A{=}B{-}Y+ \longrightarrow +A{-}B{\because}Y$$

#26-17

An allylic carbonium ion is also relatively easy to form because of the high resonance stabilization indicated by the equivalent canonical forms shown in #26-18*a* and *b*.

#26–18

$$CH_2{=}CH{-}CH_2{+} \longleftrightarrow {+}CH_2{-}CH{=}CH_2$$
(a) (b)

$$CH_2{=}CH{-}CH_2{+} \longrightarrow {+}CH_2{-}CH{\because}CH_2$$

Example 1 gives you the opportunity to apply these principles under guidance and then you may fend for yourself in the problems of §27.

*Example 1*—When acetylacetone reacts with sodium ethoxide according to equation #26–19 a resonance stabilized anion is formed. Draw the canonical forms needed to show resonance in this ion.

#26–19

$$CH_3\overset{O}{\overset{\|}{C}}CH_2\overset{O}{\overset{\|}{C}}CH_3 + {}^{-}{:}\ddot{O}CH_2CH_3 \longrightarrow CH_3\overset{O}{\overset{\|}{C}}\underset{\cdot\cdot}{C}H\overset{O}{\overset{\|}{C}}CH_3 + H\ddot{O}CH_2CH_3$$

*Step 1.* Indicate all double bonds and all nonbonding electrons by dots.

Resonance occurs only through the nonbonding and/or pi electrons, so use dots, as in #26–20, to represent these electrons and leave the uninvolved $\sigma$ bonds as lines. This practice makes it easier to identify the electrons that might be involved in resonance.

#26–20

$$CH_3{-}\overset{:\ddot{O}}{\overset{::}{C}}{-}\underset{\cdot\cdot}{\overset{H}{\overset{|}{C}}}{-}\overset{\ddot{O}:}{\overset{::}{C}}{-}CH_3 \quad (-1)$$

*Step 2.* Determine whether or not it is possible to relocate a pair of pi electrons within the molecular structure as drawn.

There are two carbon–oxygen double bonds in the molecule. A carbonyl group has both pi and non-bonding electrons and can become involved in resonance if the structure on the adjacent atom permits. In order to draw a structure which differs only in the placement of the dots representing electrons, let us first try to shift the position of one pair of pi electrons. In the electron-dot formula there is one side of each carbonyl oxygen that is "empty," and one side of each carbonyl carbon that is also "empty." "Empty" means that there are no dots. Let us redraw the formula in two ways: *a*) by shifting the pi electrons of one carbonyl group into the "empty" space on the oxygen as in #26–21; *b*) by shifting this pair of pi electrons into the "empty" space on carbon as in #26–22.

#26–21

O1 O2 (−1)

$$CH_3{-}\overset{:\ddot{O}}{\overset{::}{C}}{-}\underset{\cdot\cdot}{\overset{H}{\overset{|}{C}}}{-}\overset{:\ddot{O}}{\overset{::}{C}}{-}CH_3 \longrightarrow CH_3{-}\overset{:\ddot{O}:}{\overset{\cdot\cdot}{C}}{-}\underset{\cdot\cdot}{\overset{H}{\overset{|}{C}}}{-}\overset{\ddot{O}:}{\overset{::}{C}}{-}CH_3$$

(−1) (+1) (−1)

C2 C3 C4

#26–22

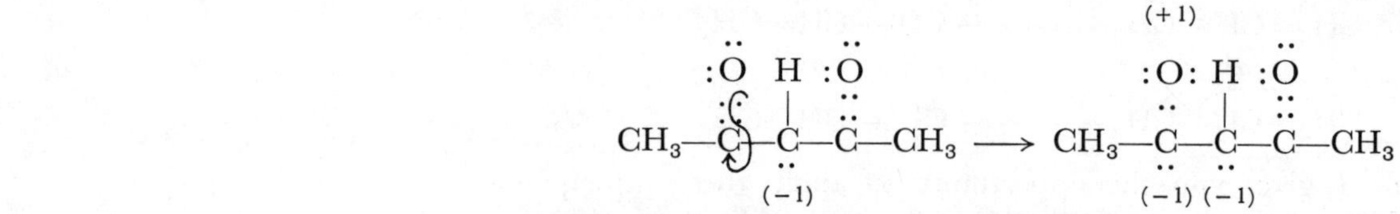

```
                                      (+1)
   :Ö  H :Ö                        :O: H :Ö
    :   |  :                        ..  |  :
CH3—C—C—C—CH3   ⟶   CH3—C—C—C—CH3
       ..                            ..  ..
      (−1)                         (−1) (−1)
```

The shift in #26–21 leaves the carbonyl carbon with an incomplete octet and generates a −1 formal charge on O*1* and a +1 formal charge on C*2*. The reverse is true of the shift in #26–22; the positive charge is on O*1* and the negative charge is on C*2*; the oxygen has the incomplete octet.

*Step 3.* Determine if there is any way to eliminate the incomplete octet on the carbon in #26–21 by a further shifting of electrons.

Whenever we are drawing canonical forms every effort must be directed to producing structures in which as many atoms as possible have a complete octet of electrons. To accomplish this we may change the position on a given atom of any pair of nonbonding or pi electrons, just as we shifted the pi electrons in Step 1.

#26–23

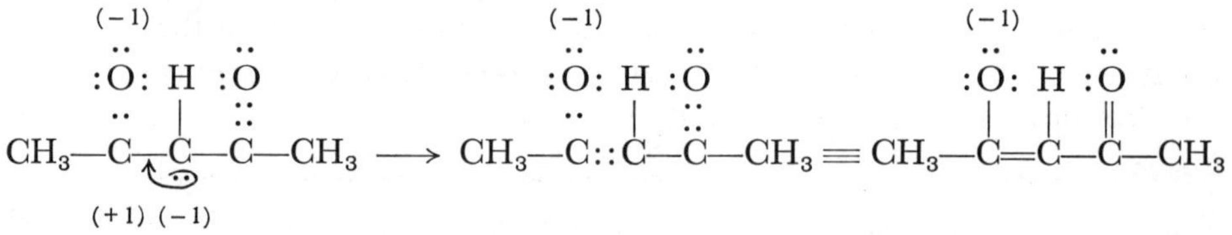

```
   (−1)                       (−1)                        (−1)
   :Ö: H :Ö                   :Ö: H :Ö                    :Ö: H :Ö
    ..  |  :                   ..  |  :                     |   |  ‖
CH3—C—C—C—CH3  ⟶  CH3—C::C—C—CH3  ≡  CH3—C=C—C—CH3
   (+1) (−1)
```

The original structure indicates a pair of nonbonding electrons on C*3*. If, as shown in #26–23, we shift these electrons from the nonbonding position into a double bond between C*2* and C*3*, we have not changed the number of electrons around C*3* but we have completed the octet for C*2*. Now both carbons are shown as having a share in eight electrons. This shift also eliminates the formal charges on the two carbons. The entire sequence of shifts results in a relocation of the −1 formal charge from its original position on C*3* to a new position on O*1*.

Remember that we are not intending to indicate that the actual electrons in the molecule move around in this manner. We are merely saying that, if we can *draw* reasonable structures with the electrons in different positions, then there can be *overlap* of the *orbitals* in the actual molecules—an overlap which may lead to charge dispersal and resonance stabilization.

*Step 4.* Determine whether there is any way to eliminate the incomplete octet on the oxygen in #26–22 (Step 2) by a further shifting of electrons.

If the nonbonding electrons on C*2* in the structure of #26–22 were shifted, the original structure would be obtained. There is no other atom on the oxygen from which an electron pair could be obtained.

We cannot take a pair of electrons away from an atom. We can merely shift positions on a given atom as we did in Steps 2 and 3. The structure in #26–22 has to be considered relatively unstable because of the incomplete octet on the oxygen and because oxygen, a highly electronegative atom, carries a positive formal charge. For these reasons we will eliminate #26–22 from further consideration.

*Step 5.* Repeat Steps 2 and 3 for the carbonyl group involving C*4* and O*2*.

This operation gives us the canonical form in #26–24.

**#26–24**

$$\mathrm{CH_3{-}\overset{:\ddot{O}}{\underset{}{C}}{-}\overset{H}{\underset{(-1)}{\ddot{C}}}{-}\overset{:\ddot{O}}{C}{-}CH_3} \longrightarrow \mathrm{CH_3{-}\overset{:\ddot{O}}{C}{-}\overset{H}{C}{::}\overset{:\ddot{O}:\,(-1)}{C}{-}CH_3} \equiv \mathrm{CH_3{-}\overset{:\ddot{O}}{\overset{\|}{C}}{-}\overset{H}{C}{=}\overset{:\ddot{O}:\,(-1)}{C}{-}CH_3}$$

*Step 6.* Conclusion

The structures in #26–23 and #26–24 are equivalent canonical forms, indicating that this anion has a high degree of resonance stabilization. These two canonical forms and a formula which combines the features of both are given in #26–25.

**#26–25**

$$\left[\ \mathrm{CH_3{-}\overset{:\ddot{O}}{\overset{\|}{C}}{-}\overset{H}{C}{=}\overset{:\ddot{O}:^{-}}{C}{-}CH_3} \longleftrightarrow \mathrm{CH_3{-}\overset{^{-}:\ddot{O}:}{C}{=}\overset{H}{C}{-}\overset{\ddot{O}:}{\overset{\|}{C}}{-}CH_3}\ \right] \equiv \mathrm{CH_3{-}\overset{O}{C}{-}\overset{H}{C}{-}\overset{O}{C}{-}CH_3}$$

(*a*) (*b*) (*c*)

These equivalent canonical forms tell us that:

- There is a high degree of resonance stabilization in the ion.
- All the bonds between C*2*, C*3*, C*4*, O*1*, and O*2* have some double bond character.
- The atoms C*2*, C*3*, C*4*, O*1* and O*2* all lie in one plane.
- There is a $\pi$ cloud that extends over all five atoms both above and below the plane in which they lie.
- The charge from the "extra" electron originally shown on C*3* is distributed over the chain of five atoms, leaving a $\delta-$ charge on each oxygen.

The molecular orbital diagram in #26–26 confirms our conclusion about resonance in the acetylacetonate ion and justifies our use of canonical forms as the basis for such conclusions. Each of the five adjacent atoms, C*2*, C*3*, C*4*, O*1* and O*2* has either a pi electron or a pair of nonbonding electrons. Thus, there is a succession of orbital overlaps from one atom to the next in the series. This continuous row of overlapped orbitals results in a $\pi$ orbital over the entire series of five atoms. Resonance stabilization of this anion means that a proton

#26–26

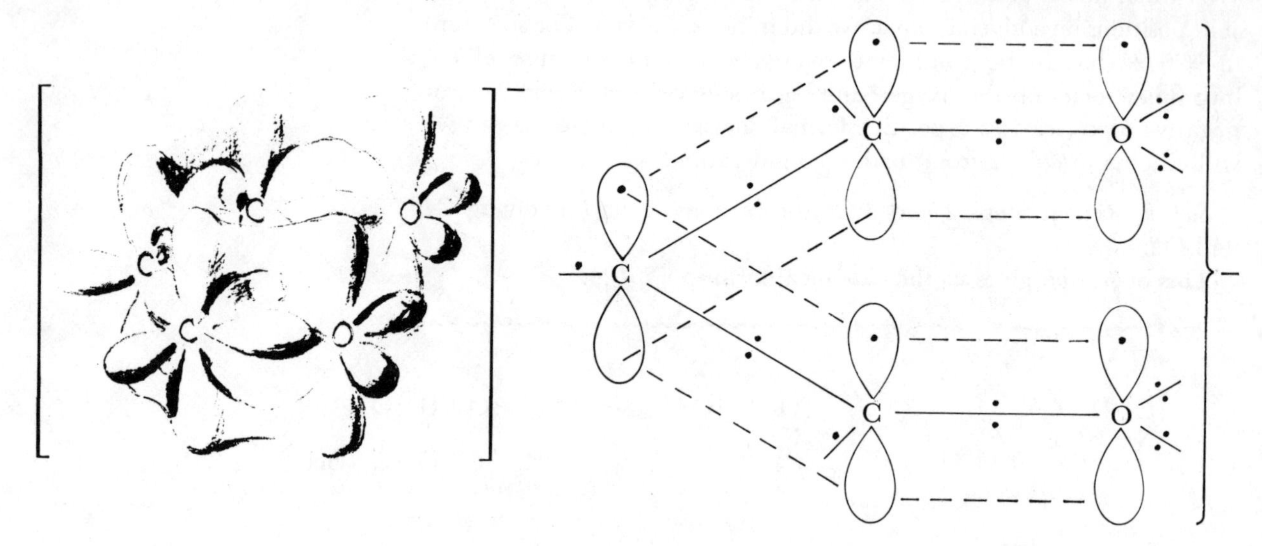


can be removed with comparative ease from C*3* and acetylacetone can function effectively as a Brönsted acid.

When you have understood Example 1, work the problems in §27.

* * * * *

A structure in which double and single bonds alternate is referred to as *conjugated*. In a conjugated chain there is one $p$ electron* on each adjacent atom as shown in #26–27. The orbitals of these $p$ electrons are parallel and each $p$ orbital overlaps the $p$ orbital on the carbon to its left and that on the carbon to its right. A $\pi$ orbital results which extends over the entire conjugated chain.

The effects of this kind of interaction between orbitals is most evident in benzene ($C_6H_6$), a six-carbon conjugated ring. The $\pi$ cloud

#26–27

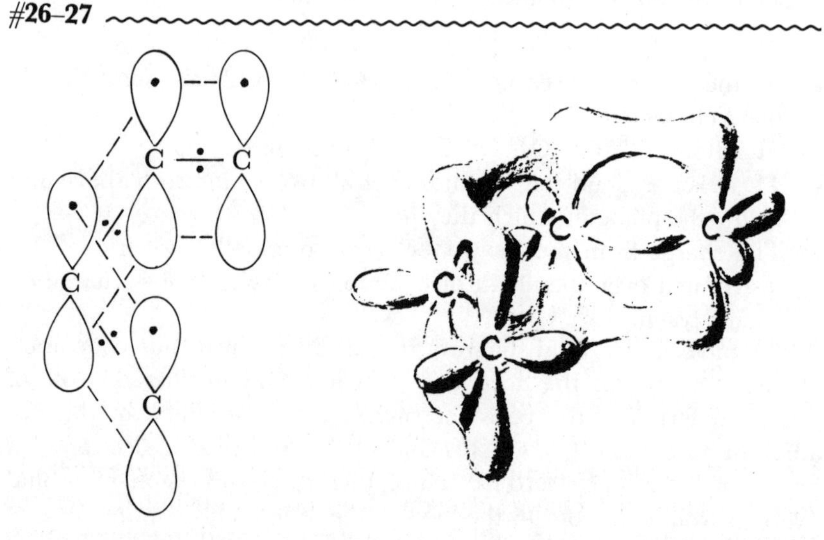


* See footnote, page 186

makes a complete circle in this molecule (#26–28) both above and below the plane in which all six carbons are held by a $\sigma$ bond framework.

**#26–28**

(*p electron on each of the six carbons*)

(*p orbitals overlap, charge distributed in a $\pi$ cloud above and below plane of ring*)

Resonance in the benzene ring can be shown by the two canonical forms in #26–29*a* and *b* or by the single formula in #26–29.

**#26–29**

[ (*a*) ⟷ (*b*) ] ≡ (*c*)

Benzene does not undergo the usual reactions of a compound with one or more double bonds. Its unique chemical reactivity has been classified under the term *aromatic properties*. We will not study aromatic properties at this time but will discuss the charge distribution in the $\pi$ cloud responsible for these properties. The charge density is the same throughout the $\pi$ cloud of benzene itself but the presence of a substituent on the ring brings about a redistribution, creating regions of greater and lesser charge density.

Consider aniline. It consists of a benzene ring with its circle of *p* orbitals forming a $\pi$ cloud and an amino group with a pair of nonbonding electrons. In the absence of any effect from the benzene ring, the bonding geometry at the amino nitrogen would be pyramidal. That is, it would be somewhat modified tetrahedral (#1–53). This hypothetical electron configuration is shown in #26–30. However, the amino group is not independent of the benzene ring. The orbital of the nonbonding electron pair can assume a position where it overlaps the $\pi$ orbital of the adjacent carbon atom. These nonbonding electrons thereby become part of the $\pi$ cloud causing it to extend over the nitrogen in addition to the six aromatic carbons, as diagrammed in #26–31. Thus, the charge of eight electrons is distributed over the seven atoms. Just as the depth of the water increased in compartment A of our analogy after the barrier was removed, so the total amount of

#26–30

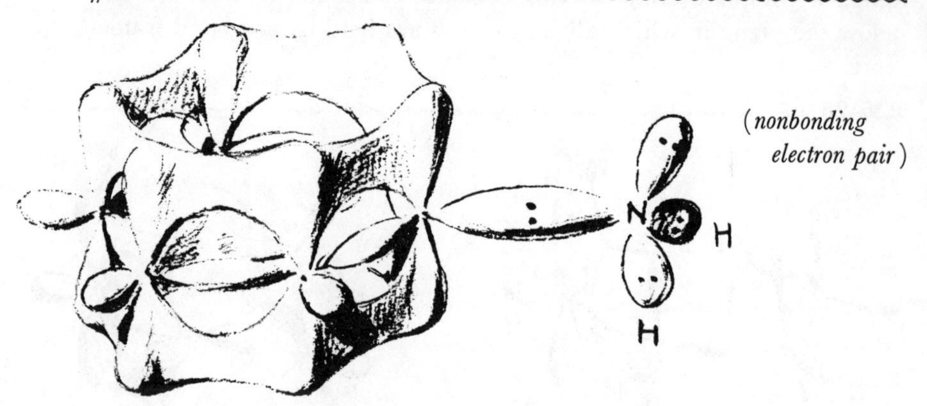


"electron stuff" has increased on the aromatic ring because of overlap between the orbital of the nonbonding electron pair from nitrogen and a $\pi$ orbital of the ring. The increase in charge on the ring is accompanied by a decrease in charge density in the region of the nitrogen. Rotation at the carbon–nitrogen bond is restricted and the two amino hydrogens lie in the plane of the ring at angles of 120° ($sp^2$ geometry).

#26–31

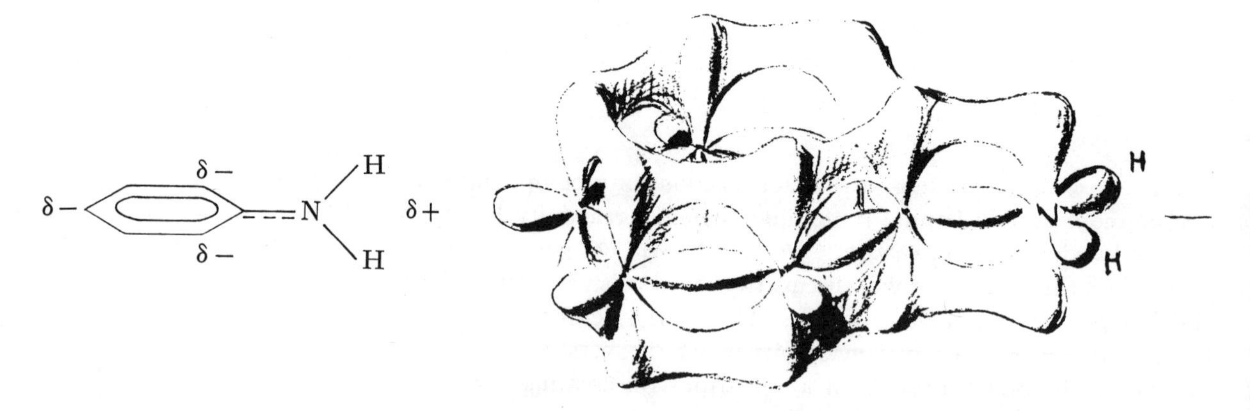


Charge distribution does not occur evenly around the ring. The $\pi$ cloud is denser in some regions than in others. Positions on a benzene ring are designated, according to their relationships to a ring substituent as *ortho*, *meta*, or *para*. These relationships are shown in #26–32, where Y represents any substituent.

#26–32

Y | Y | Y

*ortho* --- (ring) --- *ortho*

*meta* --- (ring) --- *meta*

*para*

In aniline the charge density is greater in the ortho and para positions than in the meta position.

The increased charge density at the ortho and para positions means that these carbons carry a partial negative charge. The presence of these charges can be deduced by drawing the canonical forms for aniline. Three equivalent canonical forms are shown in #26–33, wherein all the atoms have a complete octet and there is a formal negative charge on either an ortho or a para carbon. The process of drawing these forms starting with the electron-dot formulas at the left has been diagrammed. Whenever an atom has a formal charge in one of several equivalent canonical forms, that atom carries a partial charge in the actual structure. Recall your study of the —COO$^-$ group. Combining the three equivalent canonical forms #26–33*a*, *b*, and *c* gives the formula #26–33*d* for aniline.

**#26–33**

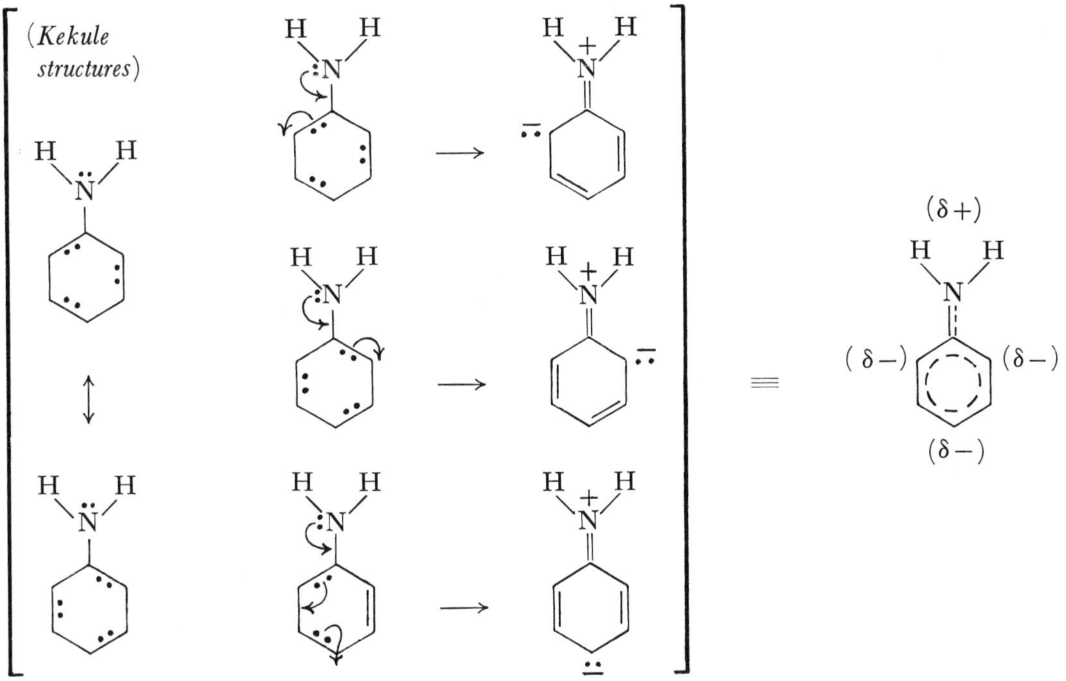

(*Kekule structures*)

($\delta+$)

($\delta-$) ($\delta-$)

($\delta-$)

To reiterate: The shifting of dots in these structures does not represent actual changes in electron configuration. These drawings are a device to help us decide whether certain electron orbitals have sufficient alignment to overlap and produce a $\pi$ cloud. Try to draw a canonical form that places a formal negative charge on a carbon meta to the amino group. You will find that you cannot devise a reasonable assignment for the electrons. No matter what you do, if you obey the rules, you end up with a carbon that has ten electrons. The impossibility of drawing for aniline a reasonable canonical form that places a formal negative charge in the meta position reflects the

Table 26–1

**Classification of Aromatic Substituents**

| *Activating ortho-para directors* | | *Effect on charge distribution* |
|---|---|---|
| strong activation | —$\ddot{N}H_2$, —$\ddot{N}HR$, —$\ddot{N}R_2$, —$\ddot{\underset{..}{O}}H$, —$\ddot{\underset{..}{O}}$:$^-$ | +**R**, −**I** |
| moderate activation | —$\ddot{\underset{..}{O}}R$, —$\ddot{\underset{..}{O}}Ar$, —NHC$\ddot{\underset{..}{O}}$R | +**R**, −**I** |
| weak activation | —Ar, —R | +**R**, +**I** hyperconjugation |

* * * * *

| *Deactivating ortho-para directors* | | |
|---|---|---|
| weak deactivation | —$\ddot{\underset{..}{F}}$:, —$\ddot{\underset{..}{Cl}}$:, —$\ddot{\underset{..}{Br}}$:, —$\ddot{\underset{..}{I}}$: | +**R**, −**I** |

* * * * *

| *Deactivating meta directors* | | |
|---|---|---|
| moderate deactivation | —$\overset{+}{N}R_3$, —$\overset{+}{S}R_2$, —$CX_3$ | −**I** |
| strong deactivation | —$NO_2$, —$SO_3H$, —COOH, —$CONH_2$, —COH, —COR, —CN | −**R**, −**I** |

physical fact that resonance in aniline increases the charge density at the ortho and para carbons over that at the meta carbons.

Substituents like —$\ddot{N}H_2$ which by resonance increase the overall charge on the benzene ring are called *activating* groups (Table 26–1) because the ring is more reactive towards aromatic substitution than is benzene itself. Groups which activate by resonance (+**R**) all have a pair of nonbonding electrons on the atom bonded to the aromatic ring carbon. These include: —$\ddot{N}H_2$, —$\ddot{N}HR$, —$\ddot{N}R_2$, —$\ddot{\underset{..}{O}}H$, —$\ddot{\underset{..}{O}}R$, etc. Alkyl groups are also activating but to a lesser extent. Their effect is due to hyperconjugation (§20). Many activating groups contain elements which because of their high electronegativity have a −**I** effect and which might be expected to reduce the charge on the ring, thereby deactivating it. However, it has been found that resonance effects dominate whenever resonance and inductive effects are at cross purposes.

Reduction of the charge density of the $\pi$ cloud and deactivation of the ring occur with substituents such as the nitro group which contain multiple bonds and highly electronegative elements.

Two canonical forms (#26–34) for a nitro group indicate double bond character for the nitrogen to oxygen bonding, a partial positive charge on the nitrogen and partial negative charges on the oxygens.

**#26–34**

A molecular orbital diagram (#26–35) shows the expected $\pi$ orbital over all three atoms. The $\pi$ orbital contains four electrons as did the $\pi$ orbital for the —$COO^-$ group.

**#26–35**

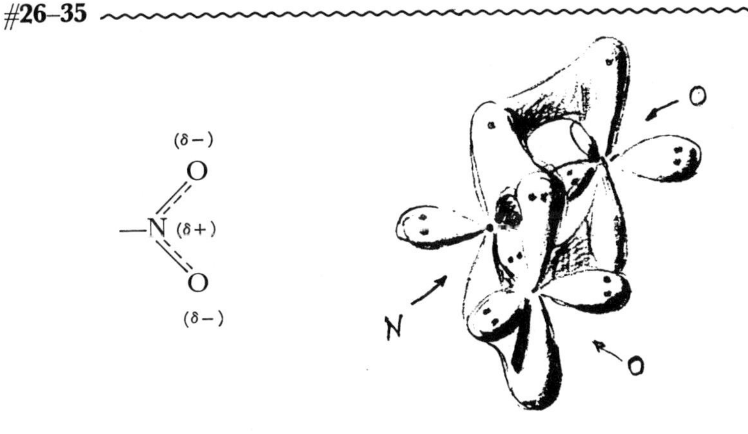


When a nitro group is substituent on a benzene ring, the $\pi$ orbital of the ring and the $\pi$ orbital of the nitro group are aligned when both oxygens lie in the plane of the ring. There is overlap between the $\pi$ orbitals at the point where nitrogen is bonded to an aromatic carbon. Thus, a single $\pi$ orbital is formed extending over the ring and the nitro group both above and below the plane of the ring as diagrammed in #26–36.

Resonance in this instance does not lead to an overall increase in charge on the ring, such as occurred with aniline. The electron deficiency (formal + charge) on the nitrogen atom and the high

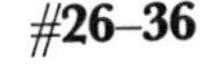

**#26–36**

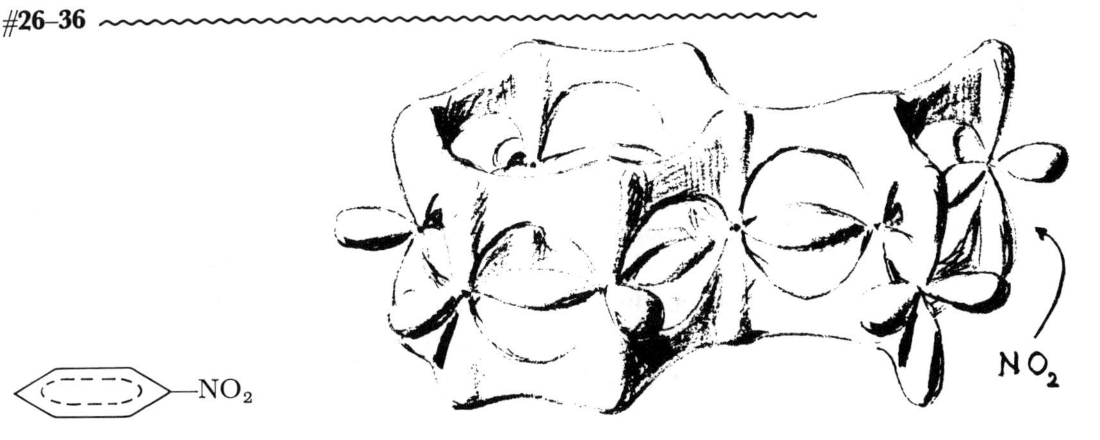

electronegativity of the oxygens tend to attract "electron stuff" away from the aromatic ring. The ring now corresponds to compartment B in our analogy, whereas in aniline it corresponded to compartment A. This electron-withdrawal leaves a deficiency in the aromatic $\pi$ cloud and creates a partial positive charge on the ortho and para carbons. Try drawing canonical forms to show the electronic configuration of nitrobenzene.

Substituents such as a nitro group that deactivate an aromatic ring by electron-withdrawal through resonance are designated as having a $-$**R** effect. Substituents such as an amino group that activate the ring by electron-release through resonance have a $+$**R** effect. Deactivating substituents in general have multiple bonds and highly electronegative atoms. Examples include: —$NO_2$, —COOH, —CN, —$SO_3H$, —CHO, —$CONH_2$, and groups with a positive charge, such as —$N(CH_3)_3^+$ (Table 26–1).

The halogens are deactivating substituents. However, this effect is due to induction rather than to resonance.

A substituent on an aromatic ring can have both an inductive and a resonance effect. In some instances the two effects supplement each other. For example, the $-$**R** and $-$**I** characteristics of a —$NO_2$ group both lead to electron-withdrawal. In other instances, the two functions produce opposite effects. An —OH($+$**R**, $-$**I**) group is electron-releasing by resonance and electron-withdrawing by induction. When the two effects oppose each other, the question arises "Will the presence of this particular group result in electron-withdrawal or electron-release?" The answer depends upon whether there are one or two substituents on the ring. If there are two or more substituents, the answer then depends upon how the substituents are positioned relative to each other.

When a group with opposing resonance and inductive effects is the only substituent on an aromatic ring, charge distribution is determined by resonance.

Resonance is a stronger effect than induction and dominates any opposing inductive effect.

There is one major exception to this statement. Halogens deactivate a ring to a moderate degree. This electron-withdrawal occurs by induction. Thus, —$\ddot{\underset{..}{O}}$H ($+$**R**, $-$**I**) activates a ring, while $:\ddot{\underset{..}{Cl}}:^-$ ($+$**R**, $-$**I**) deactivates a ring but to a lesser degree than does —$NO_2$ ($-$**R**, $-$**I**).

With two substituents on an aromatic ring, the inductive and/or the resonance effect from one substituent may modify the charge density of the electron cloud in the region of the second substituent. For the purpose of this discussion the second substituent has been designated as Y. The factors that determine the effect of a substituent on the charge density at Y can be summarized as follows:

- *Substituent* (*other than halogen*) *para to Y*

  Charge distribution is determined by both resonance and induction when these effects supplement each other. Charge distribution is determined by resonance when the inductive and resonance effects are in opposition.

- *Substituent meta to Y*
  Charge distribution is determined by induction.
- *Substituent (other than halogen) ortho to Y*
  Same relationships as with para substitution except that factors such as hydrogen bonding and the "ortho" effect* also play a role.
- *Halogens,* whether ortho, meta, or para to Y, exert a $-\mathbf{I}$ effect. This effect is strongest from the ortho position, weakest from the para position.

Thus we find that $-\ddot{\underset{\cdot\cdot}{O}}H$ in the para position is electron-releasing by resonance and $-\ddot{\underset{\cdot\cdot}{O}}H$ in the meta position is electron-withdrawing by induction.

The charge distribution patterns established by resonance play a major role in determining the relative strength of many aromatic carboxylic acids and many aryl amines. The correlation between resonance and relative acidity or basicity in the Brönsted sense can be better understood by working through the following Examples.

*Example 2*—Arrange the compounds in #26–37 in order of increasing strength as Brönsted acids.

**#26–37**

$C_6H_5COOH$ $\quad$ $HOC_6H_4COOH$ $\quad$ $NO_2C_6H_4COOH$

*Step 1.* Define the fundamental structural characteristic that determines the strength of a carboxylic acid.

The equation in #26–38 shows the type of reaction that occurs when these substances function as Brönsted acids.

**#26–38**

$$RCOOH + Base \rightleftharpoons RCOO^- + H{:}Base^+$$

The strength of a carboxylic acid can be related to the charge density on the $—COO^-$ group in the carboxylate anion in the following manner:

Charge delocalization on an ion leads to increased stability. A more stable ion is easier to form than a less stable one. The proton comes off more readily when the ion is formed more easily. The substance is, therefore, a stronger acid. Thus, a decreased charge density at the $—COO^-$ group leads to increased acidity for the parent compound.

From another point of view also, a lesser degree of negativity at the $—COO^-$ group favors an increase in the acid strength of the carboxylic acid. The proton exchange in #26–38 is an equilibrium reaction. When an acid is relatively strong, the concentration of carboxylate ion in the reaction mixture at equilibrium significantly exceeds the concentration

* Go to your textbook for an explanation of these factors.

of the unionized acid. However, when an acid is weak, the unionized acid is a major component of the reaction mixture at equilibrium.

An increased charge density at the —$COO^-$ group favors the reverse reaction in #26–38 wherein a proton ($H^+$) reacts with the —$COO^-$ group. Any factor favoring the reverse reaction reduces the strength of the acid and leads to an increase in the concentration of the unionized acid.

With these observations in mind we can evaluate the relative acidity of carboxylic acids by comparing the relative degree of charge localization on the —$COO^-$ group in their respective carboxylate ions.

*Step 2.* Examine the three compounds to determine whether one can be used as a standard of comparison for the other two.

The aromatic ring is unsubstituted in benzoic acid, whereas it contains an activating group in *p*-hydroxybenzoic acid and a deactivating group in *p*-nitrobenzoic acid. Thus, benzoic acid can serve as the standard with which the other two are compared.

*Step 3.* Analyze the electronic configuration of the three carboxylate anions to determine the factors that effect the charge density on the —$COO^-$ group.

*Benzoate anion* (#26–39)

**#26–39**

The benzoate anion is flat with a $\pi$ orbital extending over the entire structure. Resonance, therefore, determines the charge distribution. We need to analyze the variations in charge distribution caused by the presence of an activating or a deactivating substituent in the para position. The effect of such groups on charge distribution determines the acidity of the derivatives relative to the acidity of benzoic acid.

*Para-hydroxybenzoate anion* (#26–40)

**#26–40**

An —$\ddot{O}$H group has a +**R** effect that can be transmitted across the aromatic ring because it is possible for the orbital of a pair of nonbonding electrons on oxygen to overlap the $\pi$ orbital of the aromatic carbon to which it is bonded. In this way, the $\pi$ cloud of the anion is extended to include the —$\ddot{O}$H group. Thus, with a continuous $\pi$ cloud over the entire unit, the electron-releasing effect of the —$\ddot{O}$H group causes the charge density at the —$COO^-$ group in the *p*-hydroxybenzoate ion to be greater than the charge density at the —$COO^-$ group in the benzoate ion.

*Para-nitrobenzoate anion* (#26–41)

**#26–41**

A —$NO_2$ group has a −**R** effect that can be transmitted across the aromatic ring because it is possible for the orbital of the nitrogen to overlap the $\pi$ orbital of the aromatic carbon to which it is bonded. Thus, the $\pi$ cloud of the anion is extended to include the —$NO_2$

group. The continuous $\pi$ cloud over the entire unit provides the pathway for electron-withdrawal by the $—NO_2$ group. The charge density on the $—COO^-$ group in the *p*-nitrobenzoate ion is less than that of the charge density on the $—COO^-$ group in the benzoate ion.

*Step 4.* Translate the conclusions in Step 3 into terms of acid strength.

Charge localization at the $—COO^-$ group is greater in the *p*-hydroxybenzoate ion than in the benzoate ion, whereas charge localization at the $—COO^-$ group is less in the *p*-nitrobenzoate ion than in the benzoate ion. The relationship between charge localization in the ion and acid strength stated in Step 1 gives us the order in #26–42 for increasing acidity of the three carboxylic acids.

**#26–42**

$$H\ddot{\underset{\cdot\cdot}{O}}C_6H_4COOH \qquad C_6H_5COOH \qquad NO_2C_6H_4COOH$$

*Example 3*—Arrange the compounds in #26–43 in order of increasing basicity.

**#26–43**

$$\ddot{N}H_2C_6H_5 \qquad \ddot{N}H_2C_6H_4OCH_3 \qquad \ddot{N}H_2C_6H_4F$$

*Step 1.* Define the fundamental structural characteristic that determines the base strength of an amine.

The equation in #26–44 shows the type of reaction that occurs when an amine functions as a Brönsted base.

**#26–44**

$$Ar\ddot{N}H_2 + H_3\ddot{O}^+ \rightleftharpoons ArNH_3^+ + H_2\ddot{O}:$$

The base strength of the amine can be related to the charge density at the nitrogen in two ways:

Charge delocalization on an ion leads to increased stability. A more stable ion is easier to form than a less stable one. The amino nitrogen shares its electron pair more readily with a proton when the cation is easier to make. The amine is, therefore, a stronger base when the cation is more stable. Thus, an increase in charge density at the nitrogen in the cation results in increased basicity.

Increased charge density on the amino nitrogen in the neutral molecule results in an increase in the availability of the nonbonding electron pair and thereby produces an increase in basicity.

With these observations in mind we can evaluate the relative basicity of aromatic amines by comparing the relative charge localization on the nitrogen in either the amine or the corresponding anilinium ion.

*Step 2.* Examine the three compounds to determine whether one can be used as a standard of comparison for the other two.

The aromatic ring is unsubstituted in aniline, whereas it contains an activating group in *p*-methoxyaniline and a deactivating group in *p*-fluoroaniline. Thus, aniline can serve as the standard with which the other two are compared.

*Step 3.* Analyze the electronic configuration of the three amines or of the three anilinium ions to determine the factors that effect the charge concentration at the nitrogen.

It does not make any difference whether we consider the amines or the anilinium ions, all factors work in the same direction in each form.

*Aniline* (#26–45)

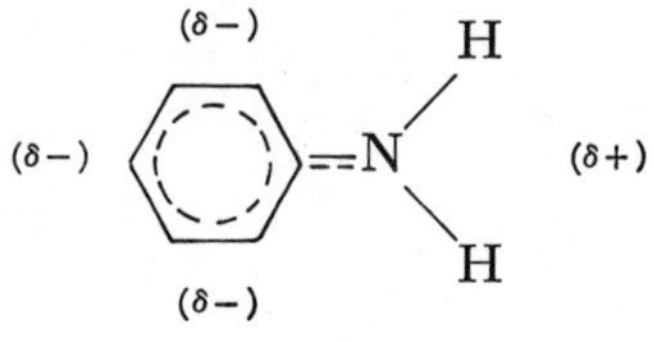

**#26–45**

A $\pi$ cloud extends over both the aromatic ring and the amino group, so resonance effects can change the charge distribution in an aryl amine.

*Para-methoxyaniline* (#26–46)

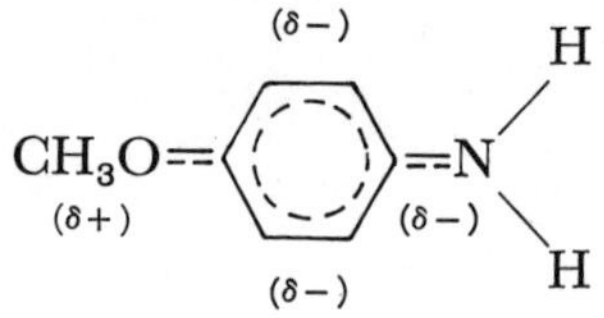

**#26–46**

A methoxy group has a $+\mathbf{R}$ effect that can be transmitted across the aromatic ring through the $\pi$ cloud to the nitrogen. This redistribution of charge can occur because of the overlap between the orbital of a nonbonding electron pair on oxygen and the $\pi$ orbital on the adjacent aromatic carbon. This overlap extends the $\pi$ orbital so that it covers the entire molecule, allowing for free charge transfer through the continuous $\pi$ cloud. Thus, the charge density is greater on the nitrogen in *p*-methoxyaniline than it is on the nitrogen in aniline.

*Para-fluoroaniline* (#26–47)

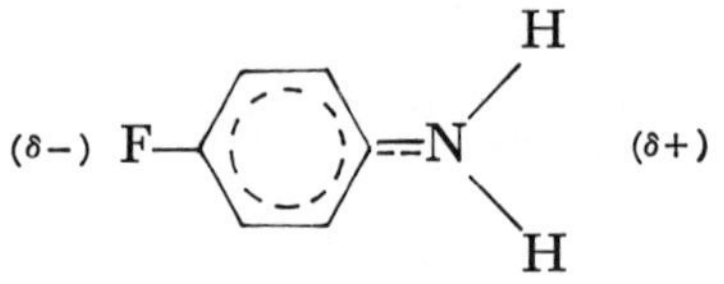

**#26–47**

A halogen exerts its major effect on the charge distribution of an aromatic ring by induction instead of by resonance. A halogen with its $-\mathbf{I}$ effect is a deactivating group. The inductive withdrawal is transmitted through the $\sigma$ bond framework of the aromatic ring. This effect is strongest with the halogen in the ortho position but is still significant with the halogen in the para position. Thus, the charge density on the nitrogen in *p*-fluoroaniline is less than the charge density on the nitrogen in aniline.

*Step 4.* Translate the conclusions in Step 3 into terms of base strength.

The increased charge density on nitrogen in *p*-methoxyaniline means that it is a stronger base than aniline. The reduced charge density on the nitrogen in *p*-fluoroaniline means that it is a weaker base than aniline. The order of increasing basicity in given in #26–48.

**#26–48**

$$FC_6H_4\ddot{N}H_2 \qquad C_6H_5\ddot{N}H_2 \qquad CH_3\ddot{\underset{..}{O}}C_6H_4\ddot{N}H_2$$

*Example 4*—Would you expect *p*-hydroxybenzoic acid or *m*-hydroxybenzoic acid to be the stronger acid?

*Step 1.* Define structural factors that help determine acidity.

See Step 1, Example 2 for reasoning leading to the conclusion that increased charge localization at the —COO⁻ group in the carboxylate anion lessens the tendency of the unionized acid to give up a proton.

*Step 2.* Identify the characteristics of the substituent on the aromatic ring.

An —ÖH exerts a +**R** effect and a −**I** effect.

*Step 3.* Determine the relative degree of charge localization on the —COO⁻ group in each anion.

*Parahydroxy benzoate ion* (#26–49)

**#26–49**

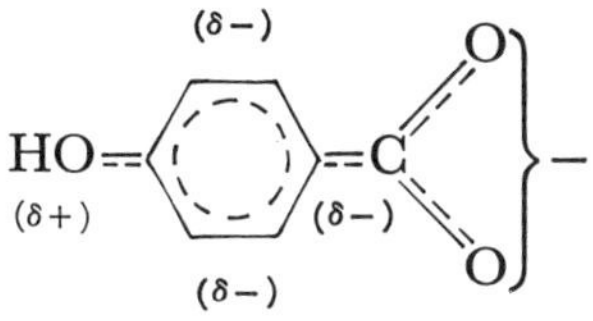


Resonance effects dominate for a substituent in the para position when the resonance and inductive effects are in opposition.

The presence of a para —ÖH group (+**R**) increases the charge localization on the —COO⁻ group.

*Meta-hydroxybenzoate ion* (#26–50)

**#26–50**

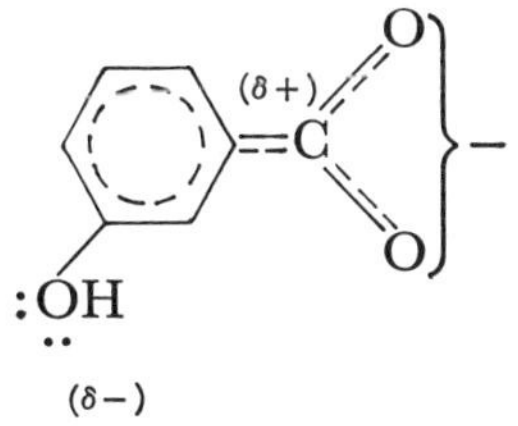


Charge distribution is determined by induction when a substituent is in the meta position.

The presence of a meta —ÖH group (−**I**) decreases charge localization in the —COO⁻ group.

*Step 4.* Express the conclusions from Step 3 in terms of relative acidity.

*m*-Hydroxybenzoic acid because of charge delocalization is a stronger acid than *p*-hydroxybenzoic acid in which there is charge localization in the region of the carboxylic acid group.

Various factors that determine the relative reactivity of Brönsted acids and bases are presented in Table 26–2. Each factor is illustrated by indicating which of two carboxylic acids is the stronger acid and which of two amines is the stronger base. Explanations for the difference in reactivity between the members of each pair are given in terms of charge distribution. Examples of electron-withdrawal (−**R**, −**I**) and of electron-release (+**R**, +**I**) by both resonance and induction have been included. Thus, when you understand this table you understand the main factors responsible for the geometry of the electron cloud in any molecule.

By definition acidity and basicity are reciprocal properties. This relationship becomes quite apparent in Table 26–2. A structural feature that leads to increased acidity in a carboxylic acid also leads to decreased basicity in an amine. The conclusions about charge distribution that can be drawn by a study of Table 26–2 are summarized in Table 26–3.

**Table 26–2**

**Comparative Acidity and Basicity**

| *Acidity* Stronger acid = 1 Weaker acid = 2 | *Reasons* Electron-withdrawal vs. Electron-release | *Basicity* Stronger base = 1 Weaker base = 2 |
|---|---|---|
| 1. $CH_3COOH$ 2. $CH_3CH_2COOH$ | $+$**I** effect of alkyl group | 1. $CH_3CH_2NH_2$ 2. $CH_3NH_2$ |
| 1. $ClCH_2COOH$ 2. $CH_3COOH$ | $-$**I** effect of chlorine atom | 1. $CH_3NH_2$ 2. $ClCH_2NH_2$ |
| 1. $FCH_2COOH$ 2. $ClCH_2COOH$ | $-$**I** effect greater for fluorine than for chlorine | 1. $ClCH_2NH_2$ 2. $FCH_2NH_2$ |
| 1. $CH_3CHClCH_2COOH$ 2. $ClCH_2CH_2CH_2COOH$ | $-$**I** greater when halogen closer to —COOH | 1. $ClCH_2CH_2CH_2NH_2$ 2. $CH_3CHClCH_2NH_2$ |
| 1. *p*-$NO_2C_6H_4COOH$ 2. $C_6H_5COOH$ | —$NO_2$ = ($-$**R**, $-$**I**) $-$**R** = main effect | 1. $C_6H_5NH_2$ 2. *p*-$NO_2C_6H_4NH_2$ |
| 1. $C_6H_5COOH$ 2. *p*-$HOC_6H_4COOH$ | $-$OH = ($+$**R**, $-$**I**) $+$**R** = main effect | 1. *p*-$HOC_6H_4NH_2$ 2. $C_6H_5NH_2$ |
| 1. *m*-$ClC_6H_4COOH$ 2. *m*-$CH_3C_6H_4COOH$ | $-$Cl = ($+$**R**, $-$**I**) $-$**I** = main effect | 1. *m*-$CH_3C_6H_4NH_2$ 2. *m*-$ClC_6H_4NH_2$ |
| 1. $C_6H_5COOH$ 2. *p*-$CH_3C_6H_4COOH$ | hyperconjugation of —$CH_3$ gives $+$**R** effect | 1. *p*-$CH_3C_6H_4NH_2$ 2. $C_6H_5NH_2$ |

**Table 26–3**

**Effect of Electron-Release and Electron-Withdrawal on Acidity and Basicity, as Exemplified by Carboxylic Acids and Amines**

| *Electron-withdrawal by induction or resonance increases acid strength* | *Electron-release by induction or resonance increases base strength* |
|---|---|
| $RC(=O)OH \underset{H^+}{\overset{-H^+}{\rightleftharpoons}} RC(=O)O^-$ | $RNH_2 \underset{-H^+}{\overset{H^+}{\rightleftharpoons}} RNH_3^+$ |
| $\leftarrow$ (weakens O—H bond); $\leftarrow$ (delocalizes negative charge) | $\rightarrow$ (increases availability of electron pair); $\rightarrow$ (delocalizes positive charge) |

Brönsted acids and bases have been used throughout these discussions to:

- illustrate the concepts of resonance and induction;
- show how these effects influence charge distribution within a molecular structure;
- relate electronic geometry with chemical reactivity.

The relationships between the configuration of the electron cloud and the relative reactivity of these Brönsted acids and bases have been studied in such detail so that you could realize that electronic geometry is in essence the ultimate factor that determines the reactivity of any chemical species.

## §27. Problems. Resonance. Molecular Orbital Diagrams. Canonical Forms

**1**) A characteristic structural feature that indicates resonance in a neutral molecule is A = B – Y = X, where A, B, Y, and X represent four atoms that may or may not be the same element. Classify each of the following structures under the heading of *resonance* or *no resonance* according to whether or not it has this characteristic feature called *conjugated double bonds.*

Draw orbital diagrams to show the relative position of the $\pi$ cloud in the structures you classify under *resonance.*

Draw the main canonical forms needed to represent the properties of the structure attributable to the configuration of the electron cloud.

*a*) $CH_3CH_2CH{=}CH_2$

*b*) $CH_2{=}CHCH{=}CH_2$

*c*) $CH_2{=}CH\overset{\overset{O}{\|}}{C}H$

*d*) $CH_2{=}CHCH_2\overset{\overset{O}{\|}}{C}H$

*e*) $CH_2{=}CH\overset{\overset{O}{\|}}{C}OH$

*f*) $CH_3CH_2\overset{\overset{O}{\|}}{C}OH$

*g*) $CH_3CH_2\overset{\overset{O}{\|}}{C}OCH_3$

*h*) $CH_2{=}CH\overset{\overset{O}{\|}}{C}OCH_3$

*i*) $CH_2{=}CH\overset{\overset{O}{\|}}{C}CH_3$

*j*) $CH_3CH{=}CHCH_2OH$

*k*) $CH_2{=}CHCH{=}CHCH{=}CH_2$

*l*) $C_6H_5{-}CH_2\overset{\overset{O}{\|}}{C}H$

*m*) $C_6H_5{-}CH_2\overset{\overset{O}{\|}}{C}OH$

*n*) $C_6H_5{-}\overset{\overset{O}{\|}}{C}OH$

*o*) $C_6H_5{-}\overset{\overset{O}{\|}}{C}OCH_3$

*p*) $C_6H_5{-}CH{=}CH_2$

*q*) $C_6H_5{-}CH_2CH{=}CH_2$

*r*) $C_6H_5{-}\overset{\overset{O}{\|}}{C}H$

**2**) Another structural characteristic leading to resonance may be summarized as A = B – Y:, where ":" is a pair of nonbonding electrons and Y may or may not have a formal charge of –1. Classify each of the following structures under the heading of *resonance* or *no resonance* according to whether or not it has this characteristic feature.

Draw orbital diagrams to show the relative position of the $\pi$ cloud in the structures you classify under *resonance.*

Draw the main canonical forms needed to represent the properties of the structure attributable to the configuration of the electron cloud.

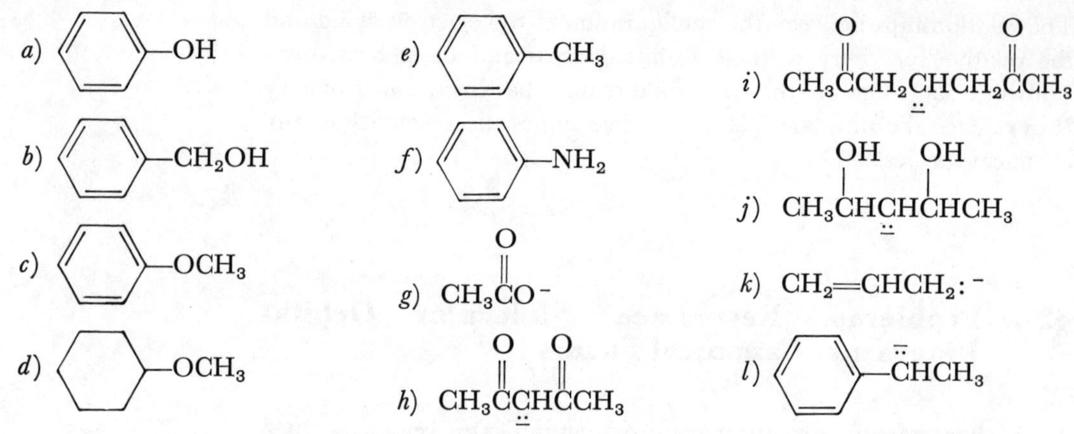


**3**) A structural characteristic that leads to resonance in a carbonium ion is A = B − Y+, where Y has an incomplete octet. Follow the directions in the two previous problems for the following list of carbonium ions.

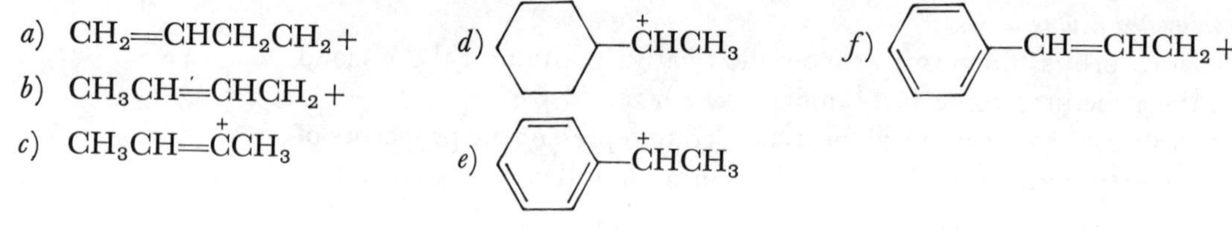


**4**) A structural characteristic that leads to resonance in a free radical is A = B − Y·. Follow the directions in problems 1 and 2 for the following list of free radicals.

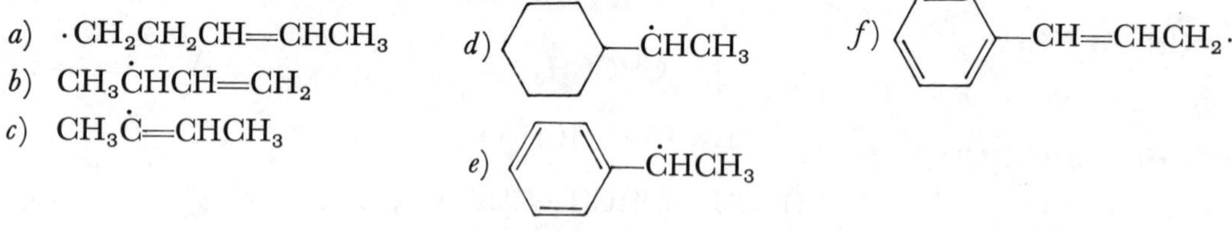


**5**) Indicate which member of each of the following pairs of substances you would expect to have a higher level of resonance energy, i.e., to be more stabilized by resonance. Explain your choice, using formulas if necessary.

*a*) $CH_3\overset{O}{\overset{\|}{C}}OH$, $CH_3\overset{O}{\overset{\|}{C}}O^-$

*b*) $CH_3\overset{O}{\overset{\|}{C}}\underset{..}{C}H\overset{O}{\overset{\|}{C}}CH_3$, $CH_3\overset{O}{\overset{\|}{C}}CH_2\overset{O}{\overset{\|}{C}}CH_3$

*c*) $CH_2{=}CHCH{=}CH_2$
$CH_2{=}CHCH_2CH{=}CH_2$

*d*) $CH_3CH_2CH_2NHCH_3$
$C_6H_5NHCH_3$

*e*) $C_6H_5OH$, $C_6H_{11}OH$

*f*) $CH_3\overset{O}{\overset{\|}{C}}O^-$, $CH_3\overset{O}{\overset{\|}{C}}Cl$

*g*) [cyclohexadiene], [benzene]

*h*) [benzonitrile (CN on benzene ring)], [cyclohexyl-CN]

**6**) Which member of each of the following pairs of substances will most easily give up the proton that is printed in boldface? To determine which of the marked hydrogens is more acidic in each pair of compounds, examine the relative stability of the two ions formed when the proton is removed. The anion that is the most stabilized by resonance comes from the stronger acid.

*a*) $(CH_3)_2CHO\mathbf{H}$

$(CH_3)_2CHCOO\mathbf{H}$

*b*) $CH_3CH{=}CHC(=O)O\mathbf{H}$

$CH_3CH_2CH_2C(=O)O\mathbf{H}$

*c*) $CH_3CH_2C(H)(\mathbf{H})CH_2CH_3$

$CH_3C(=O){-}C(H)(\mathbf{H}){-}C(=O)CH_3$

*d*) $CH_3C(=O){-}C(H)(\mathbf{H}){-}C(=O)OC_2H_5$

$CH_3CH_2C(H)(\mathbf{H}){-}C(=O)OC_2H_5$

*e*) COO**H** (benzoic acid) — COO**H** (cyclohexanecarboxylic acid)

*f*) O**H** (phenol) — O**H** (cyclohexanol)

**7**) Draw molecular orbital diagrams to show the $\pi$ cloud in each of the following substances. Draw the canonical forms needed to represent the effect of resonance on charge distribution. Draw a "combined formula" for each set of equivalent canonical forms.

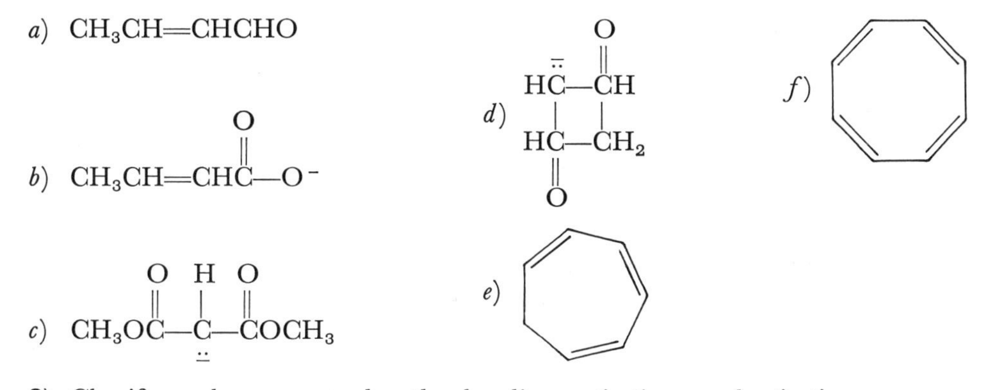


**8**) Classify each group under the heading *activating* or *deactivating* according to the effect it would have as a substituent on a benzene ring. Indicate for each group whether it is an ortho-para or a meta directing substituent.

*a*) $—NO_2$ *b*) $—NH_2$ *c*) —OH *d*) —COOH

*e*) —CN *f*) $—NHCH_3$ *g*) $—OCH_3$ *h*) —CHO

*i*) $—COOCH_3$ *j*) $—CH_2CH_3$ *k*) $—N(CH_3)_3^+$ *l*) $—SO_3H$

*m*) $—OCH_2CH_3$ *n*) —Br *o*) $—N(CH_3)_2$ *p*) $—NHC(=O)CH_3$

**9**) Arrange the members of each series in order of increasing acidity. Explain the reason for the order.

*a*) *1.* $C_6H_5COOH$ *2.* $NO_2C_6H_4COOH$ *3.* $HOC_6H_4COOH$
*b*) *1.* *p*-$BrC_6H_4COOH$ *2.* $C_6H_5COOH$ *3.* $C_6H_5OH$
*c*) *1.* *p*-$CNC_6H_4COOH$ *2.* *p*-$ClC_6H_4COOH$ *3.* *p*-$CH_3C_6H_4COOH$
*d*) *1.* *m*-$ClC_6H_4COOH$ *2.* *m*-$NO_2C_6H_4COOH$ *3.* *m*-$CH_3C_6H_4COOH$
*e*) *1.* *o*-$FC_6H_4OH$ *2.* *o*-$ClC_6H_4OH$ *3.* *o*-$IC_6H_4OH$
*f*) *1.* *p*-$CH_3C_6H_4OH$ *2.* *p*-$NO_2C_6H_4OH$ *3.* *p*-$ClC_6H_4OH$
*g*) *1.* $C_6H_{11}OH$ (cyclohexanol) *2.* $C_6H_5OH$ *3.* *p*-$BrC_6H_4OH$
*h*) *1.* $C_6H_5SO_3H$ *2.* $C_6H_5OH$ *3.* $C_6H_5COOH$
*i*) *1.* *o*-$CH_3OC_6H_4COOH$ *2.* *o*-$FC_6H_4COOH$ *3.* *o*-$CH_3C_6H_4COOH$
*j*) *1.* *p*-$CH_3OC_6H_4COOH$ *2.* *p*-$CH_3C_6H_4COOH$ *3.* *p*-$CNC_6H_4COOH$

**10**) Arrange the members of each series in order of increasing basicity. Explain the reason for the order.

*a*) *1.* $C_6H_5NH_2$ *2.* *p*-$NO_2C_6H_4NH_2$ *3.* *p*-$HOC_6H_4NH_2$
*b*) *1.* *p*-$CNC_6H_4NH_2$ *2.* *p*-$ClC_6H_4NH_2$ *3.* $C_6H_5NH_2$
*c*) *1.* $C_6H_5SH$ *2.* $C_6H_5OH$ *3.* $C_6H_5NH_2$
*d*) *1.* $C_6H_5NH_2$ *2.* $CH_3CH_2NH_2$ *3.* *p*-$NO_2C_6H_4NH_2$
*e*) *1.* *m*-$ClC_6H_4NH_2$ *2.* *m*-$NO_2C_6H_4NH_2$ *3.* *m*-$CH_3C_6H_4NH_2$
*f*) *1.* *o*-$FC_6H_4NH_2$ *2.* *o*-$BrC_6H_4NH_2$ *3.* *o*-$ClC_6H_4NH_2$
*g*) *1.* *p*-$CH_3OC_6H_4NH_2$ *2.* *p*-$BrC_6H_4NH_2$ *3.* *p*-$CNC_6H_4NH_2$
*h*) *1.* $C_6H_5NH_2$ *2.* $C_6H_5CH_2NH_2$ *3.* $C_6H_5NHC_6H_5$
*i*) *1.* *o*-$CH_3OC_6H_4NH_2$ *2.* *o*-$CH_3C_6H_4NH_2$ *3.* *o*-$BrC_6H_4NH_2$
*j*) *1.* *p*-$HOC_6H_4NH_2$ *2.* *p*-$HSC_6H_4NH_2$ *3.* *p*-$CNC_6H_4NH_2$

**11**) Draw molecular orbital diagrams for the following substances showing the relative position of the $\pi$ cloud. Draw for each the main canonical forms needed to show the effects of resonance on charge distribution.

*a*) *p*-$HOC_6H_4COO^-$ *c*) *p*-$HOC_6H_4NH_2$
*b*) *p*-$NO_2C_6H_4NH_2$ *d*) *p*-$NO_2C_6H_4COO^-$

## §28. Proton Exchange Reactions. The Acid-Base Chart

The previous sections related the variations in the strength of Brönsted acids and bases to charge distribution in the electron cloud. This section shows how a knowledge of the relative acidity and basicity of substances makes it possible to predict the extent to which proton exchange can occur between two specific substances.

The term *extent* has the following connotation: Proton exchange reactions are equilibrium reactions. The proportion of products to reactants in the reaction mixture at equilibrium is a measure of the *extent* to which a reaction occurs. A reaction proceeds to the extent of 70% if the products constitute 70% of the reaction mixture (exclusive of solvent) at equilibrium. In general chemistry you referred to acids that were 1% ionized when you were discussing the extent to which a weak acid gave up a proton in aqueous solution.

The relative acidity of two Brönsted acids can be determined from the Acid-Base Chart in Table 28–1. In this chart a variety of Brönsted acids have been listed in order of decreasing acidity: the strongest acid at the top, the weakest acid at the bottom.

**#28–1**

$$\mathrm{H{:}\underset{\cdot\cdot}{\overset{\cdot\cdot}{Cl}}{:}} + \mathrm{H{-}\underset{\cdot\cdot}{\overset{\cdot\cdot}{O}}{-}H} \rightleftharpoons \mathrm{H{-}\underset{\cdot\cdot}{\overset{\overset{\displaystyle H+}{|}}{O}}{-}H} + \mathrm{{:}\underset{\cdot\cdot}{\overset{\cdot\cdot}{Cl}}{:}^{-}}$$

Equation #28–1 shows a proton exchange between hydrogen chloride (HCl) and water ($H_2\ddot{O}{:}$). The chloride ion ($:\underset{\cdot\cdot}{\ddot{Cl}}:^-$) formed when the proton is removed from HCl satisfies the definition of a Brönsted base. A chloride ion can share a pair of nonbonding electrons with a proton. The basic species that remains after a proton has been removed from an acid is called the *conjugate base* of that acid. Thus, chloride ion ($:\underset{\cdot\cdot}{\ddot{Cl}}:^-$) is the conjugate basc of hydrogcn chloride (IICl). Water ($H_2\ddot{O}{:}$) is the conjugate base of the hydronium ion ($H_3O{:}^+$).

The Acid-Base Chart includes a second column which lists the conjugate base of each acid. A strong acid has a weak conjugate base and a weak acid has a strong conjugate base. Thus, the weakest base appears at the top of column 2 opposite the strongest acid, and the strongest base appears at the bottom of column 2 opposite the weakest acid. The relative reactivity of two Brönsted bases can be determined from column 2 of the Acid-Base Chart.

The arrangement of substances in the Acid-Base Chart makes it possible to determine from the chart whether an equilibrium mixture for a proposed proton exchange reaction would contain an excess of products or of reactants. The information you need to use this chart effectively is included in the following discussion.

The substances in equation #28–1 can each be classified as either a Brönsted acid or a Brönsted base. In the reaction that proceeds from left to right as written: HCl functions as an acid and gives up a proton; $H_2\ddot{O}{:}$ functions as a base and shares a pair of nonbonding electrons with a proton. In the reverse reaction: $H_3\ddot{O}^+$ functions as an acid and gives up a proton; $:\underset{\cdot\cdot}{\ddot{Cl}}:^-$ functions as a base and shares a pair of nonbonding electrons with a proton. These classifications are included in #28–2.

**#28–2**

$$\underset{(acid)}{\mathrm{H{:}\underset{\cdot\cdot}{\overset{\cdot\cdot}{Cl}}{:}}} + \underset{(base)}{\mathrm{H{-}\underset{\cdot\cdot}{\overset{\cdot\cdot}{O}}{-}H}} \rightleftharpoons \underset{(acid)}{\mathrm{H{-}\underset{\cdot\cdot}{\overset{\overset{\displaystyle H\ +}{|}}{O}}{-}H}} + \underset{(base)}{\mathrm{{:}\underset{\cdot\cdot}{\overset{\cdot\cdot}{Cl}}{:}^{-}}}$$

**Table 28–1**

**Acid-Base Chart**

| | Acid | Base | |
|---|---|---|---|
| (*strongest acid*) | 1. $H_2SO_4$ | $HSO_4^-$ | (*weakest base*) |
| | 2. $R-C(=OH^+)-H$ | $R-C(=O)-H$ | |
| | 3. HCl | $Cl^-$ | |
| | 4. $ArOH_2^+$ | ArOH | |
| | 5. $R-C(=OH^+)-R$ | $R-C(=O)-R$ | |
| | 6. $R-C(=OH^+)-OR$ | $R-C(=O)-OR$ | |
| | 7. $R-C(=OH^+)-OH$ | $R-C(=O)-OH$ | |
| | 8. $R-O(H^+)-R$ | $R-O-R$ | |
| | 9. $R_3COH_2^+$ | $R_3COH$ | |
| | 10. $RCH_2OH_2^+$ | $RCH_2OH$ | |
| | 11. $H_3O^+$ | $H_2O$ | |
| | 12. RCOOH | $RCOO^-$ | |
| | 13. $H_2CO_3$ | $HCO_3^-$ | |
| | 14. $NH_4^+$ | $NH_3$ | |
| | 15. ArOH | $ArO^-$ | |
| | 16. $RNH_3^+$ | $RNH_2$ | |
| | 17. $HCO_3^-$ | $CO_3^=$ | |
| | 18. $H_2O$ | $OH^-$ | |
| | 19. $RCH_2OH$ | $RCH_2O^-$ | |
| | 20. $R_3COH$ | $R_3CO^-$ | |
| | 21. $HC\equiv CH$ | $HC\equiv C^-$ | |
| | 22. $NH_3$ | $NH_2^-$ | |
| | 23. $CH_2=CH_2$ | $CH_2=CH^-$ | |
| (*weakest acid*) | 24. $CH_3CH_3$ | $CH_3CH_2^-$ | (*strongest base*) |

The reversibility of equilibrium reactions is represented by double arrows. The longer arrow points towards the major components of the reaction mixture at equilibrium. Equation #28–2 tells us that at equilibrium most of the HCl has been converted into $H_3\ddot{O}^+$ and $:\ddot{Cl}:^-$. The removal of the proton ($H^+$) from HCl by $H_2\ddot{O}:$ indicates that water has a greater affinity for the proton than does the chloride ion ($:\ddot{Cl}:^-$) from which the proton was separated. Therefore, water is a stronger base than chloride ion. The reaction can be similarly described in terms of the acids involved. The formation of $H_3\ddot{O}^+$ and the disappearance of HCl indicates that HCl gives up a proton more readily than does $H_3\ddot{O}^+$. Therefore, hydrogen chloride is a stronger acid than the hydronium ion. The acids and bases of reactions #28–2 are labelled as to their relative strengths in #28–3.

**#28–3**

$$\underset{\text{(stronger acid)}}{H:\ddot{\underset{..}{Cl}}:} + \underset{\text{(stronger base)}}{H—\ddot{\underset{..}{O}}—H} \rightleftharpoons \underset{\text{(weaker acid)}}{\left[H—\underset{..}{\overset{\overset{H}{|}}{O}}—H\right]^+} + \underset{\text{(weaker base)}}{:\ddot{\underset{..}{Cl}}:^-}$$

A useful general relationship appears in #28–3. In a proton exchange reaction, the stronger of the two acids in the equilibrium mixture has the greater tendency to give up a proton and the stronger of the two bases has the greater tendency to bond with a proton. Therefore, at equilibrium the reaction mixture contains a larger proportion of the weaker acid and the weaker base than of the stronger acid and the stronger base. Or in other words, a proton exchange reaction involves the formation of a weaker acid and a weaker base from a stronger acid and a stronger base.

**#28–4**

$$\underset{\text{(weaker acid)}}{CH_3\overset{\overset{:\ddot{O}}{\|}}{C}\ddot{\underset{..}{O}}H} + \underset{\text{(weaker base)}}{H—\ddot{\underset{..}{O}}—H} \rightleftharpoons \underset{\text{(stronger acid)}}{\left[H—\underset{..}{\overset{\overset{H}{|}}{O}}—H\right]^+} + \underset{\text{(stronger base)}}{CH_3\overset{\overset{:\ddot{O}}{\|}}{C}\ddot{\underset{..}{O}}:^-}$$

Contrast the equation in #28–4 with the one in #28–3. The reaction in #28–3 proceeds so as to produce a significant amount of product. The reaction in #28–4 reaches equilibrium while the reactants still make up a major part of the reaction mixture (exclusive of solvent). We deduce this fact from the equation because the longer arrow points towards the reactants. Although the reaction mixture at equilibrium does contain some product, we would describe the reaction in #28–4 as proceeding only to a slight extent.

The use of the Acid-Base Chart as an aid for writing equations can be illustrated by comparing the reaction between acetic acid and sodium hydroxide with the reaction between ethyl alcohol and sodium hydroxide.

*Example 1*—Write equations to show the reactions (if any) that occur between sodium hydroxide and acetic acid; between sodium hydroxide and ethyl alcohol.

*Step 1.* Identify the reactant that can function as an acid and the reactant that can function as a base for each proposed reaction.

The hydroxide ion $(:\ddot{\underset{..}{O}}H^-)$ is a strong base as shown in Table 28–1.

The other two substances can each function as a Brönsted acid.

*Step 2.* Give a skeletal equation, without arrows, showing the reactants and the products for each potential reaction.

Possible products can be determined by shifting the boldface hydrogen from each acid to a hydroxide ion, as shown in #28–5. The hydrogen comes away as a proton, leaving its electrons behind.

**#28–5**

| | (*Reactants*) acid | base | (*Products*) base | acid |
|---|---|---|---|---|
| (*a*) | $CH_3C(=\ddot{O}:)\ddot{O}\mathbf{H}$ + | $:\ddot{O}H^-$ | $CH_3C(=\ddot{O}:)\ddot{O}:^-$ + | $\mathbf{H}:\ddot{O}:H$ |
| (*b*) | $CH_3CH_2\ddot{O}\mathbf{H}$ + | $:\ddot{O}H^-$ | $CH_3CH_2\ddot{O}:^-$ + | $\mathbf{H}:\ddot{O}:H$ |

*Step 3.* Identify each acid and each base in the two reaction mixtures and determine the relative strengths of each pair of acids and each pair of bases from Table 28–1.

The two acids in #28–5*a* are acetic acid ($CH_3COOH$) and water ($H_2O$). RCOOH is above $H_2O$ in the acid column of Table 28–1. This positioning means that acetic acid is a stronger acid than water.

The two bases in #28–5*a* are hydroxide ion ($:\ddot{O}H^-$) and acetate ion ($CH_3COO^-$). $:\ddot{O}H^-$ is below $CH_3COO^-$ in the base column of Table 28–1. This positioning means that hydroxide ion is a stronger base than acetate ion. These relationships are shown in #28–6*a*.

**#28–6**

| | | | | |
|---|---|---|---|---|
| (*a*) | $CH_3C(=\ddot{O}:)\ddot{O}H$ + | $:\ddot{O}H^-$ | $CH_3C(=\ddot{O}:)\ddot{O}:^-$ + | $H—\ddot{O}—H$ |
| | (*stronger acid*) | (*stronger base*) | (*weaker base*) | (*weaker acid*) |
| (*b*) | $CH_3CH_2\ddot{O}H$ + | $:\ddot{O}H^-$ | $CH_3CH_2\ddot{O}:^-$ | $H—\ddot{O}—H$ |
| | (*weaker acid*) | (*weaker base*) | (*stronger base*) | (*stronger acid*) |

The two acids in #28–5*b* are ethanol ($CH_3CH_2OH$) and water ($H_2O$). $RCH_2OH$ is below $H_2O$ in the acid column, indicating that ethanol is a weaker acid than water.

The two bases in #28–5*b* are hydroxide ion ($:\ddot{O}H^-$) and ethoxide ion ($CH_3CH_2\ddot{O}:^-$). $:\ddot{O}H^-$ is above $CH_3CH_2\ddot{O}:^-$ in the base column, indicating that hydroxide ion is a weaker base than ethoxide ion. These relationships are shown in #28–6*b*.

*Step 4.* Insert equilibrium arrows with the longer arrow pointing towards the weaker acid and the weaker base.

The final equations including arrows are given in #28–7. They indicate that while acetic acid is a strong enough acid to react with sodium hydroxide to a reasonable extent, ethyl alcohol is not that strong an acid.

**#28–7**

$$CH_3C(=\ddot{O}:)\ddot{O}H + :\ddot{O}H^- \rightleftharpoons CH_3C(=\ddot{O}:)\ddot{O}:^- + H\ddot{O}H$$

$$CH_3CH_2\ddot{O}H + :\ddot{O}H^- \rightleftharpoons CH_3CH_2\ddot{O}:^- + H\ddot{O}H$$

Some substances can function as either an acid or a base depending upon the relative acid or base strength of the other reactant. Such substances are called *amphiprotic*. Phenol ($C_6H_5\ddot{O}H$) is such a substance, as is water. You should be familiar from general chemistry with equation #28–8. Water functions as both an acid and a base in this reaction. The amphiprotism of phenol is illustrated in Example 2.

**#28–8**

$$\underset{\text{(weaker acid)}}{H\ddot{O}H} + \underset{\text{(weaker base)}}{H\ddot{O}H} \rightleftharpoons \underset{\text{(stronger acid)}}{\left[H{-}\ddot{O}(H){-}H\right]^+} + \underset{\text{(stronger base)}}{:\ddot{O}H^-}$$

*Example 2*—Write equations for the reaction of phenol with sodium hydroxide and of phenol with hydrogen chloride.

*Step 1.* Identify the acid and the base for each set of reactants.

Phenol ($C_6H_5\ddot{O}H$) and hydroxide ion ($:\ddot{O}H^-$) each has a pair of nonbonding electrons. Thus, either of these substances could function as a base. Table 28–1 places $C_6H_5\ddot{O}H$ above $:\ddot{O}H^-$ in the base column, indicating that hydroxide ion is a stronger base than phenol. Therefore, the hydroxide ion functions as a base and the phenol functions as an acid in this reaction.

Phenol ($C_6H_5\ddot{O}H$) and hydrogen chloride (HCl) each has a proton that could be removed by a base. Thus, either of these substances could function as an acid. Table 28–1 places HCl above $C_6H_5\ddot{O}H$ in the acid column, indicating that hydrogen chloride is a stronger acid than phenol. Therefore, hydrogen chloride functions as the acid and phenol as the base in this reaction.

*Step 2.* Give skeletal equations, without arrows, showing the reactants and products for each potential reaction.

Possible products can be determined by shifting the bold face hydrogen from each acid to each base as shown in #28–9.

#28–9

| Reactants | | Products | |
|---|---|---|---|
| *(acid)* | *(base)* | *(acid)* | *(base)* |
| (*a*) $C_6H_5\ddot{\underset{..}{O}}\mathbf{H}$ | $:\ddot{\underset{..}{O}}H^-$ | $\mathbf{H}—\ddot{\underset{..}{O}}—H$ | $C_6H_5\ddot{\underset{..}{O}}:^-$ |
| (*b*) $\mathbf{H}:\ddot{\underset{..}{Cl}}:$ | $C_6H_5\ddot{O}H$ | $C_6H_5\ddot{\underset{..}{O}}H$ with $\mathbf{H+}$ | $:\ddot{\underset{..}{Cl}}:^-$ |

*Step 3.* Determine from Table 28–1 the relative strengths of each pair of acids and each pair of bases in the two reaction mixtures.

The two acids in #28–9*a* are phenol ($C_6H_5\ddot{\underset{..}{O}}H$) and water ($H_2\ddot{\underset{..}{O}}$). The chart shows phenol to be a stronger acid than water. The two bases in 28–9*a* are hydroxide ion ($:\ddot{\underset{..}{O}}H^-$) and phenoxide ion ($C_6H_5\ddot{\underset{..}{O}}:^-$). The chart shows hydroxide ion to be a stronger base than phenoxide ion. These relationships are shown in #28–10*a*.

#28–10

(*a*) $C_6H_5\ddot{\underset{..}{O}}H + :\ddot{\underset{..}{O}}H^-$ (*stronger* *acid*) (*stronger* *base*) $H—\ddot{\underset{..}{O}}—H + C_6H_5\ddot{\underset{..}{O}}:^-$ (*weaker* *acid*) (*weaker* *base*)

(*b*) $H:\ddot{\underset{..}{Cl}}: + C_6H_5\ddot{O}H$ (*stronger* *acid*) (*stronger* *base*) $C_6H_5\ddot{\underset{..}{O}}H$ (with H+) $+ :\ddot{\underset{..}{Cl}}:^-$ (*weaker* *acid*) (*weaker* *base*)

The two acids in #28–9*b* are hydrogen chloride (HCl) and protonated phenol ($C_6H_5\ddot{O}H_2^+$). The chart shows that hydrogen chloride is a stronger acid than the protonated phenol. The two bases in #28–9*b* are phenol ($C_6H_5\ddot{\underset{..}{O}}H$) and chloride ion ($:\ddot{\underset{..}{Cl}}:^-$). The chart shows that phenol is a stronger base than chloride ion. These relationships are shown in #28–10*b*.

*Step 4.* Insert equilibrium arrows with the longer arrow pointing towards the weaker acid and the weaker base in each equation.

#28–11

$$(a)\ C_6H_5\ddot{\underset{..}{O}}H + :\ddot{\underset{..}{O}}H^- \rightleftharpoons H\ddot{\underset{..}{O}}H + C_6H_5\ddot{\underset{..}{O}}:^-$$

$$(b)\ C_6H_5\ddot{\underset{..}{O}}H + H:\ddot{\underset{..}{Cl}}: \rightleftharpoons C_6H_5OH_2^+ + :\ddot{\underset{..}{Cl}}:^-$$

The equations for the two reactions are given in #28–11. Both reactions proceed far enough towards completion so that the concentration of products exceeds that of reactants at equilibrium. Yet in reaction #28–11*a* phenol has functioned as an acid, while in #28–11*b* phenol has functioned as a base.

The relationships in the chart can be summarized by the following statement:

An acid from column 1 reacts with any base from column 2 that appears at a lower level in column 2 than the level of the acid in column 1. The greater the distance between the positions of the two substances in their respective columns, the greater the extent of reaction.

This slight introduction to the Acid-Base Chart should provide you with sufficient knowledge to do the problems in §29.

## §29. Problems. Proton Exchange Reactions. The Acid-Base Chart

**1**) Each pair of substances in the following list consists of an acid and a base. Write equilibrium equations showing the extent to which proton exchange can occur between the members of each pair. Follow the procedure of Examples 1 and 2, working with the Acid-Base Chart to determine the relative strengths of both the acids and both the bases in the reaction mixture. Adjust the lengths of the equilibrium arrows to indicate whether the reactants or the products constitute the major portion of the reaction mixture at equilibrium.

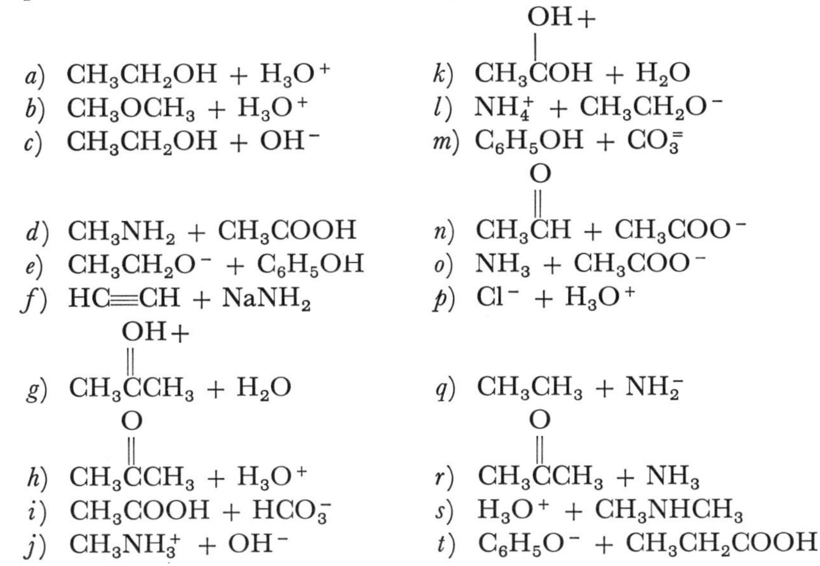

*a*) $CH_3CH_2OH + H_3O^+$

*b*) $CH_3OCH_3 + H_3O^+$

*c*) $CH_3CH_2OH + OH^-$

*d*) $CH_3NH_2 + CH_3COOH$

*e*) $CH_3CH_2O^- + C_6H_5OH$

*f*) $HC{\equiv}CH + NaNH_2$

*g*) $CH_3C(=OH^+)CH_3 + H_2O$

*h*) $CH_3C(=O)CH_3 + H_3O^+$

*i*) $CH_3COOH + HCO_3^-$

*j*) $CH_3NH_3^+ + OH^-$

*k*) $CH_3C(OH^+)OH + H_2O$

*l*) $NH_4^+ + CH_3CH_2O^-$

*m*) $C_6H_5OH + CO_3^=$

*n*) $CH_3C(=O)H + CH_3COO^-$

*o*) $NH_3 + CH_3COO^-$

*p*) $Cl^- + H_3O^+$

*q*) $CH_3CH_3 + NH_2^-$

*r*) $CH_3C(=O)CH_3 + NH_3$

*s*) $H_3O^+ + CH_3NHCH_3$

*t*) $C_6H_5O^- + CH_3CH_2COOH$

# Answers to Problems

## §2. Geometry of Sigma and Pi Bonding

All the drawings for **1**, **2**, **4**, **5**, **7**, **8**, and **9** can be made by applying the principles given in §1. **3**) Oxygen at the apex of a 105° angle. See #1–5 and #1–49. **6**) Angles of 120° at unsaturated atom. See #1–55 and #9–3. **10**) Place appropriate substituents on the legs of the tripod structure in #1–16*b*. See #1–30 for *h–j*. **11**) Use #1–52 for the basic form of these structures and #1–53 for the first part of problem 12. **12**) Diagram the overlapping orbitals as in #1–38. Add appropriate substituents in the XZ plane. **13**) $HC\equiv CH$, $HC\equiv N$, $CH_3C\equiv CCH_3$, $CH_3CH_2C\equiv CCH_2CH_3$, $CH_3C\equiv CCHClCH_3$. **14**) These compounds combine all the forms from §1. You have to figure out how best to represent them in combination.

**15**)

*a*)

|  | |  |
|---|---|---|
| C1—C2 | σ | $sp^3$—$sp^3$ |
| C1—H | σ | $sp^3$—$s$ |
| C2—H | σ | $sp^3$—$s$ |
| C2—Br | σ | $sp^3$—$p$ |
| C2—C3 | σ | $sp^3$—$sp^3$ |
| C3—H | σ | $sp^3$—$s$ |
| C3—O | σ | $sp^3$—$p$ |
| O—H | σ | $p$—$s$ |

*all angles are 109.5°*

*b*)

|  | |  |
|---|---|---|
| C1—C2 | σ | $sp^2$—$sp^2$ |
| | π | $p$—$p$ |
| C1—Br | σ | $sp^2$—$p$ |
| C1—H | σ | $sp^2$—$s$ |
| C2—C3 | σ | $sp^2$—$sp^3$ |
| C2—H | σ | $sp^2$—$s$ |
| C3—H | σ | $sp^3$—$s$ |

*angle formed by:*

| | |
|---|---|
| Br—C1—H | 120° |
| Br—C1—C2 | 120° |
| C1—C2—C3 | 120° |
| C1—C2—H | 120° |
| C2—C3—H | 109.5° |
| H—C3—H | 109.5° |

*c*)

|  | |  |
|---|---|---|
| C1—C2 | σ | $sp$—$p$ |
| | π | $p$—$p$ |
| | π | $p$—$p$ |
| C1—H | σ | $sp$—$s$ |
| C2—C3 | σ | $sp$—$sp^3$ |
| C3—H | σ | $sp^3$—$s$ |
| C3—C4 | σ | $sp^3$—$sp^3$ |
| C4—H | σ | $sp^3$—$s$ |

*angle formed by:*

| | |
|---|---|
| H—C1—C2 | 180° |
| C2—C3—H | 109.5° |
| H—C3—C4 | 109.5° |
| C3—C4—H | 109.5° |
| H—C4—H | 109.5° |

*d*)

|  | |  |
|---|---|---|
| C1—C2 | σ | $sp^2$—$sp^2$ |
| | π | $p$—$p$ |
| C1—H | σ | $sp^2$—$s$ |
| C2—C3 | σ | $sp^2$—$sp$ |
| C2—H | σ | $sp^2$—$s$ |
| C3—C4 | σ | $sp$—$sp$ |
| | π | $p$—$p$ |
| | π | $p$—$p$ |
| C4—C5 | σ | $sp$—$sp^3$ |
| C5—H | σ | $sp^3$—$s$ |

*e*)

|  | |  |
|---|---|---|
| C1—O | σ | $sp^3$—$p$ |
| C1—H | σ | $sp^3$—$s$ |
| O—C2 | σ | $p$—$sp^3$ |
| C2—H | σ | $sp^3$—$s$ |
| C2—C3 | σ | $sp^3$—$sp$ |
| C3—N | σ | $sp$—$p$ |
| | π | $p$—$p$ |
| | π | $p$—$p$ |

*f*)

|  | |  |
|---|---|---|
| C1—C2 | σ | $sp^2$—$sp^2$ |
| | π | $p$—$p$ |
| C1—H | σ | $sp^2$—$s$ |
| C2—H | σ | $sp^2$—$s$ |
| C2—C3 | σ | $sp^2$—$sp^2$ |
| C3—H | σ | $sp^2$—$s$ |
| C3—O | σ | $sp^2$—$p$ |
| | π | $p$—$p$ |

## §4. Two-Dimensional Structural Formulas. Position Isomerism

**1**) *a*) $CH_3CHClCH_2OH$ *b*) $CH_3CH_2CHCNCH(CH_3)_2$ *c*) $CHCl_2CHOHCH(CH_3)CH_2CH_3$ *d*) $CH_3C(CH_3)OHCH_2OC(CH_3)_3$ *e*) $CH_3CHBrCH(CH_3)CHO$

```
         CH3 H    H                   CH3 CH3 O                 OH Br CH3    H
          |   |   |                    |   |  ||                 |  |  |     |
2) a) CH3C---CCH2COH          b) CH3C----C---CH          c) CH3C---C-CCH2OCCl
          |   |   |                    |   |                     |  |  |     |
          H   Cl  H                   CH3  H                     H  H CH3    H

        H    CH3 H                    H    Cl CH3 OH
        |     |  |                    |    |   |  |
   d) ClCCH2C---COH              e) HOCCH2C---CCH2CCH3
        |     |  |                    |    |   |     |
        H    CH3 H                    H    Cl  H     H
```

**3**) *a*) C*1* = 1°, C*2* = 3°, methyl on C*2* = 1°, C*3* = 2°, C*4* = 1°.
*b*) C*1* = 1°, C*2* = none, methyls on C*2* = 1°, C*3* = 2°, C*4* = 2°, C*5* = 1°.
*c*) C*1* = 1°, C*2* = 2°, C*3* = 3°, methyl on C*3* = 1°, C*4* = 2°, C*5* = 1°.

**4**) prob. **3** *a*) 4; *b*) 4; *c*) 4; prob. **5** *a*) *1* = 3, *2* = 2, *3* = 3, *4* = 2, *5* = 3, *6* = 3; *b*) *1* = 6, *2* = 5, *3* = 6, *4* = 6, *5* = 6, *6* = 4; *c*) *1* = 7, *2* = 8, *3* = 7, *4* = 8, *5* = 5, *6* = 6; *d*) *1* = 4, *2* = 6, *3* = 3, *4* = 4, *5* = 4, *6* = 4; *e*) *1* = 5, *2* = 2, *3* = 5, *4* = 4, *5* = 5, *6* = 5.

**5**) *a*) 1, 3, 5, 6; 2, 4. *b*) 1, 3, 4, 5. *c*) 1, 3, 4. *d*) 1, 4, 5, 6. *e*) 1, 3; 5, 6.

**6**) These answers indicate the various carbon skeletons and the positions, on each, available for a particular substituent. The name of each compound in *a*, *b*, and *e* is given in answer to prob. 3, §6.

```
                            C
                            |
a) C—C—C—C, Br—1, 2; C—C—C, Br—1, 2. Four isomers.
   1 2                  1 2
```

```
b) C—C—C—C—C—C; C—C—C—C—C, —CH3 on 2,3; C—C—C—C, —CH3
                1 2 3                   1 2 3
```
on *2* and *3*; 2 —$CH_3$ on *2*. Five isomers.

```
c) C—C—C—C—C,   Cl on 1,1; 1,2; 1,3; 1,4; 1,5; 2,2; 2,3; 2,4; 3,3.
   1 2 3 4 5

     C  1′
     |
   C—C—C—C, Cl on 1,1; 1,1′; 1,2; 1,3; 1,4; 2,3; 2,4; 3,4; 3,3; 4,4.
   1 2 3 4

     C  1
     |
   C—C—C 1′, Cl on 1,1; 1,1′. Twenty-one isomers.
     |
     C
```

*d*) The oxygen may be present as an —OH substituent on carbon or as an ether link between two carbons (—C—O—C—). The various carbons skeletons for C*7* are shown along with the number of positions available for the —OH and the —O—.

```
C—C—C—C—C—C—C, —OH = 4 positions, —O— = 3 positions;
1 2 3 4

  C 1′
  |
C—C—C—C—C—C, —OH = 6 positions, —O— = 5 positions;
1 2 3 4 5 6
```

```
        C  7
        |
C—C—C—C—C—C, —OH = 7 positions, —O— = 6 positions;
1  2  3  4  5  6

1  C  C  6
   |  |
C—C—C—C—C, —OH = 6 positions, —O— = 5 positions;
1  2  3  4  5

1  C     C  1
   |     |
C—C—C—C—C, —OH = 3 positions, —O— = 2 positions;
1  2  3  2  1

  1  C
     |
1 C—C—C—C—C, —OH = 4 positions, —O— = 4 positions;
     |
  1  C  2  3  4

      C  3
      |
C—C—C—C—C, —OH = 3 positions, —O— = 3 positions;
      |
      C
1  2  3  2  1

  1  C  C  3
     |  |
1 C—C—C—C, —OH = 3 positions, —O— = 3 positions;
     |
     C
     1  2  3

(C—C)2—C—C—C, —OH = 3 positions, —O— = 2 positions.
 1  2     3  1  2          Seventy-two isomers.
```

*e*) C—C—C—C—C—C—C—C; C—C—C—C—C—C—C, $—CH_3$ on *2,3,4*.
*1 2 3 4 3 2 1*

C—C—C—C—C—C, $—CH_3$ on *2,3*; *2,4*; *2,5*; *3,4*; *2,2*; *3,3*; $—CH_2CH_3$ on *3*.
*1 2 3 4 5 6*

C—C—C—C—C, $—CH_3$ on *2,2,3*; *2,2,4*; *2,3,3*; *2,3,4*; $—CH_3$ and $—CH_2CH_3$ on
*1 2 3 4 5*
*3*; $—CH(CH_3)_2$ on *3*;

C—C—C—C, $—CH_3$ on *2,2,3,3*. Eighteen isomers.
*1 2 3 4*

## §6. Nomenclature of Alkanes

**1**) Answers are given here in the condensed form. Write yours so as to show clearly the branching rather than the condensed form.

*a*) $CH_3(CH_2)_6CH_3$

*b*) $(CH_3)_2CHCH_2Cl$

*c*) $CH_3CH_2C(CH_3)HCH_2CH_2CH_3$

*d*) $(CH_3)_3COH$

*f)* 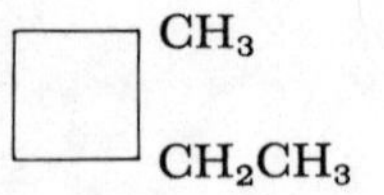 (ring) $CH_3$, $CH_2CH_3$

*g)* (ring) $CH_2CH(CH_3)_2$

*e)* $(CH_3)_2CHCH(CH_3)_2$ *h)* $CH_3CH_2OCH(CH_3)_2$
*i)* $CH_3CH_2C(CH_3)(CH_2CH_3)CH_2CH_2CH_3$
*j)* $(CH_3)_2CHCH_2NHC(CH_3)HCH_2CH_3$ *k)* $(CH_3CH_2)_2CHCH(CH_2CH_2CH_3)_2$
*l)* $(CH_3)_2CHC(CH_3)HCH(CH(CH_3)_2)CH(CH_3)—C(CH_2CH_3)_2CH_2CH(CH_3)_2$

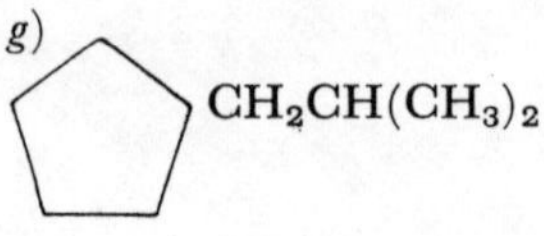

*m)* 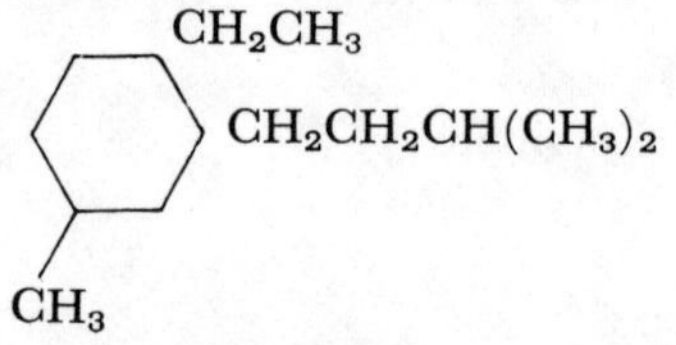 (ring) $CH_2CH_3$, $CH_2CH_2CH(CH_3)_2$, $CH_3$

*n)* $(CH_3)_2CHC(CH_2CH_3)_2C(CH_3)(CH_2CH_2CH_3)CH_2CH(C(CH_3)_3)CH(CH_3)_2$
*o)* $(CH_3)_3CCH_2CH(CH_2CH(CH_3)_2)CH(CH_2CH_3)_2$

**2)** *a)* nonane. *b)* 2,4-dimethylpentane. *c)* *tert.*-butylcyclopentane. *d)* isopentyl alcohol, 3-methylbutanol. *e)* isobutylcyclobutane. *f)* ethylisobutyl ether. *g)* 4-ethyl-2-methyl-hexane. *h)* isopropyl bromide. *i)* 2,3,4-trimethylpentane. *j)* *sec.*-butylisopropyl amine. *k)* 2,3,4,4-teramethylhexane. *l)* *sec.*-butylcyclopropane. *m)* 2,2,7,7-tetramethyloctane. *n)* 5-ethyl-2,3,3,6-tetramethyloctane. *o)* 6,7-diethyl-3-methyl-5-isopropyldecane.

**3)** *a)* *n*-, *sec*-, *tert*-, and isobutylbromide. *b)* *n*-hexane, 2-methylpentane, 3-methyl-pentane, 2,3-dimethylbutane, 2-dimethylbutane. *e)* octane, 2-, 3-, and 4-methylhep-tane; 2,2-, 2,3-, 2,4-, 2,5-, 3,4- and 3,3-dimethylhexane, 3-ethylhexane; 2,3,4-, 2,2,3-, 2,2,4-, 2,3,3-trimethylpentane, 2-methyl-3-ethylpentane (3-isopropylpentane), 3-methyl-3-ethylpentane; 2,2,3,3-tetramethylbutane.

**4)** *a)* 3-chloropentane. *b)* 3-methylhexane. *c)* 1,2-dimethylcyclohexane. *d)* 2,3,5-trimethylhexane. *e)* 2,5-dimethyl-3-isopropylhexane. *f)* 3-chloro-5-methylheptane. *g)* 2-bromomethylcyclopentane. *h)* 2-methylpentane. *i)* 2,4-diethyl-3-methyl-4-propyl-heptane. *j)* 3-ethyl-4-methylhexane.

## §8. Structural Formulas. Conformation

Figures #7 1, 7–2, 7–3, and 7–4 provide the three types of basic structures to be used in answering questions 1–4. In each instance the trans conformer is the more stable. The answers are given in terms of the substituents on the two carbons joined by the cross-piece of the structure; these are the carbons represented by the circle of the Newman structure.

**1)** Substituents on C*1* = H,H,Cl; same for C*2*.
**2)** Substituents on C*2* = $H,H,CH_3$; same for C*3*.
**3)** Substituents on C*2* = $H,H,CH_3$; C*3* = $H,H,C_4H_9$.
Substituents on C*3* = $H,H,C_2H_5$; C*4* = $H,H,C_3H_7$.
**4)** Substituents on C*2* = $H,CH_3,CH_3$; C*3* = $H,C_2H_5,C_2H_5$.
Substituents on C*3* = $H,C_2H_5,C_2H_5$; C*4* = $H,H,CH_3$.

**5)** The three basic structures for answering this question are shown here. Locate the indicated substituents in the correct positions on the forms given.

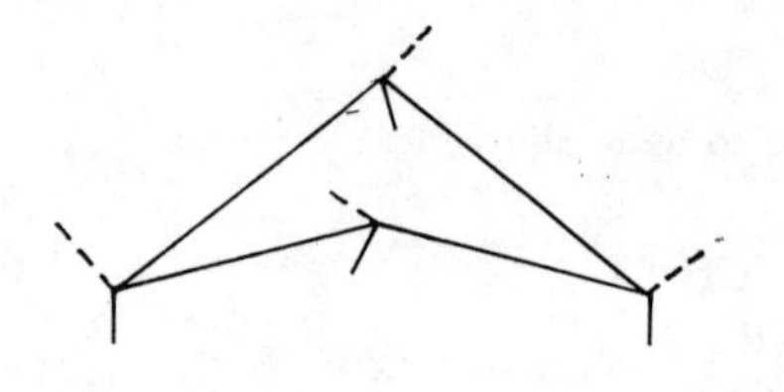
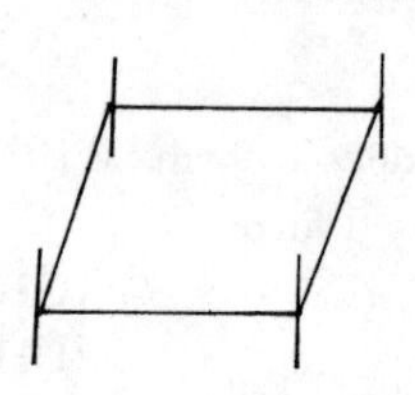
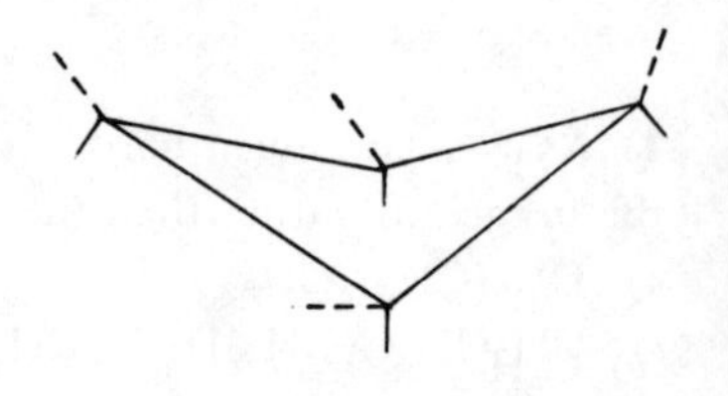

6) The three basic structures for answering this question are shown. Locate the indicated substituents in the correct positions on them.

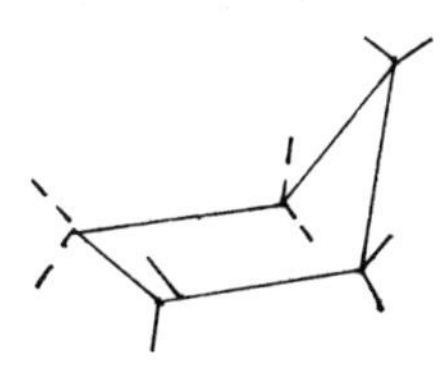

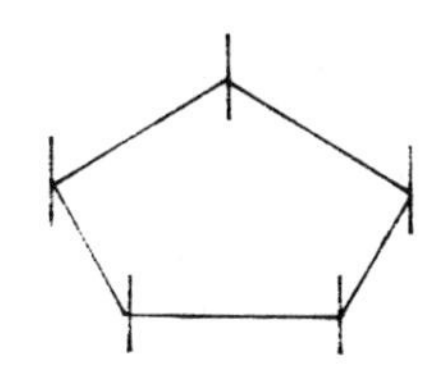

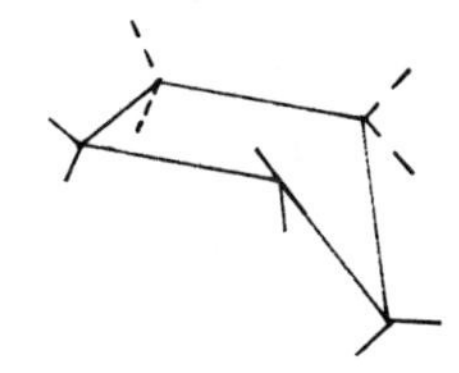

7) The three basic structures for answering this question are shown. Locate the indicated substituents in the correct positions on them.

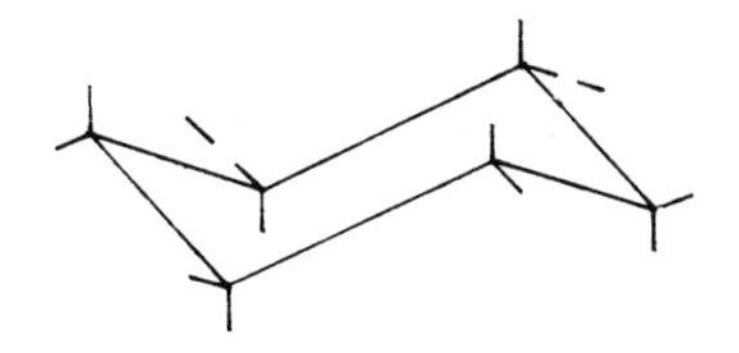

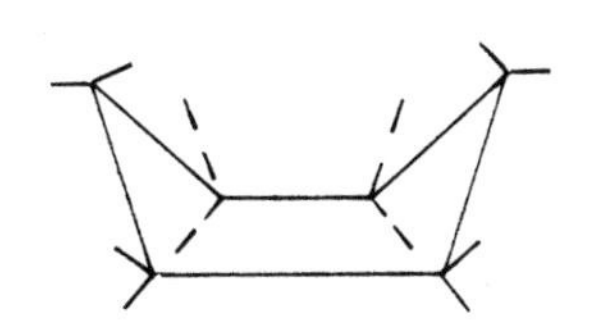

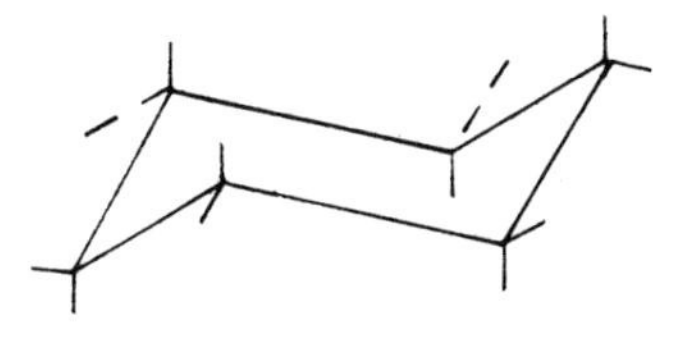

## §10. Cis-Trans Isomerism

1) Only the cis isomer is given for each compound:

*a*) 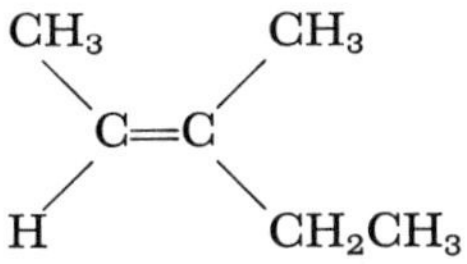 $CH_3$ and $CH_3$ / C=C / H and $CH_2CH_3$

*b*) None. *c*) None. *d*) See #9–12. *e*) None.

*f*) 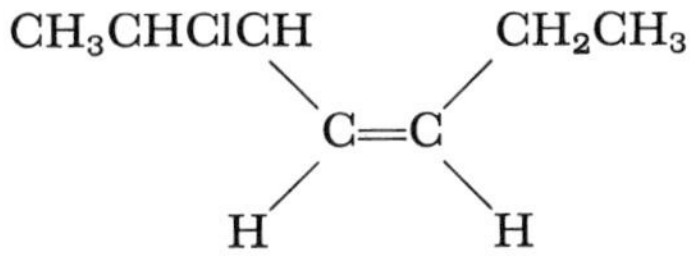 $CH_3CHClCH$ and $CH_2CH_3$ / C=C / H and H

*g*) See #11–7.

*h*) 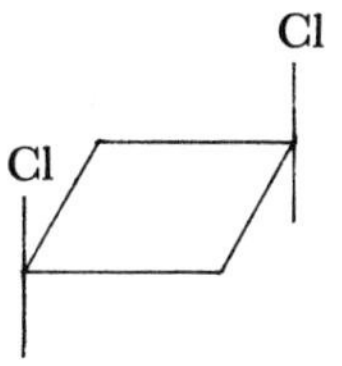 Cl, Cl

*i*) Br, Br

*j*) OH, OH, OH

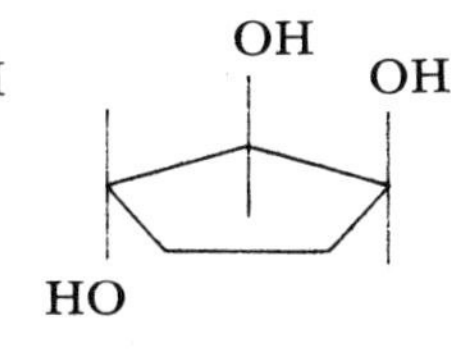

OH, OH, HO

OH, HO, OH

*k*)

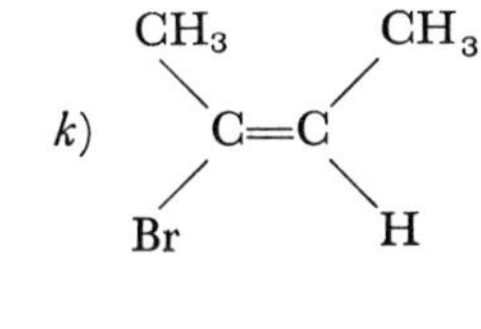



*l*) $C_6H_5$ and $CH_2CH_3$ / C=C / H and H

*m*) $CH_2Cl$, $CH_2Cl$, H, H

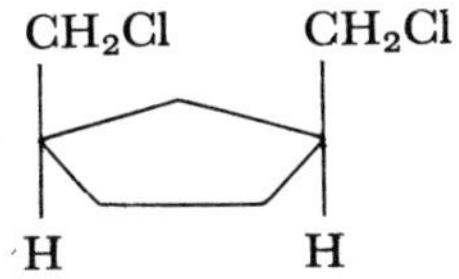

*n*) None.

*o*)

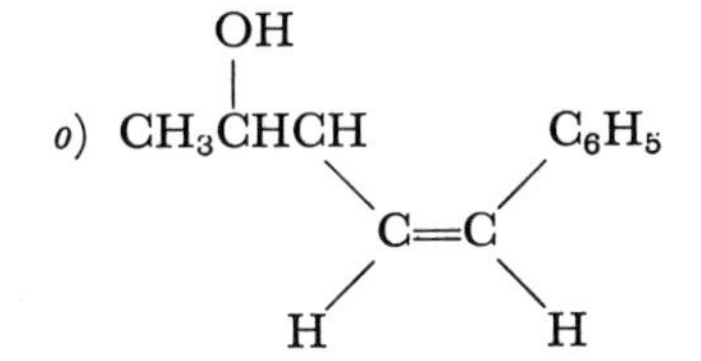



2) *a*) $C_nH_{2n}$ means one double bond or one ring.

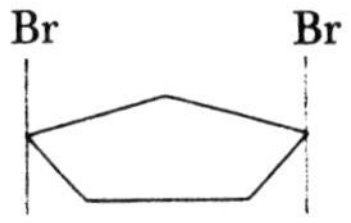

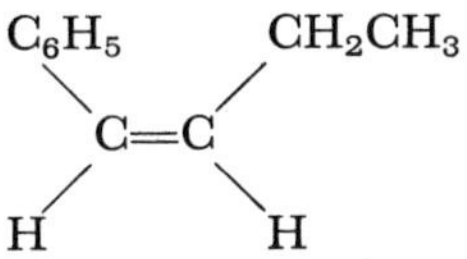

$CH_3CH_2CH{=}CH_2$, $CH_3\overset{H}{C}{=}\overset{H}{C}CH_3$, $CH_3\overset{H}{C}{=}\underset{H}{C}CH_3$, $CH_3\overset{CH_3}{C}{=}CH_2$, □, ▷ $CH_3$.

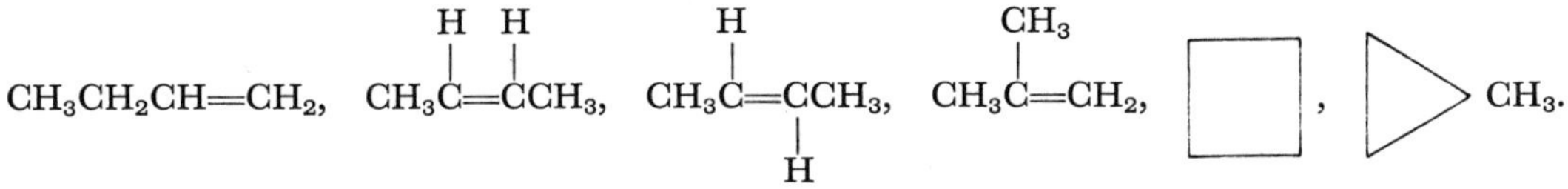

*b*) $C_nH_{2n}$ means one double bond or one ring.

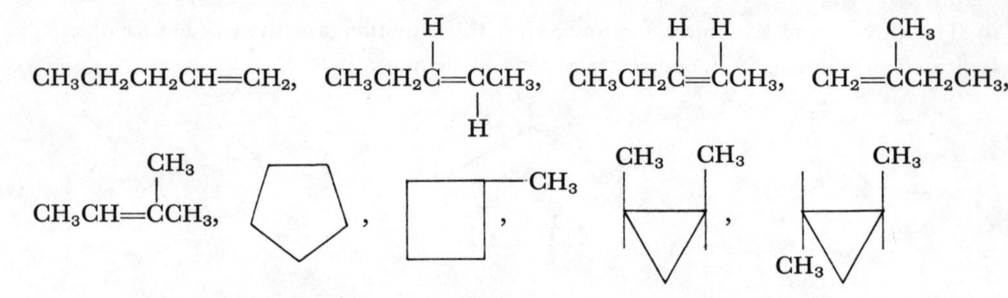


*c*) $C_nH_{2n-1}Cl$ can be considered as $C_nH_{2n-2}$ for the purpose of determining general structures. Thus this compound may have one double bond or one ring.

C=C—C—C, Cl on *1*, *2*, *3*, and *4* (cis-trans isomers for *1*-Cl);
*1* *2* *3* *4*

C—C=C—C, Cl on *1*, and *2* (both have cis-trans isomers); C—C(=C *3*)—C, Cl on *1*, and *3*.
*1* *2* *2* *1* ; *1* *2* *1*

Cyclic structures

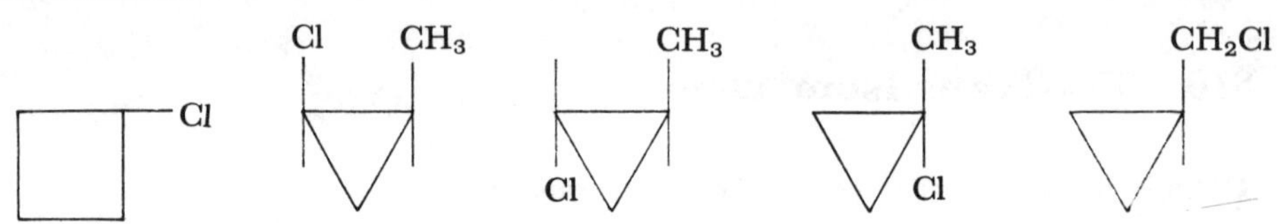


*d*) $C_nH_{2n-2}$ means one triple bond, or two double bonds. or one double bond and one ring, or two rings.

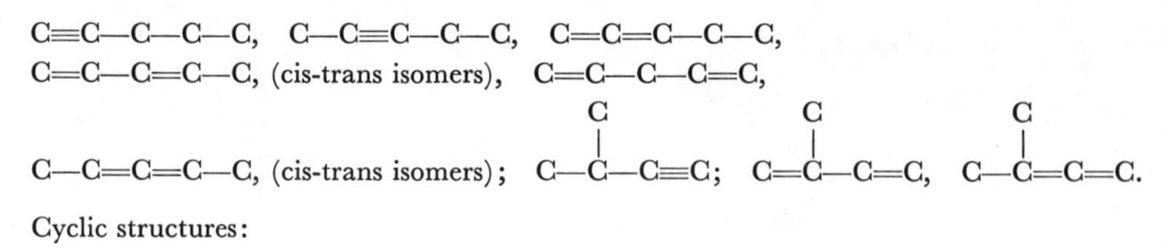


Cyclic structures:

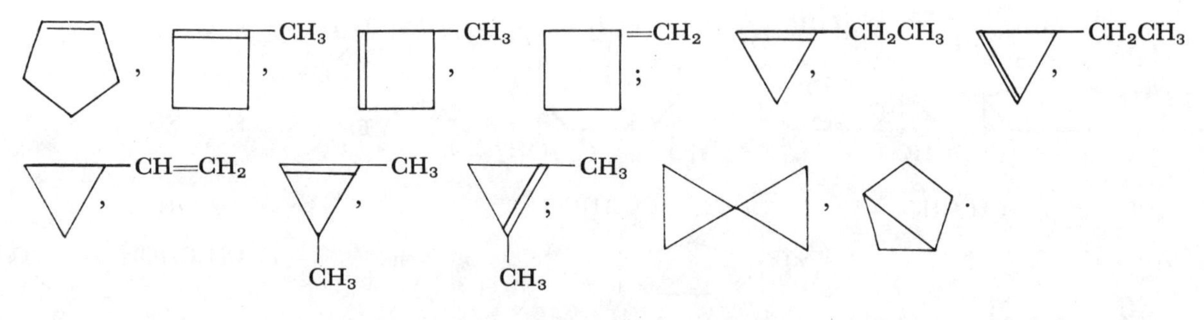


(Names are given in answer to problem 4, §12.)

*e*) Each set of structural features denoted by the formula $C_nH_{2n-6}$ is listed below along with compounds having these features. Not all of the structures that can be drawn actually exist. Cumulated double bonds tend to be unstable. Unsaturated three- and four-membered rings have a high degree of strain particularly in fused systems. You are not being asked to decide whether or not a particular structure can be made. You are to draw structures that comply with the empirical formula.

*2 triple bonds*:

C≡C—C≡C—C—C, C≡C—C—C≡C—C, C≡C—C—C—C≡C;

```
        C
        |
C≡C—C—C≡C.
```

*1 triple bond plus 2 double bonds*:

C≡C—C=C=C—C, C≡C—C=C—C=C, C≡C—C—C=C=C,
C—C≡C—C=C=C, C=C—C≡C—C=C.

*1 triple bond plus 1 double bond plus 1 ring*:

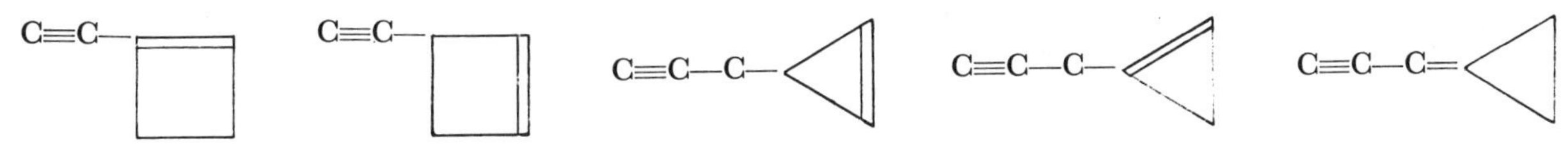


*1 triple bond plus 2 rings*:

*4 double bonds*: C=C=C=C=C—C, C=C=C=C—C=C, C=C=C—C=C=C.

These are all improbable structures.

*3 double bonds and 1 ring*:

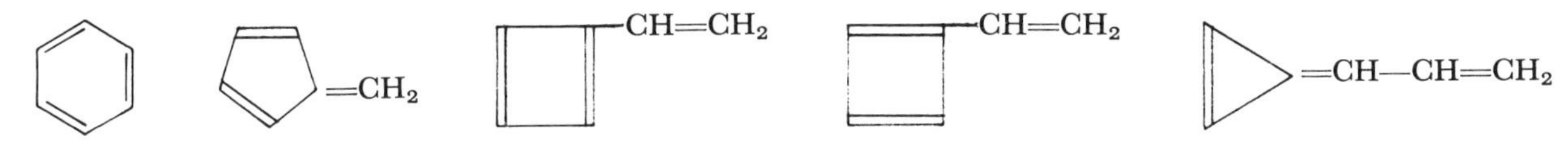


The smaller the ring gets in the above series, the less stable the structure becomes because of bond strain. The last one is improbable.

*2 double bonds and 2 rings*: There are 3 possible fused-ring systems for six carbons: two four-membered rings, a five- and a three-membered ring; two three-membered rings with a methyl substituent. However, when you start introducing two double bonds into each of these carbon skeletons the resulting structures are very strained and of low stability. You can draw 3 of the unsaturated structures for the fused four-membered rings, and 4 for the system composed of a three- and a five-membered ring. One can also draw improbable spirane ring systems with a cyclopropene ring.

*1 double bond and 3 rings*: *4 rings;* Both of these combinations lead to excessive strain when there are only six carbons in the structure.

**3**) *cis*- and *trans*-1-cyclobutylpropene, 2- and 3-cyclobutylpropene; 1- and 2-propylcyclobutene; 1- and 2-isopropylcyclobutene; *cis*- and *trans*-2-vinyl-1-methylcyclobutane, *cis*- and *trans*-3-vinyl-1-methylcyclobutane; 2-, 3-, and 4-ethyl-1-methylcyclobutene; *cis*- and *trans*-4-ethyl-3-methylcyclobutene; 3- and 4-methyl-1-ethylcyclobutene. Structures of low stability with a double bond between the ring carbon and the side chain can be drawn. There are six of them to figure out.

**4**) *a*) Fused rings composed of carbon atoms as follows: seven plus three; six plus four; five plus five. Spirane rings composed of carbons as follows: six plus three; four plus five. *b*) Fused rings composed of carbon atoms as follows: eight plus three; seven plus four; and six plus five. Spirane rings composed of carbons as follows: seven plus three; six plus four; five plus five.

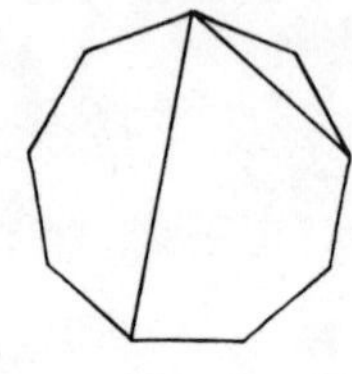

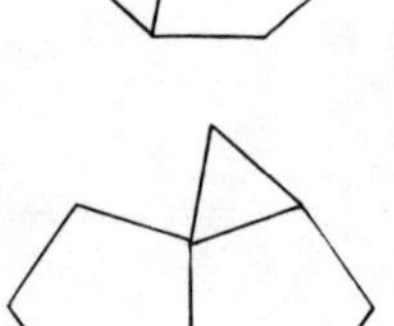

**5**) The systematic shifting of lines within the frame of a nine-membered ring will provide you with a means for identifying the various fused ring systems. However, the drawing may not readily convey to you the structural relationships. Therefore, it is a good idea to redraw the structure as shown. This flat representation does not show the three dimensional aspect of the structure, but does indicate the ring size and how the rings are fused. Draw your answers both ways.
The number of isomeric structures for each of the four different fused-ring systems are (the first three numbers indicate number of carbons in each of the three rings; fourth number indicates the number of isomers): 7,3,3; 4. 6,4,3; 6. 5,4,4; 4. 5,5,3; 3.

**6**) There are four isomers of the fused-ring system composed of a three- and a five-membered ring with a methyl substituent. There are two isomers of the fused-ring system composed of two four-membered rings with a methyl substituent. There are three positions for the methyl substituent on a spirane ring system composed of a four- and a three-membered ring. There are two positions for a methyl group on a structure composed of two cyclopropane rings joined by a single bond.

## §12. Nomenclature. Alkenes, Alkynes

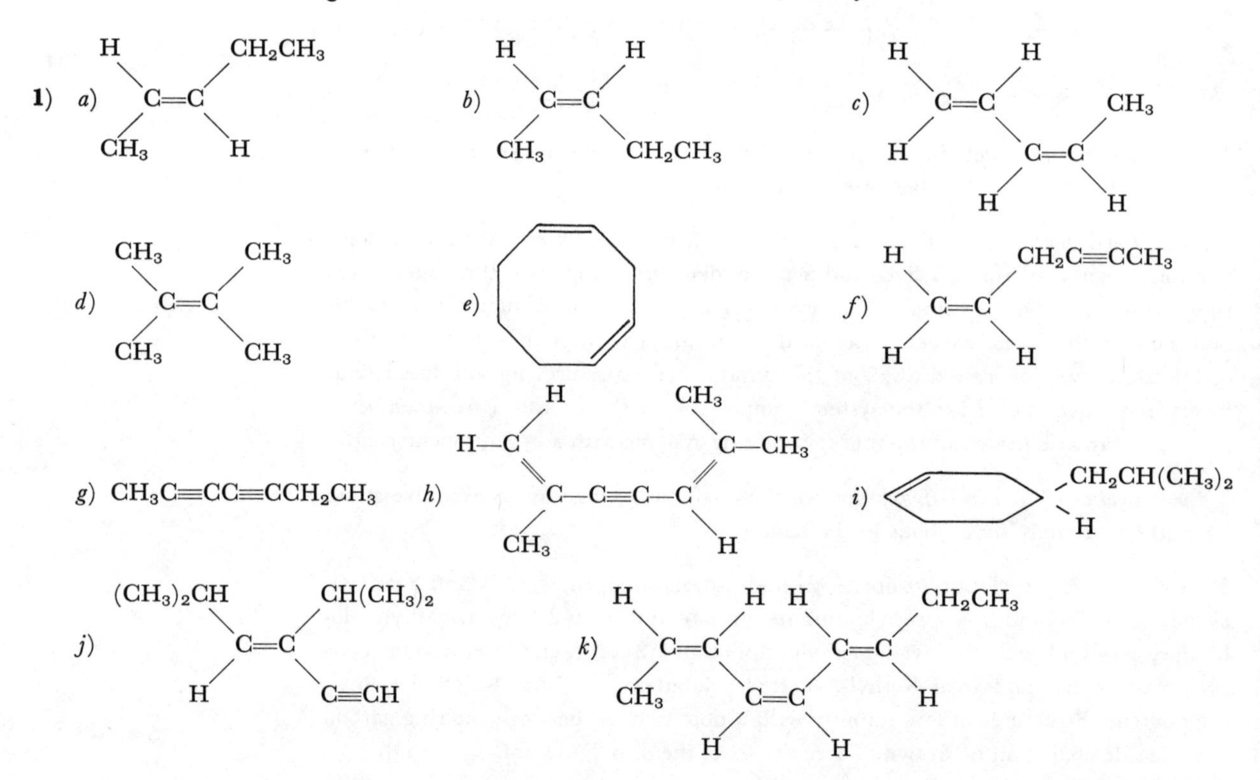


**2**) *a*) 1-pentene. *b*) 1,3-butadiene. *c*) 1-propyne. *d*) 2,5-hexadiene. *e*) 1,3-cyclopentadiene. *f*) *cis*-1,4-octadiene. *g*) 2-methyl-1-hexen-4-yne. *h*) *trans*-3,5,6-trimethyl-2,5-heptadiene. *i*) 6-ethyl-1,4-cyclooctadiene. *j*) *trans, cis, trans*-3,5,7-decatriene.

**3**) *a*) 3-methyl-2-pentene. *b*) *cis*- or *trans*-2,3-dimethyl-2-butene. *c*) 1,3-butadiene. *d*) *cis*- or *trans*-1,3-pentadiene. *e*) 4-chloro-2-methyl-2-pentene. *f*) *cis*- or *trans*-5-chloro-3-hexene. *g*) *cis, cis*-; *cis, trans*-; *trans, cis*-; *trans, trans*-2,4-hexadiene. *k*) *cis*- or *trans*-2-bromo-2-butene. *o*) 4-phenyl-3-butene-2-ol.

**4**) *b*) 1-pentene, *cis*- and *trans*-2-pentene, 2-methylbutene, 2-methyl-2-butene, cyclopentane, methylcyclobutane, *cis*- and *trans*-dimethylcyclopropane. *c*) *cis*- and *trans*-1-chlorobutene, 2-, 3-, and 4-chlorobutene; *cis*- and *trans*-2-chloro-2-butene; 1-chloro-2-methylpropene, 3-chloro-2-methylpropene, chlorocyclobutane, *cis*- and *trans*-2-chloro-2-methylcyclopropane, 1-chloro-1-methylcyclopropane, 1-chloromethylcyclopropane. *d*) 1-pentyne, 2-pentyne; *cis*- and *trans*-1,3-pentadiene, 1,2-, 1,4-, and 2,3- pentadiene; 3-methylbutyne, 2-methyl-1,3-butadiene, 3-methyl-1,2-butadiene; cyclopentene; 1-methylcyclobutene, 3-methylcyclobutene, 1-ethylcyclopropene, 2-ethylcyclopropene, 1,2-dimethylcyclopropene, 1,3-dimethylcyclopropene, cyclopropylethene.

**5**) *a*) 1-butene. *b*) 2-pentyne. *c*) 3-chloro-1-butene. *d*) 4-methylcyclopentene. *e*) 4-methylcyclohexene. *f*) hex-4-en-1-yne. *g*) pent-1-en-4-yne. *h*) 3-methyl-2-pentene. *i*) 1,3-pentadiene. *j*) 2,4,6-nonatriene. *k*) 1,1-dibromocyclohexane. *l*) 3-methylcyclopentene.

## §14. Hydrocarbon Derivatives. Nomenclature

**1**) *a*) 2-pentanone, methyl ethyl ketone. *b*) 1-methoxypropane, methyl propyl ether. *c*) propanal, propyl aldehyde. *d*) propanone, dimetyl ketone, acetone. *e*) ethanal, acetaldehyde. *f*) 2-butanol, *sec*-butyl alcohol. *g*) 2-bromobutane, *sec*-butyl bromide. *h*) 2-methylpropanol, isobutyl alcohol. *i*) 2-iodo-2-methylpropane, *tert*-butyl iodide. *j*) propanoic acid, propionic acid. *k*) butanoic acid, butyric acid. *l*) 5-methylhexanol, isohexyl alcohol. *m*) 1-chloro-2-methylpropane, isobutyl chloride. *n*) N-methylethanamine, ethylmethylamine. *o*) 1-methyletanamine, isopropylamine. *p*) 1-ethoxy-2-methylpropane, isobutyl methyl ether. *q*) 4-methyl-3-hexanone, isobutyl ethyl ketone. *r*) 2-methyl-2-propanol, *tert*-butyl alcohol. *s*) propen-2-ol, allyl alcohol. *t*) 2-methylpropanoic acid, isobutyric acid.

**3**) *a*) $CH_3CH_2CH_2CH_2CHO$. *b*) $CH_3\overset{H}{C}{=}\overset{H}{C}CH_2CHO$. *c*) $CH_3C{\equiv}CCH_2CHO$.

*d*) $CH_3\overset{\overset{O}{\|}}{C}CH_2CH_3$. *e*) $CH_2{=}CH\overset{\overset{O}{\|}}{C}CH_3$. *f*) $CH{\equiv}C\overset{\overset{O}{\|}}{C}CH_3$.

*g*) $CH_3CH_2CH_2CH(CH_3)CH_2COOH$. *h*) $CH_3CH_2CH_2\overset{\overset{CH_3}{|}}{C}{=}\underset{\underset{H}{|}}{C}COOH$.

*i*) $CH_3C{\equiv}CCH_2COOH$. *j*) $CH_3CHOHCH_2COOH$. *k*) $CH_3\overset{\overset{O}{\|}}{C}CH_2CH_2COOH$.

*l*) $CH_3\overset{\overset{O}{\|}}{C}\overset{\overset{H}{|}}{C}{=}\underset{\underset{H}{|}}{C}COOH$. *m*) $CH_3OCH_2CH_2CH_2OH$. *n*) $CH_3OCH_2CHOHCH_3$.

*o*) $CH_3OCH_2CHOHCH_2CH_2OHCH_3$. *p*) $CH_3CH_2NH_2$.

*q*) $(CH_3)_2CHCH_2NH_2$. *r*) $CH_3CH{=}CHCH_2NH_2$. *s*) $CH_3C{\equiv}CCH_2NH_2$.

*t*) $CH_3CH_2CH_2CH(CH_3)CH_2NH_2$. *u*) $CH_3CH_2C(CN){=}CHCH_2CH_2NH_2$.
*v*) $CH_3CH(NHCH_3)CHOHCH_3$. *w*) $CH_3CH_2NHCH_2CH_2OH$.
*x*) $CH_3NHCH(CH_3)_2$. *y*) $CH_3CH_2CH_2NHCH_3$
*z*) $CH_3NHCH_2CH_2CH_2NH_2$.

*a'*) $CH_3CH_2CH_2CH_2CH_2\overset{O}{\overset{\|}{C}}NH_2$. *b'*) $CH_3CH_2\overset{H}{C}{=}\underset{H}{C}CH_2\overset{O}{\overset{\|}{C}}NH_2$. *c'*) $CH_3CH_2C{\equiv}CCH_2\overset{O}{\overset{\|}{C}}NH_2$.

*d'*) $CH_3CH(COOH)CH_2CH_2COOH$. *e'*) $CH_3CH(COOH)C{\equiv}CCOOH$. *f'*) (cyclopentane ring)—OH

*g'*) (cyclopentene ring)—OH *h'*) (cyclohexane ring)=O *i'*) (cyclohexene ring, $CH_3$)=O

*j'*) $(CH_3)_2CHCH_2OCH(CH_3)CH_2CH_3$. *k'*) $CH_3CH_2OCH_2CH_2CH_2OH$.

*l'*) $CH_3OCH_2CH_2\overset{O}{\overset{\|}{C}}CH{=}CH\overset{O}{\overset{\|}{C}}H$. *m'*) $CH_3\overset{(CH_3)_2CHO}{\overset{|}{C}}H\overset{O}{\overset{\|}{C}}CH_3$.

**5**) *a*) 3,4-dichlorobutanal. *b*) 3-chlorohexane. *c*) 5-oxo-pentanoic acid. *d*) 6-oxo-3-hexanone. *e*) 2-methoxyethanol. *f*) 2-methyl-5-oxo-hexanoic acid. *g*) 4-methyl-amino-butanal. *h*) 1-amino-2-propanol. *i*) 4-cyanobutanoic acid. *j*) 3-hydroxy-2-pentanone.

**6**) *a*) 2,4-dichlorohexane. *b*) 3-hydroxybutanoic acid. *c*) 1-amino-5-hydroxy-3-hexanone. *d*) 1-methoxy-2-propanone. *e*) 3-cyano-2-methylpentane. *f*) 5-bromo-3-hexanol. *g*) 3-chlorobutanoic acid. *h*) 5-hydroxy-2-pentanone. *i*) ethoxyethanal. *j*) 3,5-dihydroxypentanol. *k*) *cis*-4-bromo-3-pentenamine. *l*) N-ethyl-3-methylpentamine. *m*) 4-methoxybutanal. *n*) 4-hydroxypentanoic acid. *o*) 2-oxo-butanal. *p*) 4-dimethylamino-2-butanol. *q*) *trans*-4-bromocyclohexanol. *s*) 4-methylcyclohexanone. *t*) *trans*-1, 3-dihydroxycyclopentane (*trans*-3-hydroxycyclopentanol).

## §16. Optical Isomerism

**1**) To answer this question determine the configuration, **R** or **S**, of each structure as drawn. Draw a tripod formula for each, keeping in mind the Fischer convention as you translate the flat formula into a three-dimensional one. After drawing the tripod with the atoms in the position given, shift the structure, either mentally or by redrawing it, so that you may view it with the hydrogen projecting away from you. Number the substituents. This process is illustrated for *a* and *h*. For *h* tip the tripod back, raising the $—CH_2CH_3$ and making the —H a leg of the tripod. Then rotate.

(*a*) (*1*) Cl; H; $CH_3$ (*3*); $CH_2CH_3$ (*2*) **R**

(*h*) H; Cl; $CH_3$; $CH_2CH_3$ → (*2*) $CH_2CH_3$; H; $CH_3$ (*3*); Cl (*1*) **S**

Structure *a* is **R**. The other **R** configurations are *b, f, i,* and *j*; **S** structures are *c, d, e, g, h,* and *k*.

**2**) The formulas represent the same substance in *a, d, g, h, j,* and *o*; mirror images, i.e., enantiomers are drawn in *b, c, f, i, k, l, m, n*; diastereoisomers in *e, p*.

*Note*: re answer to question in discussion about #15–10: All four groups are bonded through carbon. The second carbon in groups *c* and *d* have a —Cl substituent which gives them priority over *a* and *b*. The third carbon in *d* has a —Cl which places this group in position 1 and group *c* in position 2. The third carbon in *a* has one —Cl, while the third carbon in *b* has two —Cl's. Group *b* has priority over group *a*. Thus, it is not the number of —Cl's that are present, but the way in which they are bonded that determines the rating of the group.

**3**) The answer gives the name of each compound with one or more asymmetric carbons and the formulas of the possible optical isomers of that compound. Each of the compounds in the initial listing *a–e* and *g* has only *one* asymmetric carbon. Therefore, the only optical isomers are one set of enantiomers. The compounds listed subsequently, *f* and *g*, are those with *two* asymmetric carbons. Thus, there are two sets of enantiomers that are diastereoisomers of each other. There is also the possibility for a meso compound.

*a*) 1-chloro-2-propanol

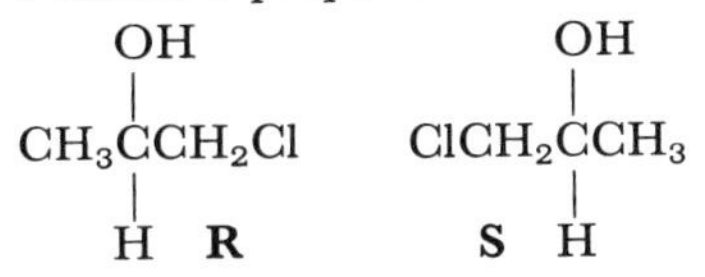

*b*) 1-bromo-1-chloroethane

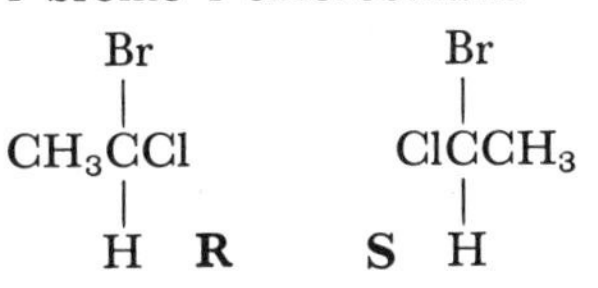

*c*) 1-deutero-1-propanol

| OH | OH |
|---|---|
| $DCCH_2CH_3$ | $CH_3CH_2CD$ |
| H **R** | **S** H |

*d*) 2-bromobutane

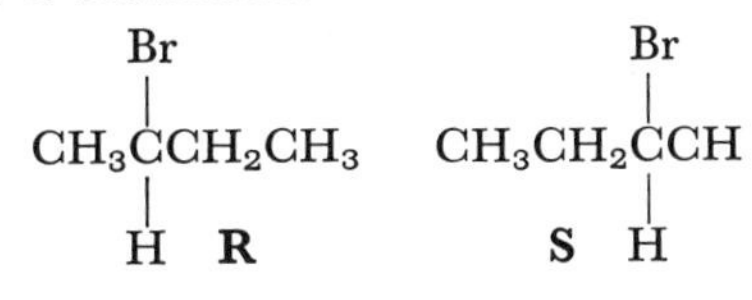

*e*) 2-chloro-N-ethylpropanamine

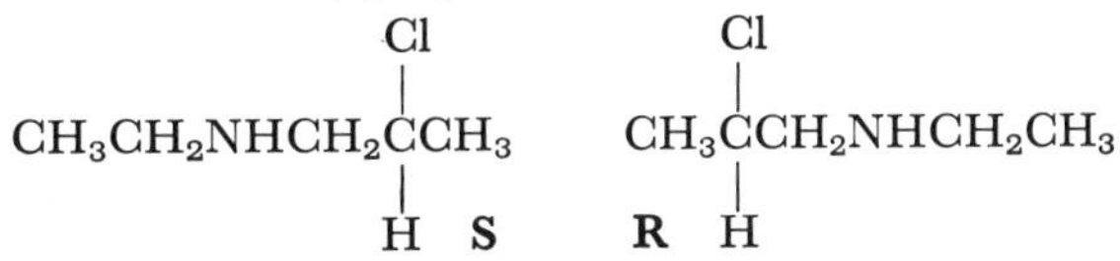

1-methoxy-2-propanol

| OH | OH |
|---|---|
| $CH_3OCH_2CCH_3$ | $CH_3CCH_2OCH_3$ |
| H **S** | **R** H |

2-cyanopentane

| CN | CN |
|---|---|
| $CH_3CH_2CH_2CCH_3$ | $CH_3CCH_2CH_2CH_3$ |
| H **S** | **R** H |

*f*) 2,4-dibromopentane

*a* and *b* are enantiomers; *c* is a meso compound and a diastereoisomer of *a* and of *b*.

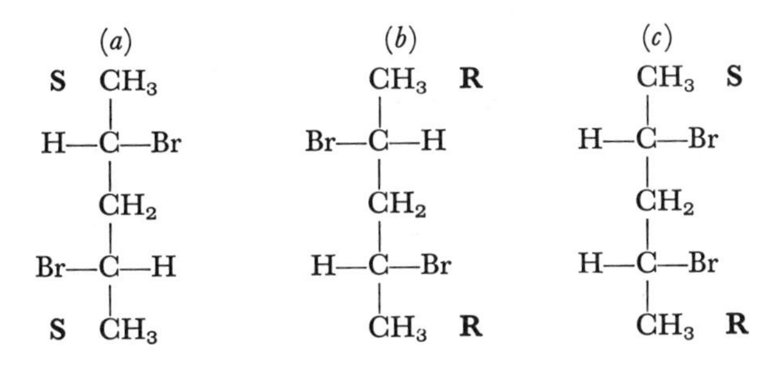

2,4-dibromohexane

*a* and *b* are enantiomers; *c* and *d* are enantiomers; either *c* or *d* is a diastereoisomer of either *a* or *b*.

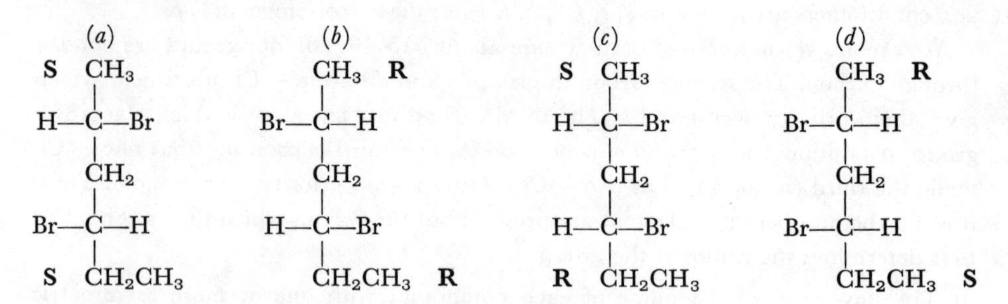

*g*) 2,3-dihydroxybutane

*a* and *b* are enantiomers; *c* is a meso compound and a diastereoisomer of *a* and of *b*.

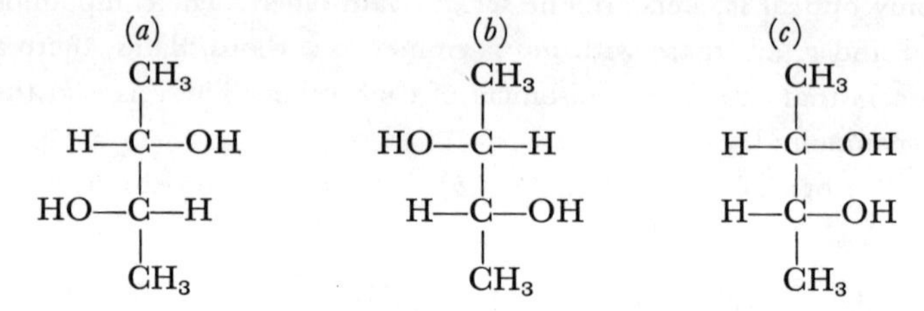

2,3-dihydroxypentane

*a* and *b* are enantiomers; *c* and *d* are enantiomers; either *c* or *d* is a diastereoisomer of either *a* or *b*.

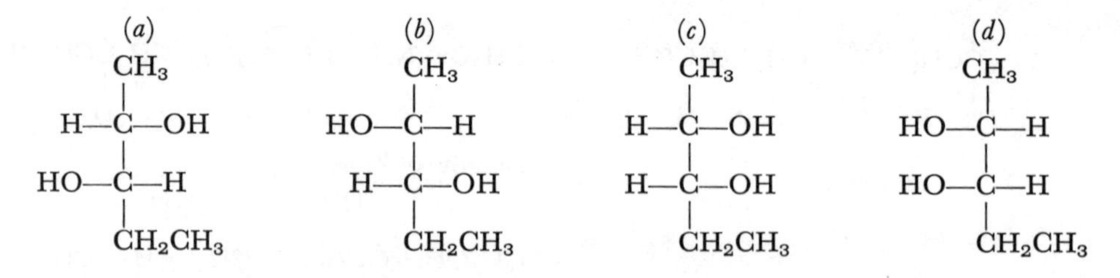

4-hydroxy-4-methyl-2-pentanol

```
    OH  OH                 OH  OH
    |   |                  |   |
CH3CCH2CCH3            CH3CCH2CCH3
    |   |                  |   |
   CH3  H   S          R   H   CH3
```

**4**) The formulas in this answer have been set up so that you can relate them directly back to the tripod structure. In each instance the hydrogen atom below the asymmetric carbon in the formula corresponds to the substituent on the arm of the tripod that extends to the right and behind the plane of the page. The substituent above the asymmetric carbon in the formula corresponds to the substituent on the upright arm of the tripod in the plane of the page. The entire group to the left of the asymmetric carbon should be attached to the left arm of the tripod in the plane of the page. The entire group to the right of the asymmetric carbon in the formula should be attached to the right arm of the tripod that extends in front of the plane of the page. The correct formulas follow:

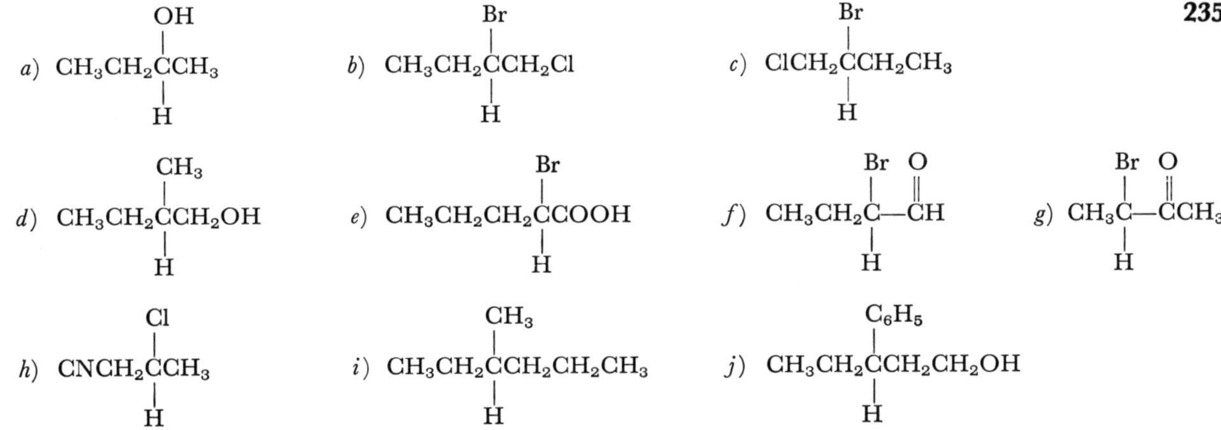

**5**) The substituents on the asymmetric carbon in each compound are listed in the priority order: 1, 2, 3, and 4. The order in which these substituents are arranged on the tripod is indicated by the notation **R** and **S**. This information is sufficient for you to determine whether your tripods have been drawn correctly. Remember the fourth substituent in the list should be on the arm that projects away from you.

*a*) **S** $NH_2$, $CH_3$, D, H.
*b*) **R** $CH_2Cl$, $CH_2CH_3$, $CH_3$, H.
*c*) **R** OH, CHO, $CH_3$, D.
*d*) **R** $CH_3$, CHO, $CH_2OH$, H.
*e*) **R** I, $CH_2CH_2Br$, $CH_2CH_2Cl$, H.
*f*) **S** $OCH_3$, $CH_2SCH_3$, $CH_2OCH_3$, H.
*g*) **R** $CH(OH)_2$, CHO, $CH_2OH$, H.
*h*) **S** COOH, CHO, $CH_2OH$, $CH_3$.
*i*) **S** $CH_2Br$, $C_6H_5$, $CH_2CH_2Cl$, $CH_3$
*j*) **S** CHO, $CH_2OH$, $CH_2{=}CH_2$, $CH_2CH_3$

**6**) Sufficient information is given in answer to problem **3** for you to check the names of the compounds.

**7**)

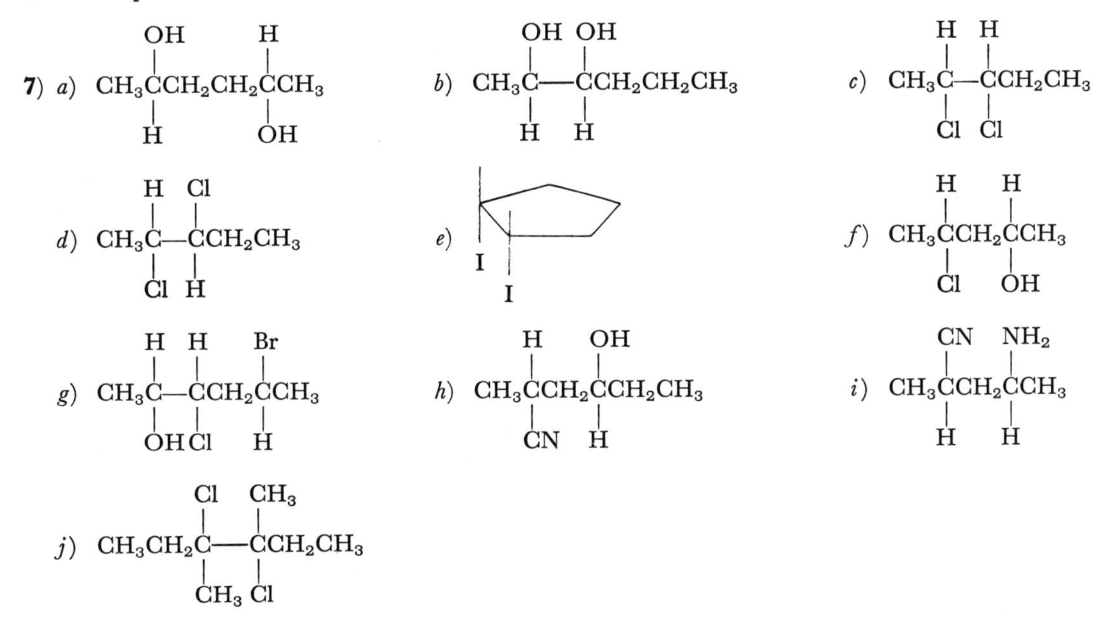

**8**) To illustrate the meaning of *enantiomer* assemble the atoms designated by the empirical formula into a molecular structure with at least one asymmetric carbon atom; to illustrate the term *diastereoisomer* arrange the atoms in a molecular structure with at least two asymmetric carbon atoms; to position the atoms so as to form a meso compound, concentrate on drawing the most symmetrical structure possible with two asymmetric carbon atoms. Then look for a plane of symmetry.

No attempt has been made to include all possible structures for each answer. Only sufficient examples are given to enable you to judge whether the structures you draw are correct and to show you a variety of molecules with the structural requirements for optical isomerism. The names are given in such a way as to emphasize the structural relationships between the compounds and may not be exact IUPAC names. Only those isomers with two or more asymmetric carbons are listed. Each potential meso compound is marked with an asterisk.

*a*) $C_6H_{12}(CN)_2$—2,3-, 2,4-, *2,5-, and *3,4-dicyanohexane; 1,3-, 1,4-, 2,3-, 2,4-, 3,4-, and 3,5- dicyano-2-methylpentane; 1,2-, 1,4-, 2,3-, *2,4-dicyano-3-methylpentane; 1,3-, *1,4-dicyano-2,3-dimethylbutane.

*b*) $C_5H_{10}O_2$—3,4-dihydroxypentene, **cis*- and *trans*- 1,2-, and 1,3-dihydroxycyclopentane; **cis*- and *trans*-1,3-dihydroxy-2-methylcyclobutane, *cis*- and *trans*-1,2-dihydroxy-3-methylcyclobutane; **cis*- and *trans*-1,2-dihydroxy-3-ethylcyclopropane, 1(1-hydroxycyclopropyl) ethanol, 1(2-hydroxycyclopropyl)ethanol. (*Note*: Only the cis isomer can be a meso compound in each case.) *c*) $C_6H_{10}Cl_2$—3,4-, 3,5-, 4,5-dichlorohexene, 4,5-dichloro-2-hexene, *2,5-dichloro-3-hexene; 3,4-, 4,5-dichloro-3-methylpentene, 4-chloro-3-chloromethylpentene, 3-chloro-4-chloromethylpentene; **cis*- and *trans*-1,2-, 1,3-, and 1,4-dichlorocyclohexane; **cis*- and *trans*-1,3-dichloro-2-methylcyclopentane, **cis*- and *trans*- 1,4-dichloro-2-methylcyclopentane, **cis*- and *trans*-1,2-dichloro-4-methylcyclopentane, *cis*- and *trans*-1,2-dichloro-3-methylcyclopentane (the cis and trans in the last two refer only to the two chloro groups. The —$CH_3$ can also be positioned cis and trans relative to the —Cl substituents.); **cis*- and *trans*-1,3-dichloro-2-ethylcyclobutane, **cis*- and *trans*-1,3-dichloro-2-propylcyclopropane.

*d*) $C_6H_8Br_2$—All of the hexenes listed under *c* can exist as hexynes; the pentenes listed under *c* can exist as pentynes; *3,4-dibromo-1,5,-hexadiene; **cis*- and *trans*-2,4-dibromocyclobutylethene, *cis*- and *trans*-2,3-dibromocyclobutylethene; **cis*- and *trans*-3(2,3-dibromocyclopropyl)propene.

*e*) $C_6H_{16}N_2$—Replace the —CN groups in problem *a* with —$NH_2$ groups in all structures: *2,3-di(N-methylamino)butane.

## §19. Electron-Dot Formulas

**1**) You will need to draw an electron-dot formula for each substance before you can answer the question.

*a*) C*1* = 2,6,9,21. C*2* and C*3* = 2,6,9,19,23. O = 4,6,11,17,22. All H's = 1,8.

*b*) C*1* and C*3* = 2,6,9,21. C*2* = 7,15. H = 1,8.

*c*) C = 2,6,9,21. F = 5,6,18. H = 1,8.

*d*) C = 2,6,9,21. P = 3,6,10,16. H = 1,8.

*e*) C = 2,6,9,21. N = 3,6,9,15,20,21. H = 1,8. Br = 5,6,13,14.

*f*) C = 2,6,9,21. O in —$OH_2^+$ = 4,6,5,15,16,20,22. 2 of O's in $NO_3^-$ = 4,6,12,14,18. 1 of O's in $NO_3^-$ = 3,6,9,15,19,23. H = 1,8.

*g*) C*1* and C*2* = 2,6,9,19,24. C*3* and C*4* = 2,6,9,21. S = 4,6,11,17,22. H = 1,8.

*h*) C*1* and C*2* = 2,6,9,21. O = 4,6,10,15,16,20,22. Cl within bracket = 5,6,12,18. $Cl^-$ = 5,6,13,14. H = 1,8.

*i*) C*1* = 2,6,10,14,19,24. C*2* = 2,6,9,19,24. C*3* = 2,6,9,21. H = 1,8. $Na^+$ = 1,7,15.

*j*) C*1* (COOH) = 2,6,9,19,23. C*2* and C*3* = 2,6,9,19,23. =O = 4,6,11,17,19,23. $O^-$ = 4,6,12,14,18. $K^+$ = 1,7,15.

**2**) For *a*, *b*, *e*, and *k*, see #18–3. *f* See #18–16*a*. *g* See #18–16*b*. *h*), *i*), and *j*), Substitute S for O in *e*, *f*, and *g* respectively. *l*) See #18–12. *m*) Extend #18–12. *r*) See #18–11*d*. *s*) See #18–20. *v*) Combine #18–11*d* and #18–12. *w*) See #22–6. *x*) Substitute Fe for Al in *w*. *y*) See #22–6. *z*) See §23, prob. 5 *n*.

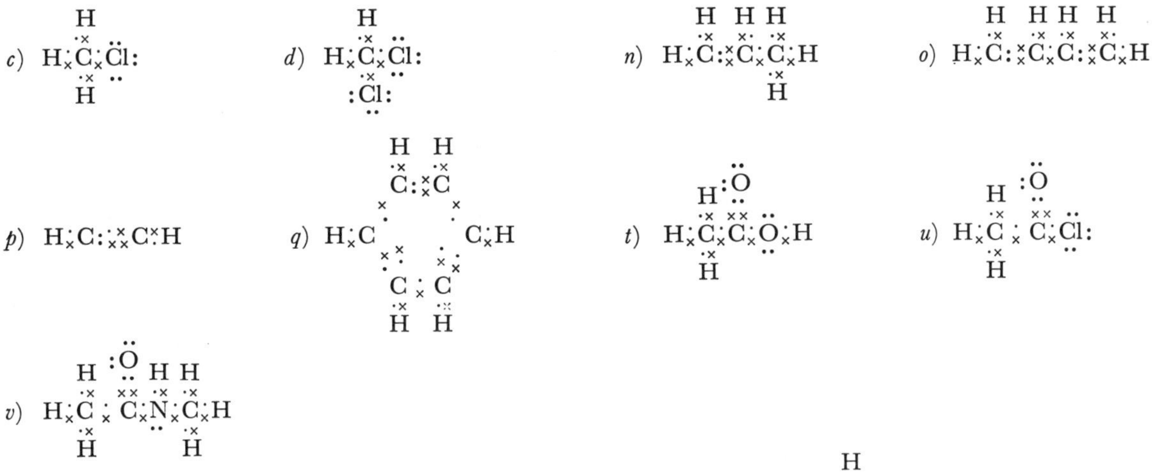

**3**) *a*) See §19, prob. 2*f*.

*b*) 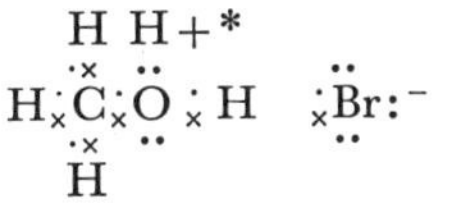

*c*) 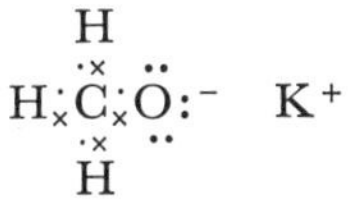

*d*) See §19, prob. 2*l*.

*e*) 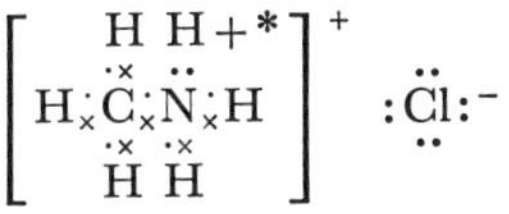

*f*) 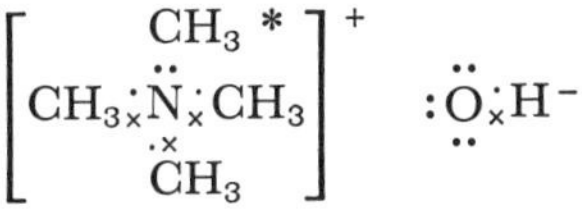

*g*) See §19, prob. 2*g*.

*h*) 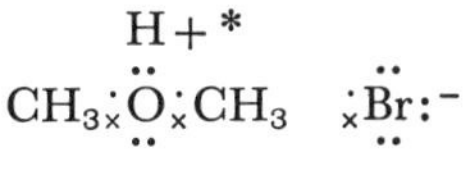

*i*) See §19, prob. 2*p*.

*j*) 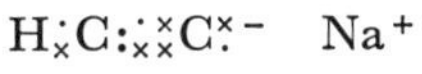

*k*) See §19, prob. 2*q*.

*l*) 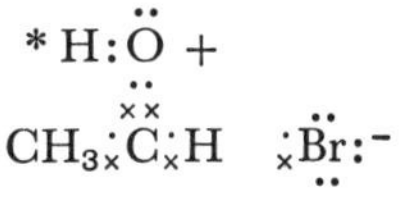

*m*) See §19, prob. 2*t*.

*n*) 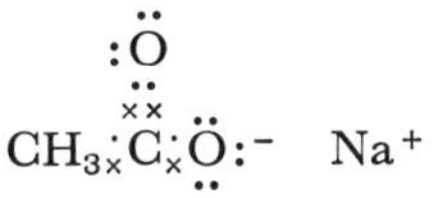

*o*) 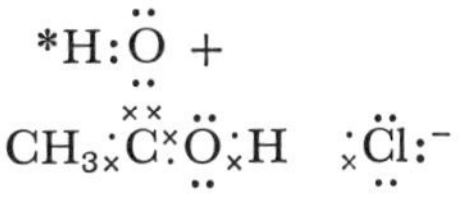

*p*) 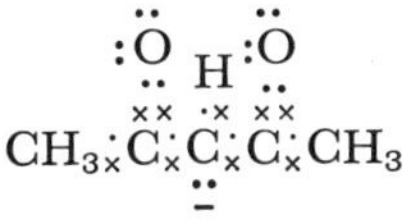

*q*) See §19, prob. 2*q*.

*r*) See §19, prob. 2*q* for ring.

**4**) *a*) 1- and 2-propanol; methyl ethyl ether. *b*) 1,1-, 1,3-, 2,3-, 3,3-dichloropropene; *cis*- and *trans*-1,2-dichloropropene. *c*) Propylamine, methylethylamine, trimethylamine. (In the subsequent answers, structures with an OH or an $NH_2$ on an unsaturated carbon have been omitted. Structures with $-\overset{|}{C}=\overset{|}{C}NHR$ and $-\overset{|}{C}=\overset{|}{C}OR$ have been included, but not the acetylenic analogs.) *d*) butanal, butanone; 3-butenol, 3-buten-2-ol; 2-butanol; *cis*- and *trans*-vinyl ethyl ether, allyl methyl ether, 2-methyl-2-propenol, 2-methoxypropene, *cis*- and *trans*-1-methoxypropene. *e*) 2- and 3-hydroxypropanal; hydroxyacetone; 1-hydroxy-2-propenol; hydroxymethyl vinyl ether. *f*) 3-butynol, 3-butyn-2-ol, 3-methoxypropyne, 3-butenal, 3-buten-2-one, 2-butenal, 2-methyl-2-propenal, $CH_2{=}C{=}CHCH_2OH$, $CH_2{=}C{=}CHOCH_3$. *g*) 4-bromobutyne, 3-bromobutyne, 4-bromo-2-butyne, 2-bromo-1,3-butadiene, 1-bromo-1,3-butadiene, $BrCH{=}C{=}CHCH_3$, $CH_2{=}C{=}CBrCH_3$, $CH_2{=}C{=}CHCH_2Br$. *h*) 3-butynamine, 2-butynamine, N-methyl-2-propynamine, 3-aminobutyne; 1-, and 2-cyanopropane; $CH_2{=}C{=}CHCH_2NH_2$, $CH_2{=}C{=}CHNHCH_3$, $CH_2{=}CHNHCH{=}CH_2$. *i*) 1-, and 2-nitropropane. *j*) Same structures as for *d*. Substitute S for O. *k*) 2-, and 3-chloropropanal; chloroacetone; 2-, and 3-chloro-2, propenol; 1-, and 2-chloromethoxyethene, chloro-methyl vinyl ether. *l*) 1-hydroxy-2-butynone, 2-hydroxy-3-propynal, 4-hydroxy-2-propynal, 2-propynyl formate, methyl-2-propynoate, 3-butynoic acid, $CH_2{=}C{=}CHCOOH$.

## §21. Carbonium Ions, Carbanions, and Free Radicals

**1**) *a*) isopropyl carbonium ion (2°),* *n*-propyl carbonium ion (1°). *b*) *tert.*-butyl free radical (3°),* isopropyl free radical (2°). *c*) methylbenzyl carbonium ion (benzyl),* isopropyl carbonium ion (2°). *d*) methylbenzyl carbanion (benzyl),* phenylethyl carbanion (1°). *e*) methyl carbonium ion (methyl), ethyl carbonium ion (1°).* *f*) allyl free radical (allyl),* methylvinyl free radical (vinyl). *g*) *tert.*-butyl carbonium ion (3°),* *n*-propyl carbonium ion (1°). *h*) *p*-methylbenzyl carbonium ion (benzyl),* *p*-methylphenylethyl carbonium ion (1°). *i*) methylbenzyl free radical (benzyl),* isopropyl free radical (2°). *j*) methyl carbanion (methyl), isopropyl carbanion (2°).* *k*) *sec.*-butyl free radical (2°), *p*-methyl(methylbenzyl) free radical (benzyl).* *l*) 1,2-dimethylvinyl carbonium ion (vinyl), 1-methylallyl carbonium ion (allyl).*

**2**) *a*) ethyl, isopropyl, *tert*-butyl. *b*) ethyl, *tert.*-butyl, allyl. *c*) ethyl, isopropyl, benzyl. *d*) methyl, ethyl, isopropyl. *e*) $C_6H_{13}\overset{+}{C}HCH_3$, $C_6H_5\overset{+}{C}HCH_3$, $(C_6H_5)_2H+$.

**3**) *a*) See #20–9 for ethyl free radical. Isopropyl free radical has six forms in addition to the one given in the problem. Those shown in *a* and *b* and two more of each form placing the free radical electron on each of the terminal hydrogens in succession.

```
    H   H   H              H   H   H
    ·   |   |              |   |   ·
H—C═C—C—H  ⟷  H—C—C═C—H
    |       |              |       |
    H       H              H       H
```

The tertiary butyl free radical has nine such structures. There are three positions for ·H on each of the three 1° carbons for each of the three positions of the C═C. One of the structures is shown.

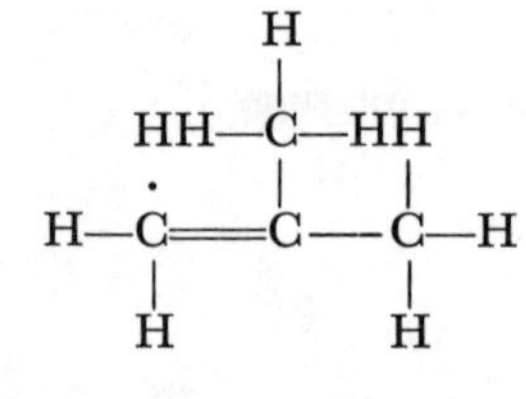

*b*) Same structures as for *a* with a + in place of the free radical electron ·.
*c*) The basic structure for the *n*-propyl radical is shown.

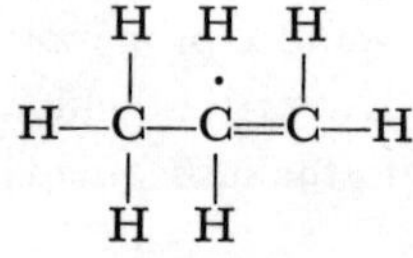

There is a second form with the ·H at the other secondary hydrogen. In $(CH_3CH_2)_3C\cdot$ allow for six structures with ·H on a secondary carbon.
*d*) Same structures as for *c* but with a + in place of the free radical electron.

* Indicates more stable species.

**4)** *a)* $CH_3CH_2\dot{C}HCH_2$, $CH_3CH_2CH_2CH_2\cdot$ *b)* $(CH_3)_3C\cdot$, $(CH_3)_2CHCH_2\cdot$

*c)* $CH_3CH_2\dot{C}HCH_2CH_3$ and $CH_3CH_2CH_2\dot{C}HCH_3$, $CH_3CH_2CH_2CH_2CH_2\cdot$

*d)* $(CH_3)_2\dot{C}CH_2CH_3$, $(CH_3)_2CH\dot{C}HCH_3$, $(CH_3)_2CHCH_2CH_2\cdot$ *e)* $(CH_3)_3CCH_2\cdot$

*f)* $(CH_3)_2\dot{C}CH_2CH(CH_3)_2$, $(CH_3)_2CH\dot{C}HCH(CH_3)_2$, $(CH_3)_2CHCH_2CH(CH_3)CH_2\cdot$

*g)* $CH_2{=}CHCH_2\cdot$, $CH_2{=}\dot{C}CH_3$ and $\dot{C}H{=}CHCH_3$

*h)* $C_6H_5\dot{C}HCH_3$, $C_6H_5CH_2CH_2\cdot$
(hydrogens on aromatic ring react only under special conditions).

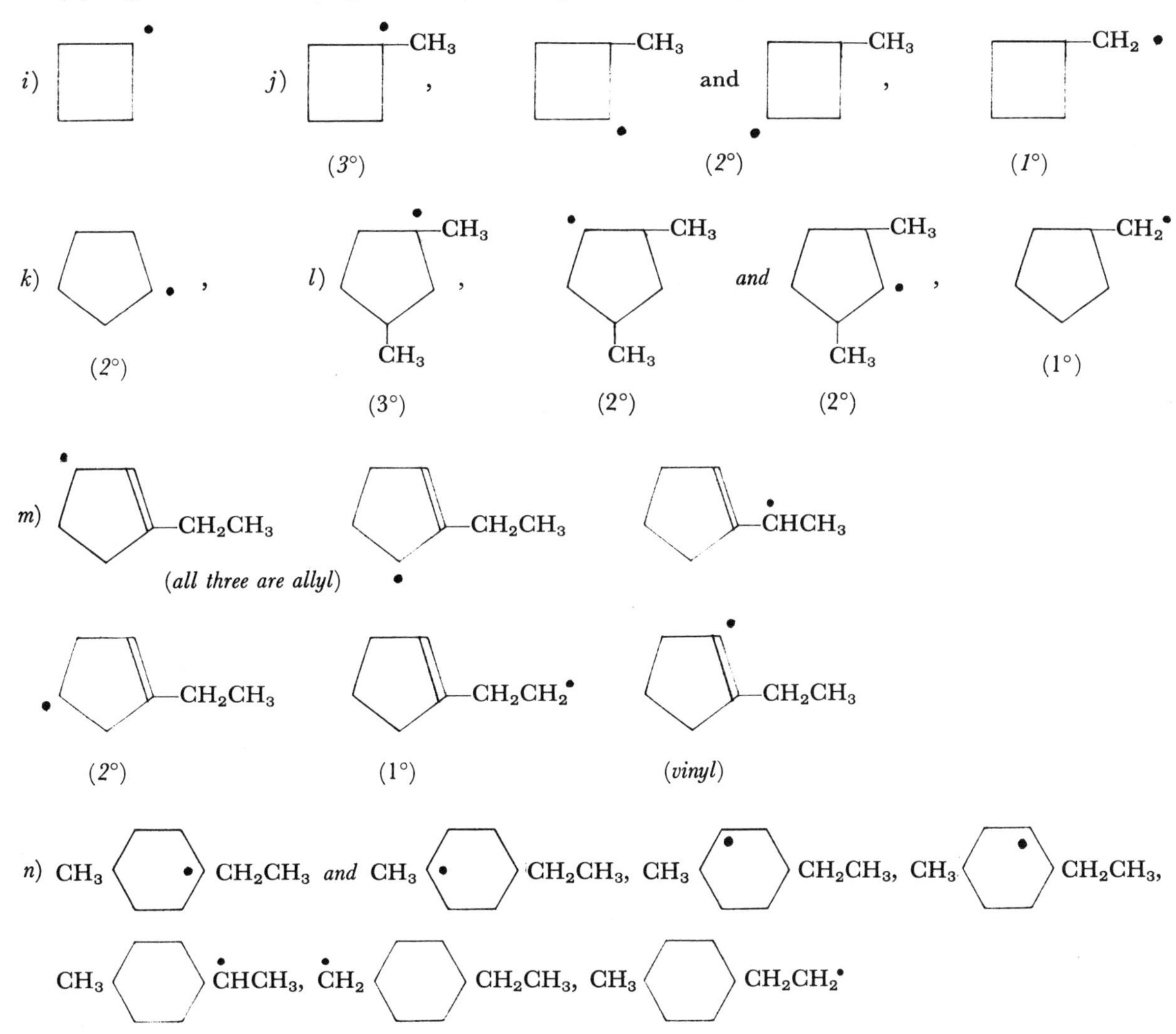


**5)** *a)* $CH_3\overset{+}{C}HCH_3^*$, $CH_3CH_2CH_2+$ *b)* $(CH_3)_2\overset{+}{C}CH_3^*$, $(CH_3)_2CHCH_2+$

*c)* $CH_3CH_2\overset{+}{C}(CH_3)CH_2CH_3^*$, $CH_3\overset{+}{C}HC(CH_3)HCH_2CH_3$

*d)* $(CH_3)_2\overset{+}{C}CH_2CH(CH_3)_2^*$, $(CH_3)_2CH\overset{+}{C}HCH(CH_3)_2$

*e)* $(CH_3)_2CHCH_2\overset{+}{C}HCH_3$, $(CH_3)_2CH\overset{+}{C}HCH_2CH_3$, both 2°. *f)* $CH_3\overset{+}{C}HCH{=}CH_2^*$, $\overset{+}{C}H_2CH_2CH{=}CH_2$

*g)* $CH_3\overset{+}{C}HC(CH_3){=}CH_2^*$, $\overset{+}{C}H_2CH_2C(CH_3){=}CH_2$, $CH_2{=}CH\overset{+}{C}(CH_3)_2^*$, $CH_2{=}CHC(CH_3)H\overset{+}{C}H_2$

*h)* $C_6H_5CH_2\overset{+}{C}HCH_3$, $C_6H_5\overset{+}{C}HCH_2CH_3^*$

*i)* $C_6H_5CH_2\overset{+}{C}(CH_3)_2$, $C_6H_5\overset{+}{C}HCH(CH_3)_2^*$

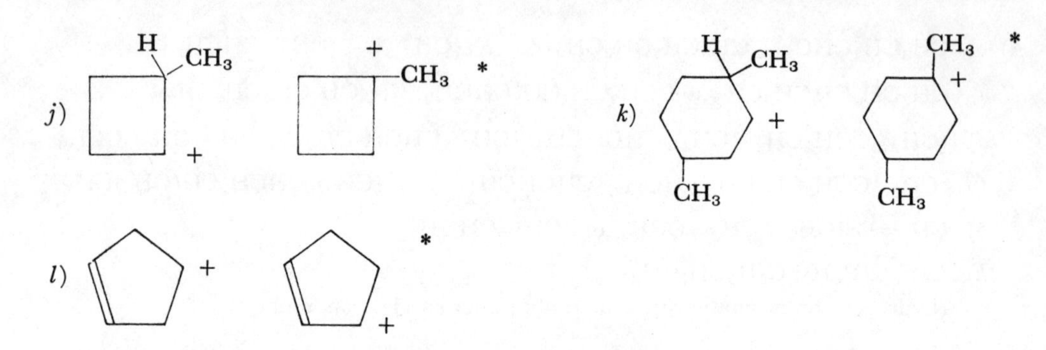

6) The answers are the same as those for problem 5 with the + of the carbonium ion replaced by the · of a free radical.

## §23. Nonbonding Electron Pairs. Proton Exchange Reactions

**1**) The following need two pairs of nonbonding electron pairs on oxygen: *a, d, e, h, i, j, k, l, n, o,* and *t*. The following need a pair of nonbonding electrons on nitrogen: *f, l, p, q,* and *r*. Each halogen needs three pairs of nonbonding electrons in addition to the bonding pair. The oxygen in the hydronium ion needs only one pair of electrons.

**2**) Wherever there is a nonbonding pair of electrons on oxygen or nitrogen, add a proton and indicate that the unit is a cation with a +1 charge.

**3**) 1 = *b, c, f, g, h, j, k, m, p, q, s, u, v,* and *w*. 3 = same as 1.
2 = *c, e, g, j, r,* and *x*. 4 = *d, i, l, m, n, o, t, y,* and *z*.

**4**) *b*) $\mathrm{NH_3} + \mathrm{H}{:}\ddot{\underset{\cdot\cdot}{\mathrm{Cl}}}{:} \rightleftharpoons \mathrm{NH_4^+} + {:}\ddot{\underset{\cdot\cdot}{\mathrm{Cl}}}{:}^-$

*c*) $\mathrm{H\underset{\cdot\cdot}{\ddot{O}}H} + \mathrm{H}{:}\ddot{\underset{\cdot\cdot}{\mathrm{Cl}}}{:} \rightleftharpoons \mathrm{H_3\ddot{O}^+} + {:}\ddot{\underset{\cdot\cdot}{\mathrm{Cl}}}{:}^-$

*f*) ${:}\underset{\cdot\cdot}{\ddot{\mathrm{O}}}\mathrm{H}^- + \mathrm{H}{:}\ddot{\underset{\cdot\cdot}{\mathrm{Cl}}}{:} \rightleftharpoons \mathrm{H\underset{\cdot\cdot}{\ddot{O}}H} + {:}\ddot{\underset{\cdot\cdot}{\mathrm{Cl}}}{:}^-$

*g*) $\mathrm{CH_3\underset{\cdot\cdot}{\ddot{O}}H} + \mathrm{H}{:}\ddot{\underset{\cdot\cdot}{\mathrm{Cl}}}{:} \rightleftharpoons \mathrm{CH_3\overset{H\,+}{\ddot{O}}H} + {:}\ddot{\underset{\cdot\cdot}{\mathrm{Cl}}}{:}^-$

*h*) $\mathrm{CH_3\ddot{N}H_2} + \mathrm{H}{:}\ddot{\underset{\cdot\cdot}{\mathrm{Cl}}}{:} \rightleftharpoons \mathrm{CH_3\overset{H\,+}{\ddot{N}}H_2} + {:}\ddot{\underset{\cdot\cdot}{\mathrm{Cl}}}{:}^-$

*j*) $\mathrm{CH_3\overset{\overset{:\ddot{O}}{\|}}{C}OH} + \mathrm{HCl} \rightleftharpoons \mathrm{CH_3\overset{\overset{H:\ddot{O}\,+}{\|}}{C}OH} + {:}\ddot{\underset{\cdot\cdot}{\mathrm{Cl}}}{:}^-$

*k*) $\mathrm{CH_3\overset{\overset{:\ddot{O}}{\|}}{C}\underset{\cdot\cdot}{\ddot{O}}{:}^-} + \mathrm{HCl} \rightleftharpoons \mathrm{CH_3\overset{\overset{:\ddot{O}}{\|}}{C}\underset{\cdot\cdot}{\ddot{O}}{:}H} + {:}\ddot{\underset{\cdot\cdot}{\mathrm{Cl}}}{:}^-$

*m*) ${:}\ddot{\underset{\cdot\cdot}{\mathrm{Cl}}}{:}^- + \mathrm{H}{:}\ddot{\underset{\cdot\cdot}{\mathrm{Cl}}}{:} \longrightarrow \mathrm{N.R.}$

*p*) ${:}\ddot{\mathrm{N}}\mathrm{H_2^-} + \mathrm{H}{:}\ddot{\underset{\cdot\cdot}{\mathrm{Cl}}}{:} \rightleftharpoons \mathrm{\ddot{N}H_3} + {:}\ddot{\underset{\cdot\cdot}{\mathrm{Cl}}}{:}^-$

*q*) $\mathrm{CH_3\overset{\overset{:\ddot{O}}{\|}}{C}H} + \mathrm{H}{:}\ddot{\underset{\cdot\cdot}{\mathrm{Cl}}}{:} \rightleftharpoons \mathrm{CH_3\overset{\overset{H:\ddot{O}\,+}{\|}}{C}H} + {:}\ddot{\underset{\cdot\cdot}{\mathrm{Cl}}}{:}^-$

*s*) $\mathrm{CH_3\overset{\overset{:\ddot{O}}{\|}}{C}CH_3} + \mathrm{HCl} \rightleftharpoons \mathrm{CH_3\overset{\overset{H:\ddot{O}\,+}{\|}}{C}CH_3} + {:}\ddot{\underset{\cdot\cdot}{\mathrm{Cl}}}{:}^-$

*u*) $\mathrm{CH_3\underset{\cdot\cdot}{\ddot{O}}CH_3} + \mathrm{HCl} \rightleftharpoons \mathrm{CH_3\overset{H}{\underset{\cdot\cdot}{O}}CH_3}$

*v*) $\mathrm{HC{\equiv}C{:}^-} + \mathrm{H}{:}\ddot{\underset{\cdot\cdot}{\mathrm{Cl}}}{:} \rightleftharpoons \mathrm{HC{\equiv}CH} + {:}\ddot{\underset{\cdot\cdot}{\mathrm{Cl}}}{:}^-$

*w*) $\mathrm{CH_2{=}CHCH_2{:}^-} + \mathrm{H}{:}\ddot{\underset{\cdot\cdot}{\mathrm{Cl}}}{:} \rightleftharpoons \mathrm{CH_2{=}CHCH_3} + {:}\ddot{\underset{\cdot\cdot}{\mathrm{Cl}}}{:}^-$

5) *c*) $H\ddot{O}H + :\ddot{O}H^- \rightleftharpoons H\ddot{O}:^- + H\ddot{O}H$

*e*) $H{-}\overset{H\,+}{\overset{|}{\ddot{O}}}{-}H + :\ddot{O}H^- \rightleftharpoons H\ddot{O}H + H\ddot{O}H$

*g*) $CH_3\ddot{O}H + :\ddot{O}H^- \rightleftharpoons CH_3\ddot{O}:^- + :\ddot{O}H^-$

*j*) $CH_3\overset{:\ddot{O}}{\overset{\|}{C}}\ddot{O}H + :\ddot{O}H^- \rightleftharpoons CH_3\overset{:\ddot{O}}{\overset{\|}{C}}\ddot{O}:^- + H\ddot{O}H$

*r*) $CH_3CH_2\overset{H\,+}{\ddot{O}}H + :\ddot{O}H^- \rightleftharpoons CH_3CH_2\ddot{O}H + H\ddot{O}H$

*x*) $CH_3\overset{H\,+}{\ddot{O}}CH_3 + :\ddot{O}H^- \rightleftharpoons CH_3\ddot{O}CH_3 + H\ddot{O}H$

6) *a*) $H:\ddot{Cl}: + H\ddot{O}H \rightleftharpoons :\ddot{Cl}:^- + H:\overset{H\,+}{\ddot{O}}:H$
(*B. base*)

*c*) $H\ddot{O}H + H\ddot{O}H \rightleftharpoons H\overset{H\,+}{\ddot{O}}H + :\ddot{O}H^-$
(*B. acid*) (*B. base*)

*d*) $H\ddot{O}H + H^+ \rightleftharpoons H_3\ddot{O}^+$
(*B. base*)

*j*) $CH_3\overset{:\ddot{O}}{\overset{\|}{C}}\ddot{O}H + H\ddot{O}H \rightleftharpoons CH_3\overset{:\ddot{O}}{\overset{\|}{C}}\ddot{O}:^- + H_3O:^+$
(*B. base*)

*l*) $H{-}\overset{H}{\overset{|}{\underset{H}{\underset{|}{C}}}}{+} + H\ddot{O}H \rightleftharpoons H{-}\overset{H}{\overset{|}{\underset{H}{\underset{|}{C}}}}\ddot{O}H_2^+$
(*L. base*)

*n*) $\overset{:\ddot{O}\cdot\ \ \cdot\ddot{O}:}{\overset{\diagdown\ \diagup}{\underset{:\ddot{O}:}{\underset{|}{S}}}} + H\ddot{O}H \rightleftharpoons :\ddot{O}{-}\overset{:\ddot{O}:H}{\overset{|}{\underset{:\ddot{O}:}{\underset{|}{S}}}}{-}\ddot{O}:H$
(*L. base*)

*o*) $C_6H_5CH_2{+} + H\ddot{O}H \rightleftharpoons C_6H_5\overset{H}{\overset{|}{\underset{H}{\underset{|}{C}}}}:\ddot{O}H_2^+$
(*L. base*)

*r*) $CH_3CH_2\ddot{O}H_2^+ + H\ddot{O}H \rightleftharpoons CH_3CH_2\ddot{O}H + H_3\ddot{O}^+$
(*L. base*)

*t*) $CH_3\overset{+}{\underset{H}{\underset{|}{C}}}CH_3 + H\ddot{O}H \rightleftharpoons CH_3\overset{CH_3}{\overset{|}{C}}\ddot{O}H_2^+$
(*L. base*)

*x*) $CH_3\overset{H+}{\ddot{O}}CH_3 + H\ddot{O}H \rightleftharpoons CH_3\ddot{O}H + H_3\ddot{O}^+$
(*B. base*)

*z*) $CH_2{=}CHCH_2{+} + H\ddot{O}H \rightleftharpoons CH_2{=}CH\overset{H}{\overset{|}{\underset{H}{\underset{|}{C}}}}:\ddot{O}H_2^+$
(*L. base*)

## §25. Inductive Effect. Partial Charge

**1**) Each halogen, each oxygen, and each nitrogen has a partial negative charge, $\delta-$. Each carbon adjacent to one of these hetero atoms has a partial positive charge, $\delta+$. Only the first carbon need be marked although it should be kept in mind that this effect is transmitted through the $\sigma$ bond framework and may function as far as three carbons from the hetero atom. If there are two carbons adjacent to the hetero atom as in an ether, both should be marked with $\delta+$.

**2)** *a*) $\overset{\delta+}{CH_3}\overset{\delta-}{\ddot{\underset{..}{O}}H} + \overset{\delta+}{H}{\times}\overset{\delta-}{\ddot{\underset{..}{Br}}}: \rightleftharpoons CH_3\overset{H^+}{\ddot{\underset{..}{O}}}H + {\times}\ddot{\underset{..}{Br}}:^-$

*b*) $\overset{\delta+}{CH_3}\overset{\delta-}{\ddot{\underset{..}{O}}H} + H{\times}H \longrightarrow$ N.R.

*c*) $\overset{\delta+}{CH_3}\overset{\delta-}{\ddot{N}H_2} + Na^+:\ddot{\underset{..}{O}}{\times}H^- \longrightarrow$ N.R.

*d*) $\overset{\delta+}{CH_3}\overset{\delta-}{\ddot{N}H_2} + \overset{\delta+}{H}{\times}\overset{\delta-}{\ddot{\underset{..}{Cl}}}: \rightleftharpoons CH_3NH_3^+:\ddot{\underset{..}{Cl}}{\times}^-$

*e*) $CH_3CH_2CH_3 + H{\times}\ddot{\underset{..}{F}}: \longrightarrow$ N.R.

*f*) $\overset{\delta+}{CH_3}\overset{\delta-}{\ddot{S}H} + \overset{\delta+}{H}{\times}\overset{\delta-}{\ddot{\underset{..}{F}}}: \rightleftharpoons CH_3\overset{H+}{\ddot{S}}H + {\times}\ddot{\underset{..}{F}}:^-$

*g*) $\overset{\delta+}{CH_3}\overset{\delta-}{\ddot{\underset{..}{O}}}\overset{\delta+}{CH_3} + H{\times}\ddot{\underset{..}{Cl}}: \rightleftharpoons CH_3\overset{H}{\ddot{\underset{..}{O}}}CH_3 + {\times}\ddot{\underset{..}{Cl}}:^-$

*h*) $\overset{\delta+}{CH_3}\overset{\delta-}{\ddot{\underset{..}{O}}}\overset{\delta+}{CH} + H{\times}H \longrightarrow$ N.R.

*i*) $(CH_3)_2CHCH_3 + \overset{\delta+}{H}{\times}\overset{\delta-}{\ddot{\underset{..}{F}}}: \longrightarrow$ N.R.

*j*) $\underset{\delta+}{CH_3}\overset{\overset{\delta-}{:\ddot{O}}}{\overset{\|}{C}}\underset{\delta-}{\ddot{\underset{..}{O}}H} + H{\times}\ddot{\underset{..}{I}}: \rightleftharpoons CH_3\overset{H:\ddot{O}^+}{\overset{\|}{C}}\ddot{\underset{..}{O}}H + {\times}\ddot{\underset{..}{I}}:^-$

*k*) $\overset{\delta+}{CH_3CH_2}\overset{\delta-}{\ddot{\underset{..}{Cl}}}: + \overset{\delta+}{H}{\times}\overset{\delta-}{\ddot{\underset{..}{Br}}}: \longrightarrow$ N.R.

*l*) $\underset{\delta+}{CH_3}\overset{\overset{\delta-}{:\ddot{O}}}{\overset{\|}{C}}\underset{\delta-}{\ddot{\underset{..}{O}}H} + Na^+ + :\ddot{\underset{..}{O}}H^- \rightleftharpoons CH_3\overset{:\ddot{O}}{\overset{\|}{C}}\ddot{\underset{..}{O}}:^- + H\ddot{\underset{..}{O}}H$

**3)** *a*) 1,2,3; *b*) 1,3,2; *c*) 2,1,3; *d*) 2,1,3; *e*) 3,2,1.

**4)** *a*) 3,2,1; *b*) 2,3,1; *c*) 3,1,2; *d*) 3,1,2; *e*) 3,1,2.

**5)** *a*) 1,3,2; *b*) 1,3,2; *c*) 2,1,3; *d*) 3,2,1; *e*) 3,1,2.

**6)** *a*) 2,3,1; *b*) 2,3,1; *c*) 3,1,2; *d*) 1,2,3; *e*) 2,1,3.

**7)** *a*) 1,2,3; *b*) 2,1,3; *c*) 2,1,3; *d*) 3,2,1; *e*) 3,1,2.

**8)** *a*) 2,3,1; *b*) 3,1,2; *c*) 3,2,1; *d*) 3,2,1; *e*) 2,1,3; *f*) 1,3,2; *g*) 3,1,2; *h*) 3,2,1.

**9)** *a*) 2,1,3; *b*) 2,1,3; *c*) 1,3,2; *d*) 2,3,1; *e*) 2,3,1; *f*) 2,1,3; *g*) 1,3,2; *h*) 2,1,3.

## §27. Resonance. Molecular Orbital Diagrams. Canonical Forms

**1)** The molecular orbital diagrams used to answer these questions do not attempt to draw the $\pi$ cloud. They merely indicate the orbitals that overlap. This is a useful device that enables you to represent the electron configurations that lead to resonance and a $\pi$ cloud.

Compounds that do not have conjugated bonds are *a*, *d*, *f*, *g*, and *j*. Compounds in which there is conjugation because of the aromatic ring but not because of the side chain are *l*, *m*, and *q*. The effects of conjugation may be diagramed as follows:

*b*)

H H C-C-C-C H H

[ H H H H / H—C∸C—C∸C—H ↔ H H H H / H—Ċ—C∸C—Ċ—H ↔

H H H H / H—C(:)—C∸C—C(+)—H ↔ H H / H—C(+)—C∸C—C(:)—H ]

*c*)

H H C-C-C-O

[ H H H / H—C∸C—C∸Ö: ↔ H H H / H—C(+)—C∸C—Ö:(:) ↔ H H H / H—C(:)—C∸C—Ö:(+) ]

*e*)

H H C-C-C-O-O H

[ H H :Ö / H—C∸C—C(:)—ÖH ↔ H H :Ö:$^{-1}$ / H—C∸C—C∸ÖH$^{+1}$ ↔

H H :Ö:$^{-}$ / H—C(+)—C∸C—ÖH ↔ H H :Ö+ / H—C(:)—C∸C—ÖH ]

*h*) is the same as *e* except that —OH is replaced by $—OCH_3$.

*i*) is the same as *c* except that —H on the carbonyl carbon is replaced by $—CH_3$.

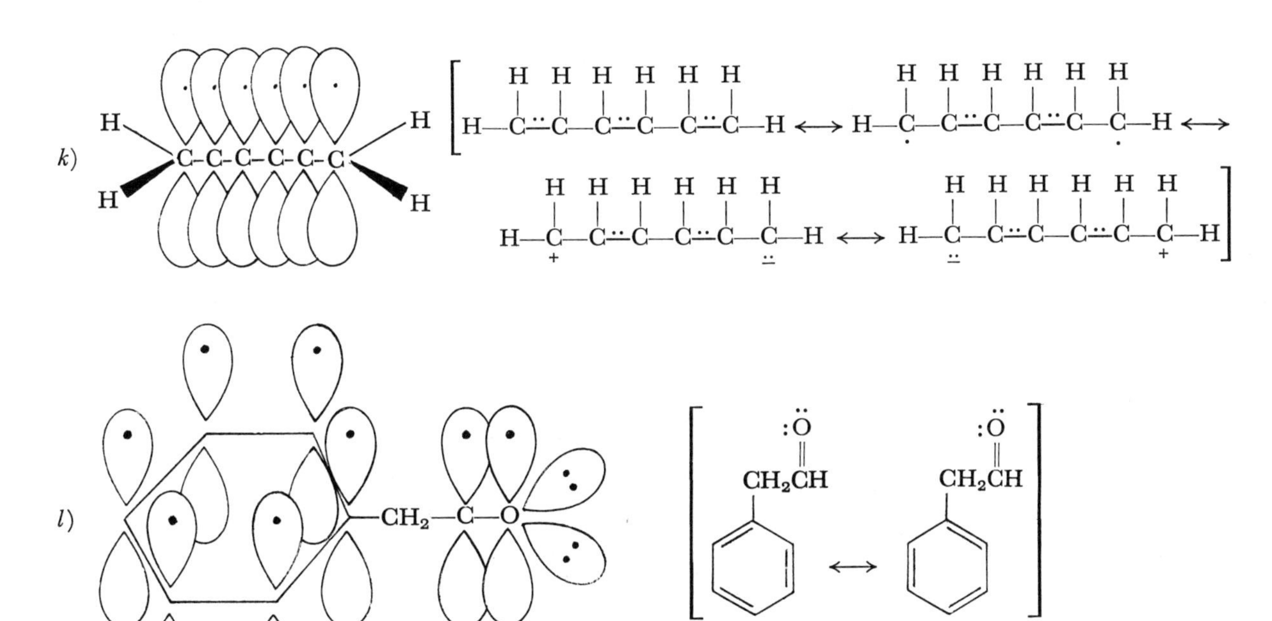

*m*)

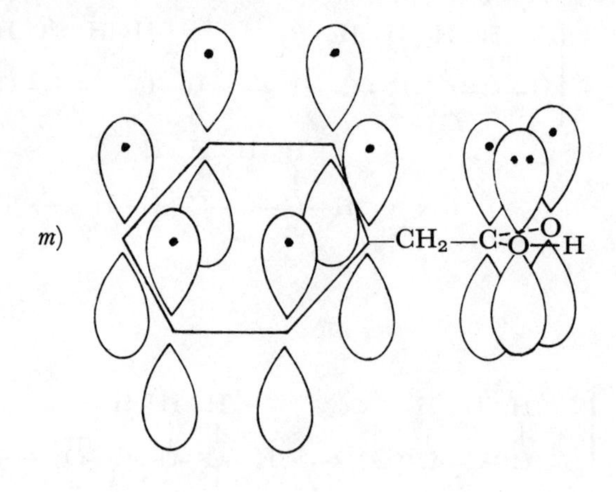


:Ö: $CH_2C$ÖH ⟷ :Ö: $CH_2C$ÖH ⟷ :Ö:$^-$ $CH_2C$=ÖH$^+$ ⟷ :Ö:$^-$ $CH_2C$=ÖH$^+$

*n*

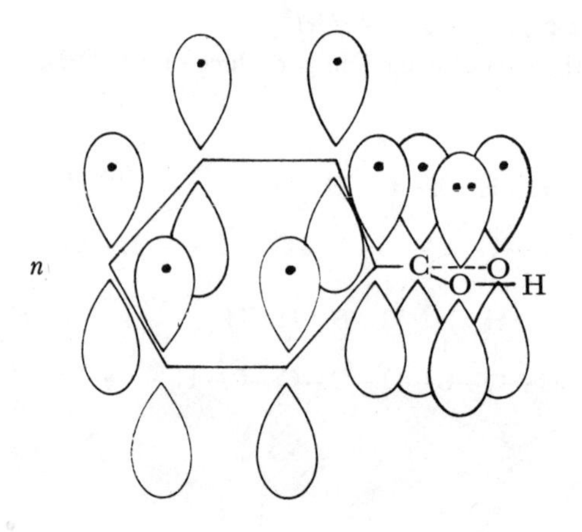


$^-$:Ö: ÖH C + ⟷ $^-$:Ö ÖH C + ⟷ $^-$:Ö ÖH C + ≡ O C OH δ$-$ δ+ δ+ δ+

*o*) is the same as *n* except that —OH is replaced by —$OCH_3$.

p)

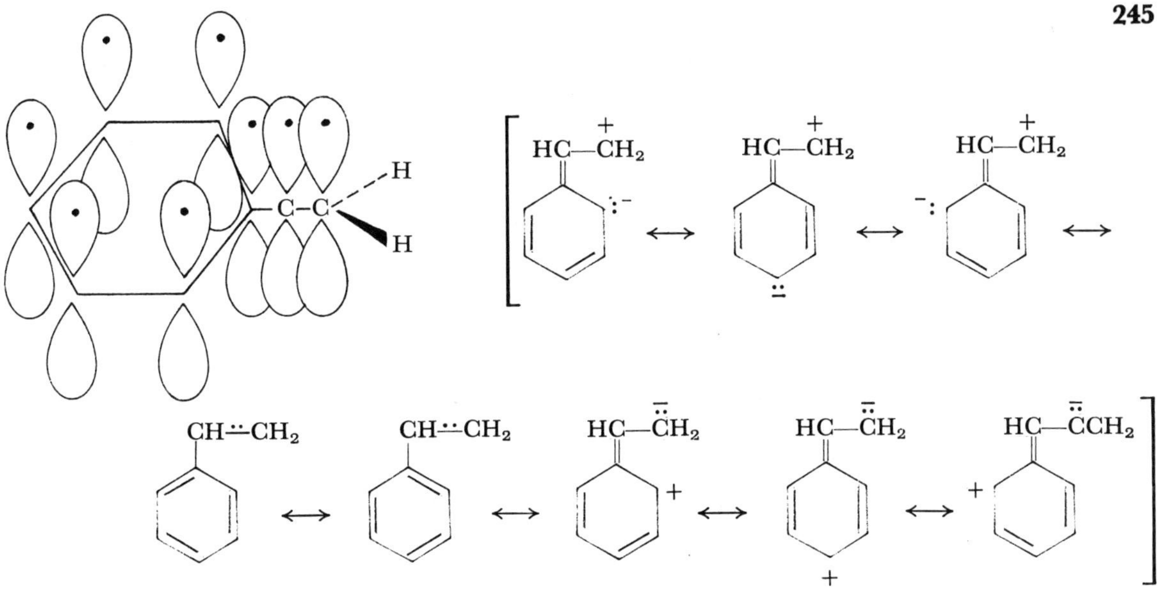

$C{-}C$ H H

$HC{-}\overset{+}{C}H_2$ ⟷ $HC{-}\overset{+}{C}H_2$ ⟷ $HC{-}\overset{+}{C}H_2$ ⟷

$CH{\cdots}CH_2$ ⟷ $CH{\cdots}CH_2$ ⟷ $HC{-}\ddot{C}H_2$ + ⟷ $HC{-}\ddot{C}H_2$ + ⟷ + $HC{-}\ddot{C}CH_2$

q)

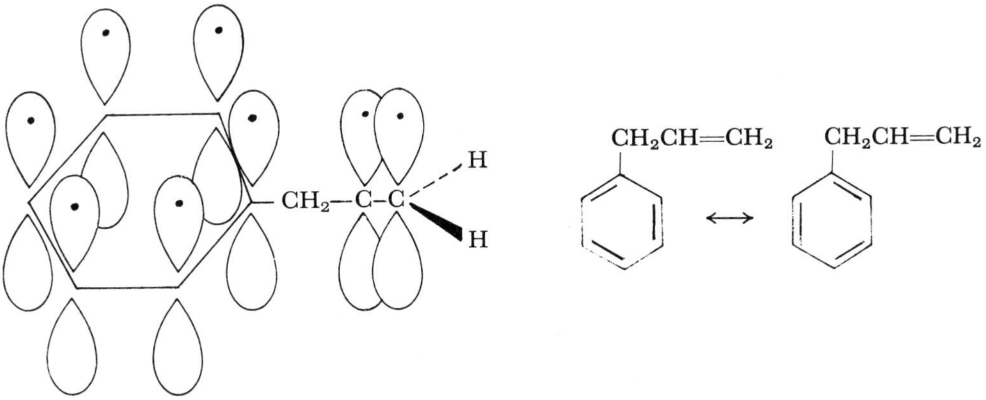

$CH_2{-}C{-}C$ H H

$CH_2CH{=}CH_2$ ⟷ $CH_2CH{=}CH_2$

r)

$C{-}O$

$\ddot{O}$ H C + ⟷ $\ddot{O}$ H C + ⟷ + $\ddot{O}$ H C ≡ O H C $\delta-$ $\delta+$ $\delta+$ $\delta+$

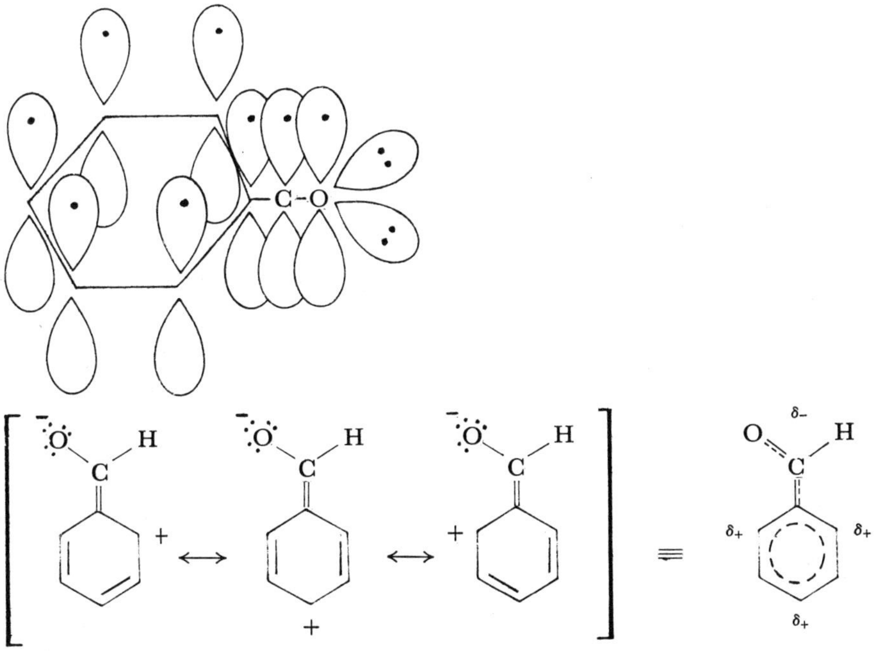

2) Compounds *b*, *d*, *e*, *i*, and *j* do not have a nonbonding pair of electrons on an atom adjacent to an unsaturated atom. There is ring resonance and hyperconjugation in *b* and *e* but not the type of resonance under consideration in this problem. The effects of a nonbonding electron pair on an atom adjacent to a doubly bonded atom can be diagrammed as follows:

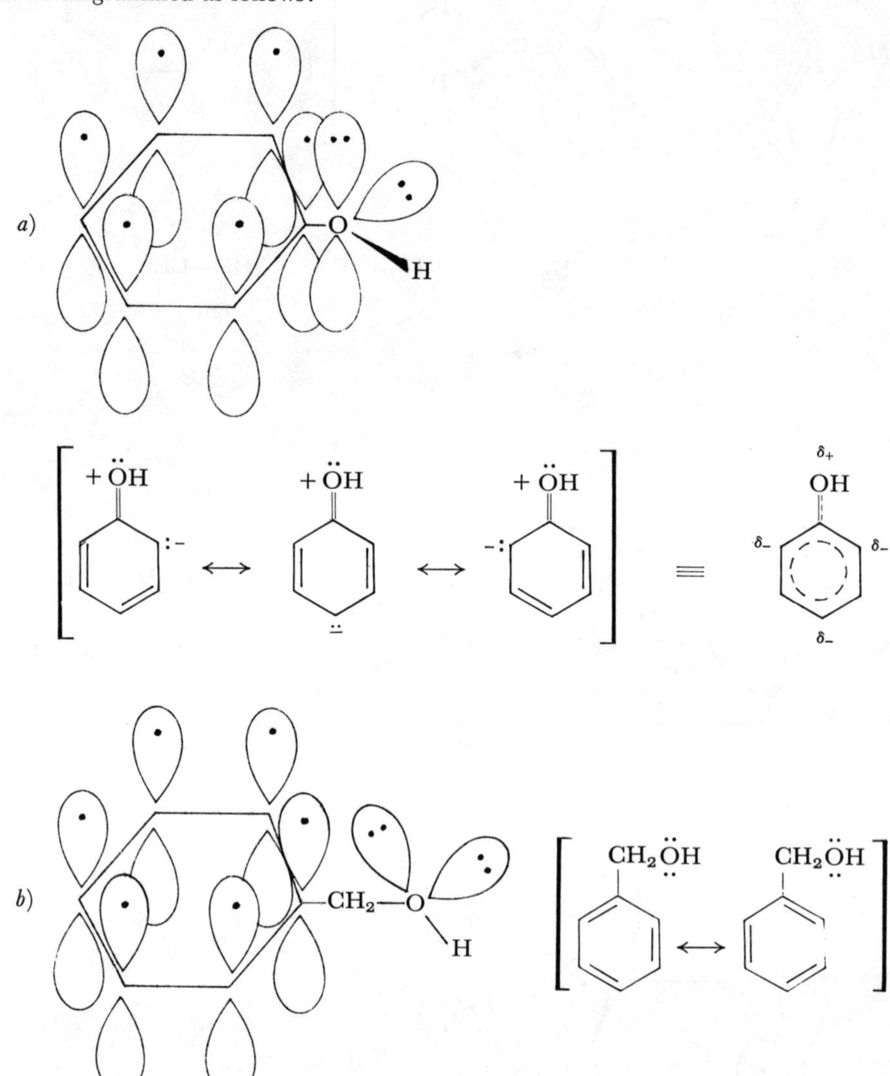


*c*) is the same as a with an $—OCH_3$ in place of an —OH.
*f*) is given in #26–31 and #26–33.
*g*) is given in #26–27 and #26–28.
*h*) is given in #26–25 and #26–26.

*k*) $[CH_2{\cdots}CH—\ddot{C}H_2 \longleftrightarrow \ddot{C}H_2—CH{\cdots}CH_2] \equiv \overbrace{CH_2{\cdots}CH{\cdots}CH_2}^{\cdot\cdot}$

*l*)

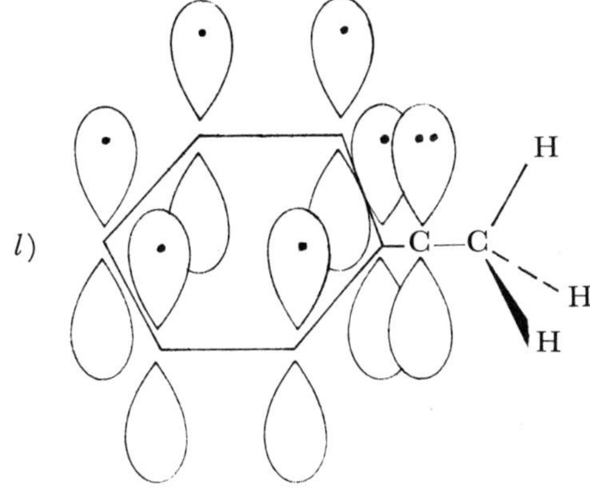

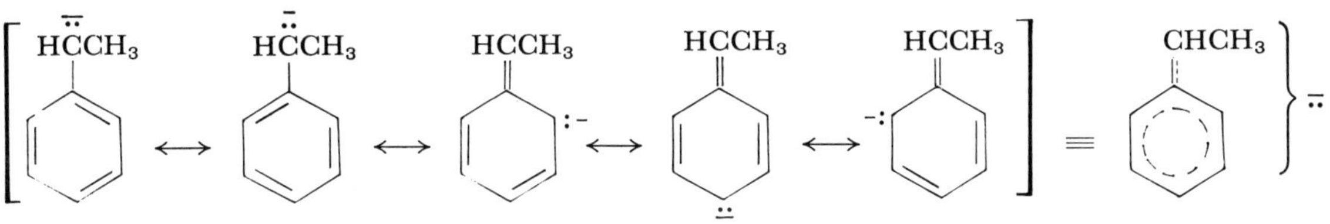

HCCH3 ⟷ HCCH3 ⟷ HCCH3 ⟷ HCCH3 ⟷ HCCH3 ≡ CHCH3

**3**) Only *b*, *e*, and *f* have structures that lead to resonance. The effect can be diagrammed as follows:

*b*)

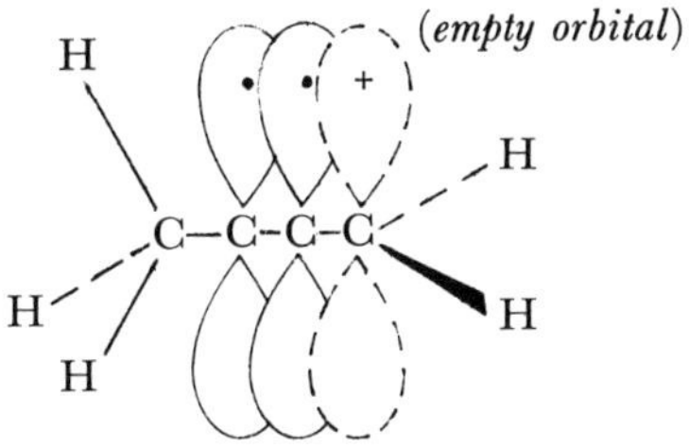

(*empty orbital*)

$$\left[CH_3\text{—}CH\overset{..}{=}CH\text{—}\overset{+}{C}H_2 \longleftrightarrow CH_3\overset{+}{C}H\text{—}CH\overset{..}{=}CH_2\right] \equiv CH_3\overbrace{CH\overset{..}{=}CH\overset{..}{=}CH_2}^{+}$$

*e*)

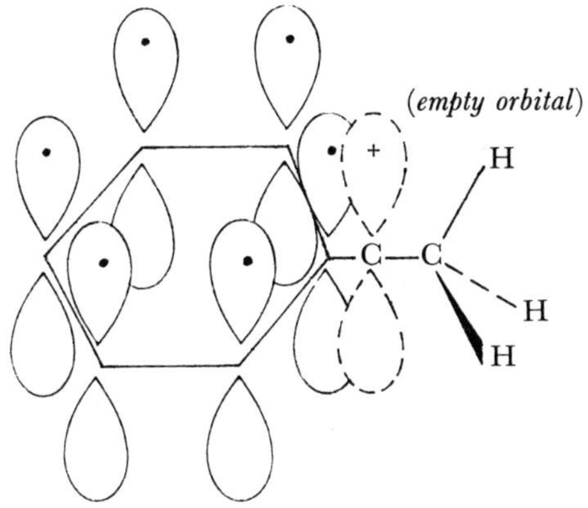

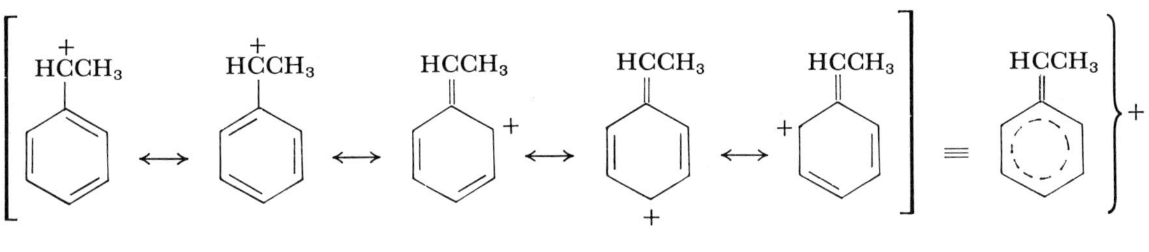

(*empty orbital*)

HCCH3 ⟷ HCCH3 ⟷ HCCH3 ⟷ HCCH3 ⟷ HCCH3 ≡ HCCH3

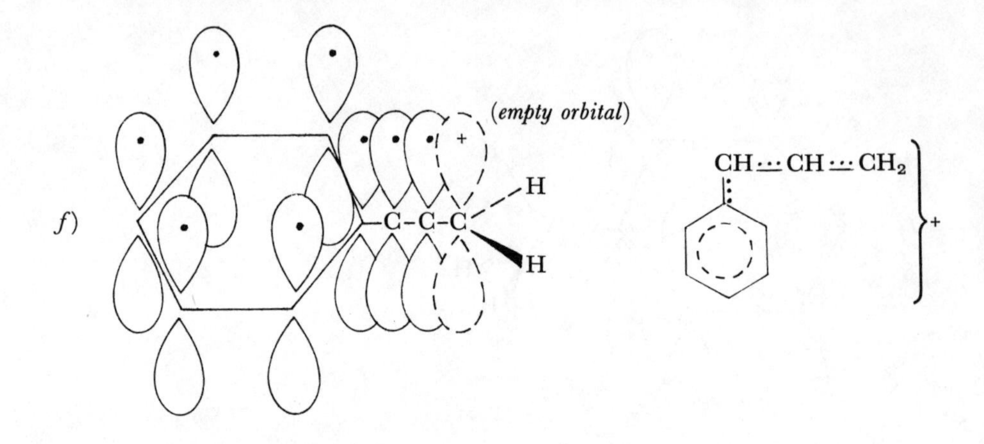


**4**) In this problem *b*, *e*, and *f* correspond to *b*, *e*, and *f* in problem 3. The diagrams are the same except that in each formula the + incomplete octet is replaced by a single nonbonding electron. Compare *b* and *e* with *k* and *l* in problem 2.

**5**) *a*) acetate ion. *b*) anion. *c*) 1,3-butadiene. *d*) N-methylaniline. *e*) phenol. *f*) acetate ion. *g*) benzene. *h*) cyanobenzene.

**6**) *a*) 2-methylpropanoic acid. *b*) 2-butenoic acid. *c*) 4-oxo-2-pentanone. *d*) ethyl acetoacetate. *e*) benzoic acid. *f*) phenol.

**7**) Only the canonical forms are included in these answers. You have enough experience with diagrams by now to know whether you have drawn it correctly. *a*) See problem 1*c*.

*b*) $\left[ CH_3CH{\cdots}CH\overset{:\ddot{O}}{\overset{\vdots}{C}}{-}\ddot{O}{:}^- \longleftrightarrow CH_3CH{\cdots}CH\overset{:\ddot{O}:^-}{\overset{|}{C}}{\cdots}O{:} \longleftrightarrow CH_3\underset{+}{C}H{-}CH{\cdots}\overset{:\ddot{O}:^-}{\overset{|}{C}}{-}\ddot{O}{:}^- \right] \equiv CH_3CH{\cdots}CH{\cdots}\overset{O}{\overset{\vdots}{C}}{\cdots}O$ (with $^-$ spread over both O)

*c*) $\left[ CH_3\ddot{O}\overset{:\ddot{O}}{\overset{\vdots}{C}}{-}\underset{..}{\overset{H}{\overset{|}{C}}}{-}\overset{\ddot{O}:}{\overset{\vdots}{C}}\ddot{O}CH_3 \longleftrightarrow CH_3\ddot{O}\overset{:\ddot{O}:^-}{\overset{|}{C}}{\cdots}\overset{H}{\overset{|}{C}}{-}\overset{\ddot{O}:}{\overset{\vdots}{C}}\ddot{O}CH_3 \longleftrightarrow CH_3\ddot{O}\overset{:\ddot{O}}{\overset{\vdots}{C}}{-}\overset{H}{\overset{|}{C}}{\cdots}\overset{:\ddot{O}:^-}{\overset{|}{C}}\ddot{O}CH_3 \right] \equiv CH_3\ddot{O}\overset{O}{\overset{\vdots}{C}}{\cdots}\overset{H}{\overset{|}{C}}{\cdots}\overset{O}{\overset{\vdots}{C}}\ddot{O}CH_3$ (with $^-$ spread over both O)

*d*) [ four-membered ring canonical forms: H–C(ring, with −: carbanion)–C(=O:), C–CH₂, C=O ↔ H–C⋯C–O:⁻, C–CH₂, C=O ↔ H–C–C(⋯O:), C⋯C–CH₂, C–O:⁻ ] ≡ H–C⋯C–O, C–CH₂, C–O (with $^-$ delocalized)

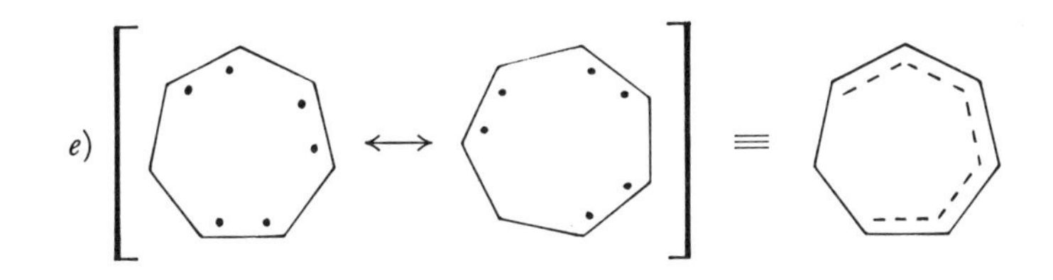

f)

8) Check your answers by comparing them with the lists in Table 26–1.

9) Check your reasons by comparing them with the information in Table 26–2.
*a*) 3, 1, 2. *b*) 3, 2, 1. *c*) 3, 2, 1. *d*) 3, 1, 2.
*e*) 3, 2, 1. *f*) 1, 3, 2. *g*) 1, 2, 3. *h*) 3, 2, 1.
*i*) 1, 3, 2. *j*) 1, 2, 3.

**10**) *a*) 2, 1, 3. *b*) 1, 2, 3. *c*) 1, 2, 3. *d*) 3, 1, 2.
*e*) 2, 1, 3. *f*) 1, 3, 2. *g*) 3, 2, 1. *h*) 3, 1, 2.
*i*) 3, 2, 1. *j*) 3, 2, 1.

**11**) You have worked out the diagrams for each individual ring substituent. You now merely have to combine two of these para to each other on a ring. The canonical forms for *a* and *d* are given in #26–40 and #26–41. Draw *b* to show the nitro group withdrawing electrons from the ring and reducing the basicity of the amino group. Draw *c* to show the hydroxy group releasing electrons into the ring and increasing the basicity of the amino group.

## §29. Proton Exchange Reactions

*a*) $CH_3CH_2\ddot{\underset{\cdot\cdot}{O}}H + H_3\ddot{O}^+ \rightleftharpoons CH_3CH_2\ddot{O}H_2^+ + H_2\ddot{O}:$

(*weaker base*) (*weaker acid*) (*stronger acid*) (*stronger base*)

*b*) $CH_3\ddot{\underset{\cdot\cdot}{O}}CH_3 + H_3\ddot{O}^+ \rightleftharpoons CH_3\overset{H^+}{\ddot{\underset{\cdot\cdot}{O}}}CH_3 + H_2\ddot{O}:$

(*w.b.*) (*w.a.*) (*s.a.*) (*s.b.*)

*c*) $CH_3CH_2\ddot{\underset{\cdot\cdot}{O}}H + :\ddot{\underset{\cdot\cdot}{O}}H^- \rightleftharpoons CH_3CH_2\ddot{\underset{\cdot\cdot}{O}}:^- + H\ddot{\underset{\cdot\cdot}{O}}H$

(*w.a.*) (*w.b.*) (*s.b.*) (*s.a.*)

*d*) $CH_3NH_2 + CH_3COOH \rightleftharpoons CH_3NH_3^+ + CH_3COO^-$

(*s.b.*) (*s.a.*) (*w.a.*) (*w.b.*)

*e*) $CH_3CH_2\ddot{\underset{\cdot\cdot}{O}}{:}^- + C_6H_5\ddot{\underset{\cdot\cdot}{O}}H \rightleftharpoons CH_3CH_2\ddot{\underset{\cdot\cdot}{O}}H + C_6H_5\ddot{\underset{\cdot\cdot}{O}}{:}^-$

(*s.b.*) (*s.a.*) (*w.a.*) (*w.b.*)

*f*) $HC{\equiv}CH + NaNH_2 \rightleftharpoons HC{\equiv}C{:}^-Na^+ + NH_3$

(*s.a.*) (*s.b.*) (*w.b.*) (*w.a.*)

*g*) $CH_3\overset{\overset{OH^+}{\|}}{C}CH_3 + H_2O \rightleftharpoons CH_3\overset{\overset{:\ddot{O}}{\|}}{C}CH_3 + H_3O^+$

(*s.a.*) (*s.b.*) (*w.b.*) (*w.a.*)

*h*) $CH_3\overset{\overset{O}{\|}}{C}CH_3 + H_3O^+ \rightleftharpoons CH_3\overset{\overset{OH^+}{\|}}{C}CH_3 + H_2O$

(*w.b.*) (*w.a.*) (*s.a.*) (*s.b.*)

*i*) $CH_3COOH + HCO_3^- \rightleftharpoons CH_3\overset{\overset{:\ddot{O}}{\|}}{C}\ddot{\underset{\cdot\cdot}{O}}{:}^- + (H_2O + CO_2)$

(*s.a.*) (*s.b.*) (*w.b.*) (*w.a.*)

*j*) $CH_3NH_3^+ + {:}\ddot{\underset{\cdot\cdot}{O}}H^- \rightleftharpoons CH_3NH_2 + H\ddot{\underset{\cdot\cdot}{O}}H$

(*s.a.*) (*s.b.*) (*w.b.*) (*w.a.*)

*k*) $CH_3\overset{\overset{:O:H^+}{\|}}{C}\ddot{\underset{\cdot\cdot}{O}}H + H_2\ddot{O}{:} \rightleftharpoons CH_3\overset{\overset{:\ddot{O}}{\|}}{C}\ddot{\underset{\cdot\cdot}{O}}H + H_3\ddot{O}^+$

(*s.a.*) (*s.b.*) (*w.b.*) (*w.a.*)

*l*) $NH_4^+ + CH_3CH_2\ddot{\underset{\cdot\cdot}{O}}{:}^- \rightleftharpoons CH_3CH_2\ddot{\underset{\cdot\cdot}{O}}H + NH_3$

(*s.a.*) (*s.b.*) (*w.a.*) (*w.b.*)

*m*) $C_6H_5OH + CO_3^= \rightleftharpoons C_6H_5O^- + HCO_3^-$

(*s.a.*) (*s.b.*) (*w.b.*) (*w.a.*)

*n*) $CH_3\overset{\overset{:\ddot{O}}{\|}}{C}H + CH_3COOH \rightleftharpoons CH_3\overset{\overset{:O:H^+}{\|}}{C}H + CH_3COO^-$

(*w.b.*) (*w.a.*) (*s.a.*) (*s.b.*)

*o*) $NH_3 + CH_3COO^- \rightleftharpoons {:}\ddot{N}H_2^- + CH_3COOH$

(*w.a.*) (*w.b.*) (*s.b.*) (*s.a.*)

*p*) ${:}\ddot{\underset{\cdot\cdot}{Cl}}{:}^- + H_3\ddot{\underset{\cdot\cdot}{O}}^+ \rightleftharpoons H{:}\ddot{\underset{\cdot\cdot}{Cl}}{:} + H_2\ddot{O}{:}$

(*w.b.*) (*w.a.*) (*s.a.*) (*s.b.*)

*q*) $CH_3CH_3 + {:}\ddot{N}H_2^- \rightleftharpoons CH_3CH_2{:}^- + \ddot{N}H_3$

(*w.a.*) (*w.b.*) (*s.b.*) (*s.a.*)

$$r)\quad \underset{(w.b.)}{CH_3\overset{\overset{:\ddot{O}}{\|}}{C}CH_3} + \underset{(w.a.)}{NH_3} \rightleftharpoons \underset{(s.a.)}{CH_3\overset{\overset{:O:H^+}{\|}}{C}CH_3} + \underset{(s.b.)}{:\ddot{N}H_2^-}$$

$$s)\quad \underset{(s.a.)}{H_3\ddot{O}^+} + \underset{(s.b.)}{CH_3\ddot{N}HCH_3} \rightleftharpoons \underset{(w.b.)}{H_2\ddot{O}:} + \underset{(w.a.)}{CH_3\overset{+}{N}H_2CH_3}$$

$$t)\quad \underset{(s.b.)}{C_6H_5\ddot{\underset{..}{O}}:^-} + \underset{(s.a.)}{CH_3CH_2COOH} \rightleftharpoons \underset{(w.a.)}{C_6H_5\ddot{\underset{..}{O}}H} + \underset{(w.b.)}{CH_3CH_2COO^-}$$